PCHEM

# Advances in
# Physical Organic Chemistry

# Advances in
# Physical Organic Chemistry

## Volume 20

*Edited by*

V. GOLD

Department of Chemistry
King's College London
Strand, London WC2R 2LS

*and*

D. BETHELL

The Robert Robinson Laboratories
University of Liverpool
P.O. Box 147, Liverpool

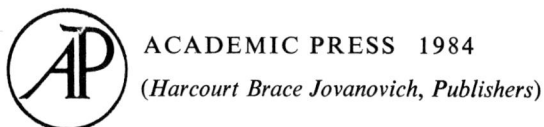

ACADEMIC PRESS  1984
(*Harcourt Brace Jovanovich, Publishers*)

London   Orlando
San Diego   San Francisco   New York
Toronto   Montreal   Sydney   Tokyo

ACADEMIC PRESS INC. (LONDON) LTD
24/28 Oval Road
London NW1 7DX

*United States Edition published by*
ACADEMIC PRESS INC.
(Harcourt Brace Jovanovich, Inc.)
Orlando, Florida 32887

Copyright © 1984 by
ACADEMIC PRESS INC. (LONDON) LTD

*All Rights Reserved*

No part of this book may be reproduced in any form by photostat, microfilm, or any other means, without written permission from the publishers

*British Library Cataloguing in Publication Data*
Advances in physical organic chemistry.
  Vol. 20
  1. Chemistry, Physical organic
  547.1'3'05  QD476

ISBN 0-12-033520-4
ISSN 0065-3160

TYPESET BY BATH TYPESETTING LTD., BATH, U.K.
AND PRINTED IN GREAT BRITAIN BY THOMSON LITHO LTD.,
EAST KILBRIDE, SCOTLAND

# Contributors to Volume 20

**R. Stephen Davidson**  Department of Chemistry, The City University, London EC1V 0H8, U.K.

**Jonathan W. Goodin**  Plastics Division, Ciba-Geigy Plastics and Additives Co., Duxford, Cambridge CB2 4QA, U.K.

**Ian R. Gould**  Department of Chemistry, Columbia University, Havemeyer Hall, New York, N.Y. 10027, U.S.A.

**Ole Hammerich**  Department of General and Organic Chemistry, University of Copenhagen, The H.C. Ørsted Institute, DK-2100 Copenhagen, Denmark

**Graham Kemp**  Plastics Division, Ciba-Geigy Plastics and Additives Co., Duxford, Cambridge CB2 4QA, U.K.

**Vernon D. Parker**  Laboratory of Organic Chemistry, Norwegian Institute of Technology, University of Trondheim, N-7034 Trondheim-NTH, Norway

**Nicholas J. Turro**  Department of Chemistry, Columbia University, Havemeyer Hall, New York, N.Y. 10027, U.S.A.

**Matthew B. Zimmt**  Department of Chemistry, Columbia University, Havemeyer Hall, New York, N.Y. 10027, U.S.A.

# Contents

Contributors to Volume 20   v

**Magnetic Field and Magnetic Isotope Effects on the Products of Organic Reactions**   1

IAN R. GOULD, NICHOLAS J. TURRO and MATTHEW B. ZIMMT

1  Introduction   1
2  Experimental examples   16
3  Conclusion   50
   Acknowledgements   51

**Kinetics and Mechanisms of Reactions of Organic Cation Radicals in Solution**   55

OLE HAMMERICH and VERNON D. PARKER

1  Introduction   56
2  Dimerization and cyclization reactions of cation radicals   56
3  Mechanisms of the reactions of cation radicals with nucleophiles   68
4  Electron-transfer reactions initiated by cation radicals   94
5  Fragmentation reactions of cation radicals   123
6  Cation radicals as intermediates in conventional organic reactions   151
7  Concluding remarks   180

## The Photochemistry of Aryl Halides and Related Compounds 191

R. STEPHEN DAVIDSON, JONATHAN W. GOODIN and GRAHAM KEMP

1 Introduction 192
2 Chloroaromatics 196
3 Bromo- and iodoaromatics 212
4 Assisted dehalogenation of halogenoaromatics 219
5 Photoinduced nucleophilic substitution 222

**Author Index** 235

**Cumulative Index of Authors** 245

**Cumulative Index of Titles** 247

# Magnetic Field and Magnetic Isotope Effects on the Products of Organic Reactions

IAN R. GOULD, NICHOLAS J. TURRO and MATTHEW B. ZIMMT

*Department of Chemistry, Columbia University, New York, U.S.A.*

1 Introduction 1
    Origin of magnetic field effects 1
    Radical pair systems 2
    Important interactions in radical systems 5
    Mechanisms of intersystem crossing 8
    Classification of magnetic field effects 14
2 Experimental examples 16
    Class I effects: Reactions influenced by the Zeeman effect 16
    Class II effects: Reactions for which the singlet-triplet splitting is non-zero 27
    Class III effects: Reactions for which the $\Delta g$-mechanism for intersystem crossing is dominant 29
    Class IV effects: Magnetic isotope effects 33
3 Conclusion 50
Acknowledgements 51
References 51

## 1 Introduction

ORIGIN OF MAGNETIC FIELD EFFECTS

The effects of magnetic fields on chemical reactions have been the subject of several review articles in recent years (Atkins, 1976; Atkins and Lambert, 1975; Sagdeev *et al.*, 1977b; Lawler and Evans, 1971; Turro and Kraeutler, 1980; Turro, 1983). It is the object of the present review to describe the current theory and to illustrate, with examples, the changes that arise in the formation of products of organic reactions as a function of applied field and of nuclear magnetic spins.

Sporadic reports of magnetic field effects on chemical reactions have appeared for almost 80 years. However, early work was plagued by non-reproducible effects and retractions. The early work is summarized by Atkins and Lambert (1975). Recently, reports of magnetic field effects on chemical reactions have become more widespread and better documented. These reports have been coincident with the observation and subsequent theoretical development of chemically induced magnetic polarization, CIDNP (Kaptein, 1975; Closs, 1971; Buchachenko and Zhidomirov, 1971; Frith and McLauchlan, 1975; Lawler, 1972). Prior to these recent developments, it had been thought that there was no possibility of observing significant magnetic effects on the products of chemical reactions. Based on a simple thermodynamic argument, it can be shown that a very strong magnetic field (100,000 G) can induce an energy change of at most $c$. 0.03 kcal mol$^{-1}$ even in paramagnetic molecules (Atkins, 1976). Since this energy is negligible compared to commonly encountered activation energies ($\sim 10$ kcal mol$^{-1}$) it could be concluded that no effect should be observed. However, the rates of chemical reactions depend upon both energetic and entropic factors. Substantial magnetic field effects on reaction probabilities are possible if the effect of the field is to change the number of degrees of freedom open to a chemical system, i.e. affect the factor $A$ in (1) and thus $k$ (Turro and Kraeutler, 1980; Turro, 1983). Specifically, observation of magnetic field effects depends

$$k = (Ae^{-\Delta E/RT}) \qquad (1)$$

upon the system possessing states which have different magnetic properties (e.g. singlet and triplet states). If the application of a magnetic field can either shut off or enhance selected reaction channels for these states, then an effect on the product distribution becomes possible.

A deeper insight into the origin of magnetic effects lies in a description of the electronic and nuclear spin states of a system, and in an understanding of possible magnetic interactions. The bulk of the observations described in the present review are concerned with the reactions of radical pairs and thus we will examine the spin states of radical pair systems as archetypes.

RADICAL PAIR SYSTEMS

An introductory account of this topic has been given by Turro (1978a). Homolytic cleavage of a single bond leads to two one-electron fragments, i.e. radicals; the two radicals constitute a radical pair. If the molecule cleaves from the singlet state (most common for thermal reactions), then the resulting radical pair will be in a singlet state, since the process of bond cleavage is three to four orders of magnitude faster than any mechanism which interconverts the radical pair spin states. Similarly, if the molecule cleaves from a triplet state (commonly encountered in photochemical processes), then the radical pair will initially be in a triplet state.

*Vector spin description of radical pairs*

The simultaneous occurrence of charge and spin motion imbues an electron with a magnetic moment. In the presence of a laboratory magnetic field, which is strong compared to molecular magnetic fields, the electron moment of a radical will interact with the applied field. The electron spin can be represented by an angular momentum vector, with the direction of applied field, $H$, as the positive $z$-axis. Then, the electron spin vector has a component of angular momentum along the field direction that is either $+1/2\hbar$ ($\alpha$, parallel) or $-1/2\hbar$ ($\beta$, antiparallel) (Fig. 1). The Uncertainty Principle dictates that the spin vector also has a component that is perpendicular to and that precesses about the field direction.

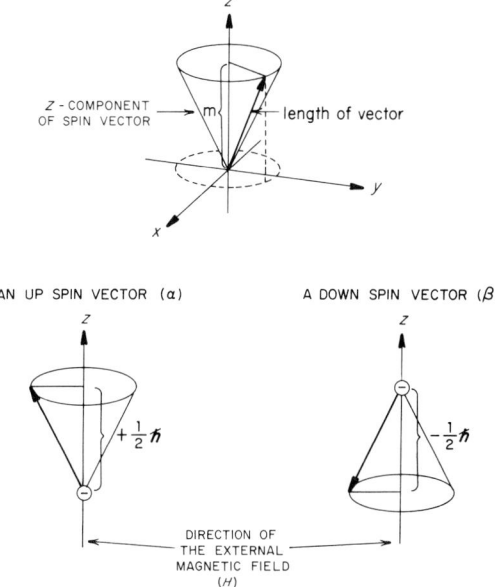

FIG. 1 Vector representation of an electron's spin moment. In the presence of an external magnetic field ($H$, $z$-axis) the moment precesses around the axis, which corresponds to the direction of that field. Two orientations of the vector are allowed; an up ($\alpha$) orientation and a down ($\beta$) orientation

In a radical pair the vectors corresponding to the spin of each electron exhibit the same behaviour as those of a single electron. Quantum mechanics dictates, however, that only specific interactions of the two spin vectors may occur in the initially formed pair (Fig. 2). If formed from a singlet state, the spin vectors of electrons 1 and 2 ($\mathscr{S}_1$ and $\mathscr{S}_2$) must be aligned so that one spin is α, one spin is β, ($M_s = 0$), and their components perpendicular to the field must be 180° out of phase ($S = 0$): a state of zero total spin angular momentum. If formed from a triplet state, three possible interactions may occur, each of which has the spin vectors in phase (0° separation of the perpendicular components, $S = 1$). The electron spins may be aligned with both moments parallel to the field $T_+(\alpha\alpha)$, with spin angular momentum $+\hbar$ along the field axis ($M_s = 1$); both antiparallel $T_-(\beta\beta)$ with spin angular momentum $-\hbar$ along the field axis ($M_s = -1$); or one spin parallel (α) and one spin antiparallel (β), $T_0$ (αβ) with zero spin angular momentum along the field axis ($M_s = 0$).

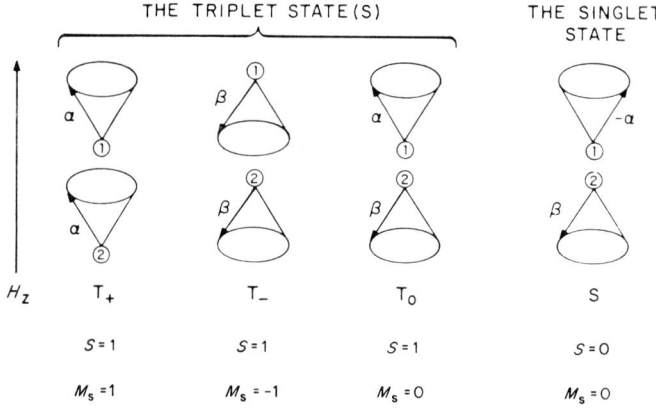

FIG. 2  Vector model of the singlet and triplet states of a radical pair

Thus, a transformation from S to $T_0$ requires a rephasing of the two spin vectors with respect to each other, whereas S to $T_\pm$ requires a spin flip, i.e. a change in total electronic spin momentum. Transitions between the triplet and singlet states (intersystem crossing) are forbidden in the absence of other magnetic interactions. However, in the next section, we shall discuss the magnetic interactions which cause the breakdown in the "forbiddenness" of such state-changing processes.

*Importance of spin states in radical pair chemistry*

If the two radicals which are produced from a singlet molecule should reencounter, then a molecular product from radical combination or disproportionation may result. Such a product is described as resulting from a geminate recombination. A bond will be formed during the encounter only if the electron spins correspond to a singlet state. Encounters involving triplet state radical pairs are unproductive. Processes which transform initially formed singlet radical pairs into triplet radical pairs prevent geminate re-encounters from forming products. When this occurs, escape of either radical from the initial solvent cage becomes more likely, and thus nongeminate, or random, encounters become more probable. Products which arise from geminate encounters are termed cage products; those from random encounters are escape products. Hence, an intersystem crossing process in an initially formed singlet radical pair will tend to increase the formation of escape products (and increase cage products for initially triplet radical pairs). Furthermore, if the intersystem crossing transformation is modulated by the magnetic field, then the origin of magnetic field effects on chemical reactions may be realized. The crucial condition for observation of magnetic field effects is a *competition* between two processes: one of the processes must be magnetic field dependent, the other is magnetic field independent[1]. For the example of radical pair chemistry, the processes are intersystem crossing (field dependent) and diffusive escape from the solvent cage or reaction to form a chemically modified radical pair (field independent).

IMPORTANT INTERACTIONS IN RADICAL SYSTEMS

The mechanisms of intersystem crossing depend intimately upon the magnetic interactions that occur in radical systems both from internal molecular and external fields. These interactions are described in this section.

*Single radicals*

The interactions which are characteristic of single radicals are those which are probed in electron spin resonance (esr) experiments (Carrington and McLaughlan, 1967). Interaction of an electron spin with an applied magnetic field results in an energy difference between the parallel ($\alpha$) and antiparallel ($\beta$) spin states. The energy gap is defined as $\Delta E = g\beta H$ (Fig. 3), where $\beta$ is the electron Bohr magneton, $H$ is the applied field, and $g$ is a proportionality constant related to the total angular momentum of the electron. Just as the spin motion of a charged electron results in a magnetic moment, the orbital motion of an electron results in a contribution to the overall electronic angular momentum. Coupling of spin and orbital angular momenta causes a change in the total angular momentum for a given electron. Thus electrons in different radicals, which possess different orbital angular momenta, will interact to varying extents with an external field. The $g$-value is the experimental measure of the extent of interaction and, in effect, is analogous to the familiar chemical shift measured in nmr spectroscopy. The value of $g$ is 2.0023 for a "free" electron and ranges from 2.0000 to 2.003

---

[1]More generally, the two processes should have *different* magnetic field dependences.

for typical carbon-based organic free radicals. Such values are determined by irradiating the radical with microwave radiation and varying an external magnetic field until the spin state energy gap equals the microwave photon energy and resonance occurs.

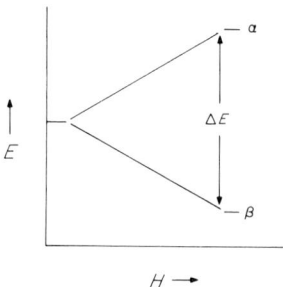

FIG. 3 Diagram showing the effect of an external magnetic field on the energy difference between the α and β spin states of a single electron

Additional intramolecular magnetic fields which may couple with the electronic magnetic moments arise from the spin moments of magnetic nuclei. These couplings lead to splittings in the absorption lines observed in esr spectra and are termed hyperfine couplings. Since the esr spectrum is recorded by varying the magnetic field, the splittings are observed at different values of the applied field, and the separation (called the hyperfine splitting constant, $a$) is usually expressed in gauss (G). The size of a particular hyperfine interaction is directly related to the electron density of the unpaired electron on the coupling nucleus. Typical values of $a$ for protons range from 25 G to less than 1 G ($c.$ $10^{-6}$ kcal mol$^{-1}$), but can reach 200 G for $^{13}$C nuclei. Esr parameters for the organic radicals produced by homolytic cleavage of dibenzyl ketone are shown in Fig. 4.

*Radical pairs*

A non-magnetic electron exchange phenomenon occurs between interacting unpaired electrons. Interacting electrons are not allowed to be specifically defined as being located on one particular radical centre. In effect, the exchange processes, which are distinguished from Coulombic interactions by their dependence on spin, serve to preserve electron indistinguishability. As a result of these exchange interactions, the energies of the singlet and triplet states differ by $2 \times J$, where $J$ is the electron exchange integral. The size of the exchange integral is determined in part by the separation of the two radical pair centres (Fig. 5). As the components of a radical pair begin to diffuse apart, the exchange interaction falls off rapidly. At a certain distance ($c.$ 10 Å for typical organic radical pairs), $J$ becomes negligible and the singlet and triplet states become energetically degenerate. At this point, intersystem crossing processes become energetically feasible.

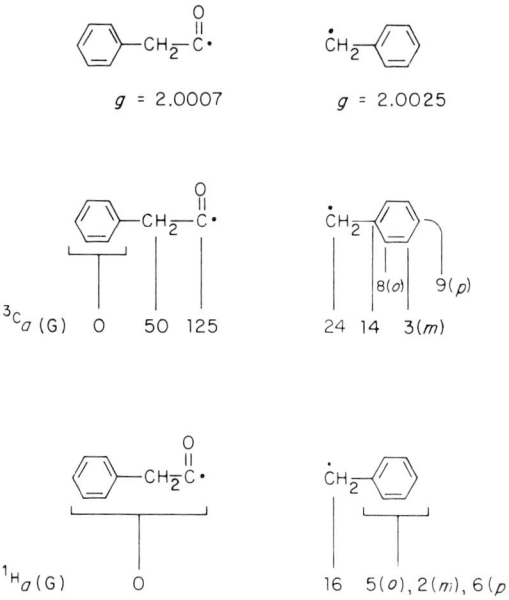

FIG. 4 Esr parameters of the $C_6H_5CH_2^{\cdot}$ and $C_6H_5CH_2CO\cdot$ radicals which are formed upon photolysis of dibenzyl ketone. (Hyperfine couplings, $a$, in gauss)

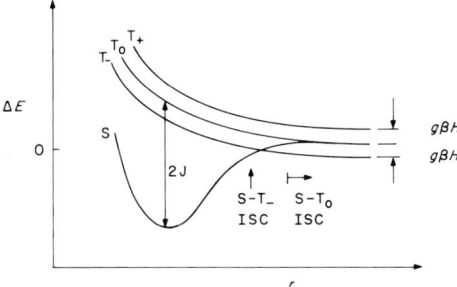

FIG. 5 Potential energy diagram showing the effect of separation of the partners in a radical pair ($r$), on the energy difference between the singlet and triplet states ($\Delta E$). As $r$ increases, the exchange integral ($J$) decreases. When $J$, and thus $\Delta E$ become negligible, intersystem crossing becomes energetically allowed. The application of a magnetic field causes a splitting in the energy levels of the $T_+$, $T_0$ and $T_-$ states (Zeeman effect). The size of the splitting is given by $g\beta H$, where $g = (g_A + g_B)/2$

Finally, an important interaction in the context of the present discussion is that with an external magnetic field. As previously described, the application of a magnetic field to a radical pair system results in the alignment of the spin vectors for each electron along the axis of that field (Fig. 1). Because two of the triplet substates have non-zero projections of their spin momenta along the axis of the field, the three substates become non-degenerate. Those radical pairs with "up" or $\alpha$ electron spin, $M_s = 1$ ($T_+$), are raised in energy and those with $\beta$ spin, $M_s = -1$ ($T_-$), are lowered in energy. The energies of the $T_0$ and S states remain unchanged. This splitting of the three sublevels upon application of an external magnetic field is commonly called the Zeeman effect. The extent of the splitting is given by the Zeeman energies which are $(g_A + g_B)\beta H/2$, $0$, $-(g_A + g_B)\beta H/2$ for $T_+$, $T_0$ and $T_-$, respectively. The effect is shown diagrammatically in Fig. 5.

MECHANISMS OF INTERSYSTEM CROSSING

As mentioned previously, the theory of magnetic field effects on radical pair reactions comes directly from the theory of CIDNP, which considers two main mechanisms for intersystem crossing. These intersystem crossing processes, and the magnetic field effects upon them, arise as a direct consequence of the interactions described in the previous section.

As the two radical centres in a radical pair begin to separate, the exchange interaction ($J$) which both couples the precessional motion of the two electron spins, and maintains a difference in the energies of the S and $T_0$ states, becomes negligible relative to the interactions of the individual electrons with external fields, and with nearby magnetic nuclei ($a$). Instead of precessing strictly in phase, as when coupled by $J$, each electron spin now precesses at a different rate determined by the magnetic interaction to which it is coupled. Actually, $J$ need not become zero for intersystem crossing to occur (*vide infra*); however, for the following discussion it will be assumed that $J$ is zero, i.e. $T_0$ and S are degenerate.

An initially formed, singlet radical pair remains in a singlet state if the precessional frequencies remain the same for the two electrons, i.e. the difference in precessional rates ($\Delta\omega$) remains zero. This is the case if the magnetic fields experienced by the two electrons on the radical pair centres are the same. However, if these magnetic fields should differ, then the precessional rates may become unequal ($\Delta\omega \neq 0$), and the electrons will lose their singlet phasing. This implies acquisition of the triplet phasing, as shown diagrammatically in Fig. 6. In fact, because the precessional frequencies remain different, given sufficient time, the spin state will pass through pure triplet ($T_0$) and continue back to singlet (S). The radical pair spin state will be a sinusoidally varying mixture of singlet (S) and triplet ($T_0$) states.

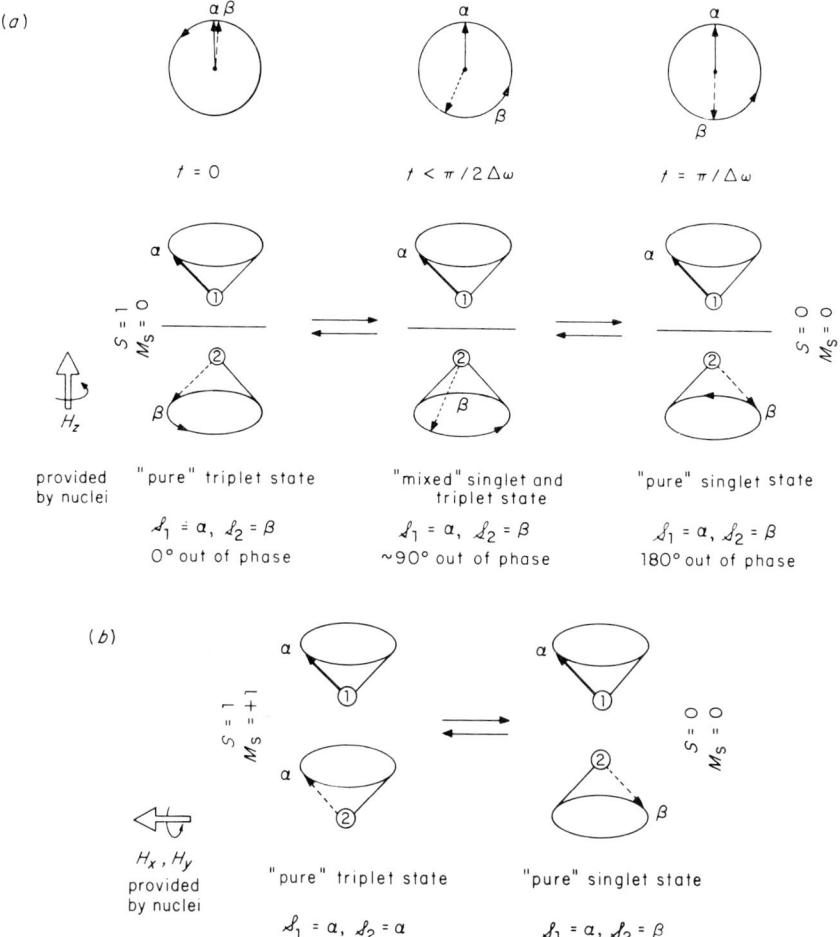

FIG. 6 (a) A spin clock and vectorial representation of the $T_0 \rightleftharpoons S$ intersystem crossing mechanism in radical pairs. (b) A vectorial representation of a $T_+ \rightleftharpoons S$ intersystem crossing showing the required change in electron spin

An alternative and useful illustration of the $S \rightleftharpoons T_0$ intersystem crossing can be made in terms of a "clock" representation (Kaptein, 1977). If the $z$-axis corresponds to the direction of the applied field, then the projection of the spins in the $xy$-plane produces a vector description of the $T_0$ and $S$ states that resembles the hands of a clock. When electron exchange is large relative to other interactions, then the clock can read only 12 or 6, i.e. $T_0$ and $S$ do not mix. When electron exchange is small, then rephasing of the two spin vectors can occur and the clock can read any arbitrary time (Fig. 6).

The time $\tau_{ST}$ it takes to convert S to $T_0$ is the time required to rephase the two vectors by $\pi$ radians. If $\Delta\omega$ equals the differential precessional rate in radians s$^{-1}$, then $\tau_{ST} = \pi/\Delta\omega$. The rate constant for intersystem crossing (the inverse of $\tau_{ST}$) is thus given by (2). Hence, the mechanism of intersystem crossing from S to $T_0$ depends on the methods by which $\Delta\omega$ is caused to change from zero.

$$k_{ST} = \Delta\omega/\pi \qquad (2)$$

Intersystem crossing from $T_+$ or $T_-$ to S requires a change in the z-component of electronic spin angular momentum from $1\hbar$ or $-1\hbar$ to $0\hbar$, respectively.[2] As is evident from the previous discussion, and from Fig. 2, the rephase mechanism which is used for $T_0 \to S$ intersystem crossing is incapable of promoting a change in the z-component of angular momentum. Instead, a mechanism is required which can "flip" the spin of one of the electrons i.e. an exchange of electron spin angular momentum with another source of angular momentum is required. Although the vector model implies that both a spin flip and rephase are required, actually only the former is necessary. The mechanism whereby the electron spin angular momentum is allowed to change makes use of the hyperfine coupling phenomenon. Thus, an electronic spin may exchange momentum with a nuclear spin to which it is coupled by means of a "double flip" in which each spin state, nuclear and electronic, is changed and thus total angular momentum is conserved. The effect is shown diagramatically in Fig. 6 in which a $T_+$ state is transformed into an S state *via* a flip in one spin vector.

*Hyperfine coupling mechanisms*

The electron-nuclear hyperfine coupling promotes intersystem crossing by two independent mechanisms. The presence of local nuclear magnetic fields, as transmitted to the electron spin *via* the hyperfine coupling, provide the torque required to rephase the electron spin vectors and promote $T_0 \rightleftharpoons S$ intersystem crossing (for radical pairs and other systems with $J \to 0$). In Fig. 6 it is shown how this torque, oriented around the z-axis, enables the rephasing of a $T_0$ to an S state to occur. The rate of intersystem crossing in terms of the $\Delta\omega$ factor is given in (3), where $a_n$ is the hyperfine constant,

$$\Delta\omega \sim |\Sigma_i a_{1_i} m_{1_i} - \Sigma_j a_{2_j} m_{2_j}| \qquad (3)$$

and $m_{ni}$ the spin states ($\pm 1/2$ for $^{13}$C and $^1$H) of nucleus $i$ on radical $n$.

---

[2]Strictly this represents an oversimplification since at zero field a dynamic equilibrium exists between the three triplet substates and in effect only one mixed triplet state exists (see Kaptein, 1977).

For a radical pair consisting of identical radicals ($a_1 = a_2$)[3], $\Delta\omega$ can be non-zero provided the nuclear spin states on the two radicals differ.[4] For non-identical radical pairs ($a_1 \neq a_2$), $\Delta\omega$ will be non-zero independent of the nuclear spin states, although some combinations of nuclear spin states will be more effective than others in promoting $T_0 \rightleftharpoons S$ intersystem crossing.

The hyperfine coupling also promotes $T_\pm \rightleftharpoons S$ intersystem crossing. In this case the local nuclear magnetic fields which act in the $x$- and $y$-axes can provide the torque necessary to flip the direction of an electron spin vector (Fig. 6). The hyperfine interaction causes transitions between radical electron-nuclear spin states which have the same total angular momentum and have similar energies. For example, an electron-nuclear spin system in the spin state $\alpha_e \beta_N$ will be transformed into the spin state $\beta_e \alpha_N$ (Fig. 7).

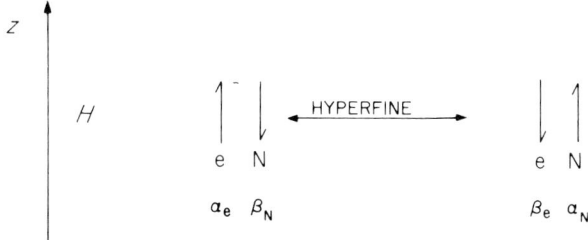

FIG. 7 Schematic representation of the mechanism by which hyperfine coupling causes a change in the $z$-component of the angular momentum of an electron (e). The simultaneous change in direction for both the electron and nuclear (N) spin vectors preserves the overall momentum in the system

In the triplet radical pair $[\alpha_{e1}, \alpha_{e2} \beta_N]$, $T_+$, this hyperfine transition produces the radical pair $[\alpha_{e1}, \beta_{e2} \alpha_N]$, S, a singlet radical pair. Hence, at zero applied magnetic field, hyperfine mechanisms exist which can induce intersystem crossing from each of the triplet levels to and from the singlet level (assuming $J = 0$). Application of a magnetic field causes splitting of $T_+$ and $T_-$ from S due to the Zeeman effect, and destroys the degeneracy needed for the coupled nuclear-electronic spin flips. At applied fields about one order of magnitude greater than the hyperfine coupling, the processes of $T_+ \rightleftharpoons S$ intersystem crossing are shut off. Since the magnitude of hyperfine couplings are less than 100 G, at applied fields of $> 1000$ G only radical pairs in the $T_0$ state can intersystem cross to S (by the rephasing mechanism).

[3] For simplicity, we consider the case of each electron being coupled to only one nucleus.
[4] Herein lies the basis for the observation of CIDNP effects (Kaptein, 1977) and of the unusually large, magnetically induced, ratio of reaction of $^{12}C$ vs $^{13}C$ in radical pairs compared to the ratio expected from mass isotope effects (see p. 33).

In terms of radical pair chemistry this means that the extent of cage reaction from a *singlet* radical pair will *increase* at fields > 1000 G relative to the zero field situation, because intersystem crossing is slowed down. In turn, the extent of cage reaction of a *triplet* radical pair will *decrease* at fields > 1000 G relative to the zero field situation.

In certain systems, such as radical pairs constrained to remain as proximate collision pairs and short chain diradicals, the electronic exchange interactions are of the order of $10^5$G, so that even at zero applied field, the singlet and triplet states of the radical pair (or diradical) are not degenerate. In such circumstances, intersystem crossing cannot be effectively induced by hyperfine interactions. However, application of an external field of sufficient strength that the Zeeman splitting ($g\beta H$) matches the singlet-triplet splitting due to exchange ($2J$), will create a degeneracy between $T_-$ and S (for singlet ground state radical pairs) (Fig. 5) and $T_+$ and S (for triplet ground state radical pairs). Thus, under these conditions, hyperfine induced spin-flip mechanisms become feasible because of the magnetic field imposed level crossing, i. e. at fields substantially lower or higher than that field required to cause a degeneracy of singlet and triplet states, intersystem crossing is inhibited. Numerous examples of these magnetic effects are known in diradical CIDNP research. In terms of diradical chemistry, a triplet diradical which can be scavenged intermolecularly would be expected to yield more "cage" or intramolecular products (due to coupling or disproportionation) at fields which induce level crossing.

The rate of hyperfine induced intersystem crossing ($k_{ISC}$) by the rephasing mechanism is given approximately by (4), where $a$ is in G (Buchachenko,

$$k_{ISC} \sim 3 \times 10^6 a \qquad (4)$$

1976). Since typical values of $a$ (for a single nucleus) fall in the range 10–100 G, $k_{ISC} \sim 3 \times 10^7$ to $3 \times 10^8 s^{-1}$. Comparable rates are expected for the spin-flip mechanism ($T_\pm \rightleftharpoons S$) in cases where the triplet-singlet gap is small compared to hyperfine interactions.

*The $\Delta g$ mechanism*

If the individual components of a radical pair possess different g-factors ($\Delta g \neq 0$), a hyperfine independent mechanism (the "$\Delta g$" mechanism) for inducing intersystem crossing occurs. This mechanism operates only to induce $S \rightleftharpoons T_0$ intersystem crossing, and results from the relationships between the g-factor, the applied field, and the rate of spin-vector precession about an applied field. The rate of precession ($\omega$) of a spin vector about a field direction is proportional to the g-factor and the strength of the field to which the vector is coupled as in (5).

$$\omega \sim g\beta H \qquad (5)$$

The difference in precessional rate ($\Delta\omega$) for the spin vectors of a radical pair is proportional to the difference of $g$-factors if the field is applied homogeneously across both radicals (6).

$$\Delta\omega \sim (g_1 - g_2)\beta H = \Delta g \beta H \qquad (6)$$

An order of magnitude estimate for the rate constant for intersystem crossing is given by (7), where $H$ is in gauss, and $k_{ISC}$ is in $s^{-1}$. For typical

$$k_{ISC} \sim 3 \times 10^6 \Delta g H \qquad (7)$$

carbon-centred organic radicals, $\Delta g$ varies from $0.5 \times 10^{-3}$ to $5 \times 10^{-3}$. Thus, in the earth's field ($\sim 1$ G), maximal values of $k_{ISC}$ for carbon-centred radical pairs fall in the range of $10^3$–$10^4 s^{-1}$. In strong laboratory magnetic fields ($\sim 10^5$G), the values of $k_{ISC}$ for carbon-centred radical pairs fall in the range $10^8$–$10^9 s^{-1}$. In the upper range, the value of $k_{ISC}$ is comparable to or greater than that for hyperfine-induced $T_0 \rightleftharpoons S$ intersystem crossing.

In summary, both $a$- and $\Delta g$-mechanisms can contribute to $T_0 \rightleftharpoons S$ intersystem crossing of carbon-centred radical pairs. A more general relationship for $k_{ISC}$, for the case of $T_0 \rightleftharpoons S$ intersystem crossing, is therefore given by (8). The contributions of $\Delta g$ and $a$ to $k_{ISC}$ can be constructive or destruc-

$$k_{ISC} \sim 3 \times 10^6 [\Delta g H + \sum_i a_{1i} M_{1i} - \sum_j a_{2j} M_{2j}] \qquad (8)$$

tive depending upon the relative signs of the pertinent constants and spin states.

*The golden rule*

The previous discussion describes the mechanisms of intersystem crossing using a convenient and easily understandable model, the vector spin description. A more complete description of the intersystem crossing process arises from the postulates of quantum dynamics. A "golden rule" expression (9) is available for understanding the rates of transitions between states of differing energies (non-degenerate) in terms of quantum dynamics, where ME represents the matrix element coupling the two states, and $\Delta E$ their

$$k_{ST} \approx \frac{1}{\hbar} \frac{<ME>^2}{\Delta E} \qquad (9)$$

energy gap (Atkins, 1970). The matrix elements for the transitions for each triplet substate are given in (10), (11) and (12) (Buchachenko, 1976) where $I_i$ is the spin quantum number of nucleus $i$, and $M_i$ is the projection of the spin along the magnetic field direction. For the case of interactions with a

$$\langle T_0|\mathcal{H}|S\rangle = 1/2\Delta gH + a_i m_i - a_j m_j \tag{10}$$

$$\langle T_+|\mathcal{H}|S\rangle = -(1/8)^{1/2}\, a_i[I_i(I_i + 1) - m_i(m_i - 1)] \tag{11}$$

$$\langle T_-|\mathcal{H}|S\rangle = (1/8)^{1/2}\, a_i[I_i(I_i + 1) - m_i(m_i + 1)] \tag{12}$$

single magnetic nucleus of spin 1/2, e.g. $^1H$ or $^{13}C$, $I_i = 1/2$, (11) and (12) reduce to (13) and (14). The matrix elements reveal the different behaviour of the triplet substrates at low field. Equation (9) shows that as the energy gap

$$\langle T_+\beta|\mathcal{H}|S\alpha\rangle = -(1/8)^{1/2}a_i \qquad \langle T_+\alpha|\mathcal{H}|S\alpha\rangle = 0 \tag{13}$$

$$\langle T_-\alpha|\mathcal{H}|S\alpha\rangle = 0 \qquad \langle T_-\alpha|\mathcal{H}|S\beta\rangle = -(1/8)^{1/2}a_i \tag{14}$$

between the states increases, the transition rate between the states decreases toward a limiting value of zero.

## Environmental effects

Many magnetic field effects on the chemistry of radical pairs have their origin in the competition between intersystem crossing and escape from the solvent cage. If escape is very fast relative to intersystem crossing and is also irreversible, then the number of radical pairs undergoing intersystem crossing will be small. In this case, magnetic field effects on the reactions of the pair will also be small. If an environment can be produced which allows diffusive separation such that $J$ is allowed to diminish to such an extent as to allow intersystem crossing, but which slows irreversible diffusive separation, then the competition between intersystem crossing and diffusion may be controlled so that the effect of an external magnetic field may be increased. Evidently, the observation of significant magnetic field effects requires a proper balance of the rates for diffusion, intersystem crossing, and electron exchange interactions. Certain examples in the next section illustrate the usefulness of changing solvent viscosity and of utilizing microheterogeneous environments such as micelles to influence the diffusion term, so that magnetic field effects may be magnified.

## CLASSIFICATION OF MAGNETIC FIELD EFFECTS

An overview of the literature allows division of the magnetic field effects into four classes. Of these, three are observed upon the application of an external laboratory magnetic field and can be distinguished in terms of the intersystem crossing mechanisms discussed previously. The fourth class of effect is attributable entirely to the local magnetic fields which are characteristic of magnetic nuclei.

## Class I effects

Class I effects are defined as those in which $J \approx 0$ and for which intersystem crossing occurs predominantly by the hyperfine mechanisms. The effects are characterized by a decrease in intersystem crossing as a function of field (Fig. 8), a result of the Zeeman splitting of the triplet sublevels. Since the $T_+$ and $T_-$ states become non-degenerate with the S state, $T_\pm \rightleftharpoons S$ intersystem crossing is prevented and thus the overall rate of intersystem crossing is slowed down. The maximum effect is observed when the external field exceeds the magnitude of the hyperfine couplings. In Fig. 8 this field is shown as $H_I$ and is typically $c$. 1000 G.

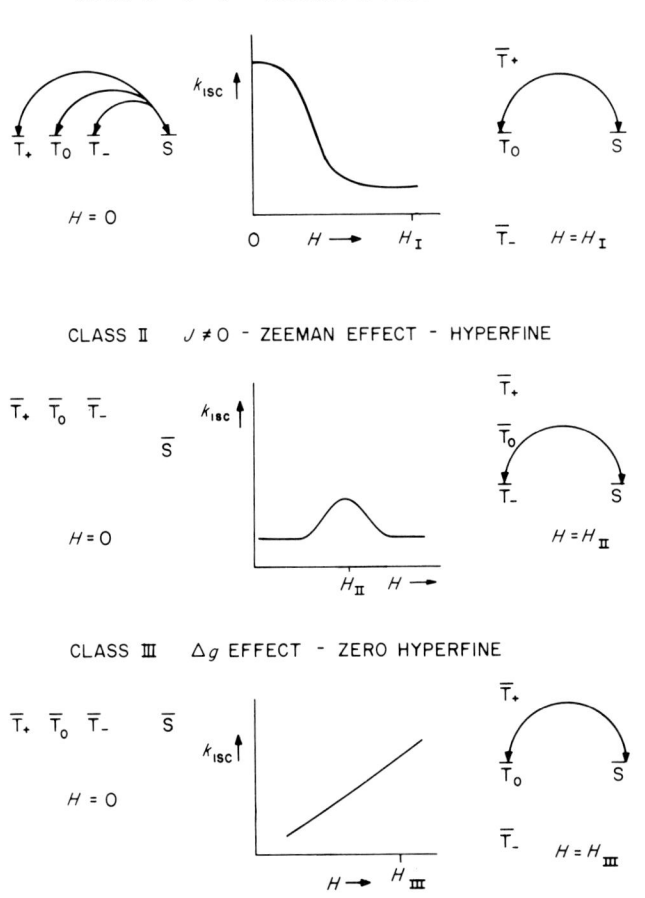

FIG. 8 Diagram showing the three types of behaviour of $k_{ISC}$ as a function of applied field which allows classification of the external magnetic field effects

*Class II effects*

Class II effects are defined as those in which $J \neq 0$ and for which intersystem crossing is hyperfine induced. These effects are characterized by an increase in intersystem crossing around an external field strength at which the Zeeman splitting is approximately equal to $J$ ($H_{II}$ in Fig. 8). At this field the Zeeman splitting of the triplet sublevels causes a degeneracy between one triplet state and the singlet state. Intersystem crossing can thus occur at this external field strength. Above or below this value of the applied field, the appropriate degeneracy does not exist, and thus intersystem crossing is inefficient.

*Class III effects*

Class III effects are observed for reactions in which the $\Delta g$-mechanism is responsible for intersystem crossing. These effects are characterized by an increase in intersystem crossing rate as a function of applied field, since the rate of precession of an electron spin vector about an applied field is directly proportional to the strength of that field. These effects are commonly only observed at very high external magnetic fields (i.e. $H_{III}$ in Fig. 8 > $10^5$G).

*Class IV effects*

Class IV effects are those for which $J \approx 0$, $\Delta gH \approx 0$, and $a \neq 0$. This class differs from the previous three because the magnetic field which dominates the effect is due to the magnetic moments of nuclei; hence, class IV effects are termed *magnetic isotope effects*. These effects allow a differentiation in the behaviour of radical pairs containing magnetic nuclei from radical pairs containing nonmagnetic nuclei or, more generally, allows a differentiation in the behaviour of radical pairs possessing nuclei of differing magnetic moments.

## 2 Experimental examples

CLASS I EFFECTS: REACTIONS INFLUENCED BY THE ZEEMAN EFFECT

Experimental examples of class I effects are the most common and best documented of all known magnetic field effects on organic chemical reactions. They involve both initially singlet and triplet radical pairs.

One of the first magnetic field effects was reported by Sagdeev *et al.* (1973a,b) who noted a substantial magnetic field effect on the product ratio from the reaction of n-butyllithium with benzyl chloride and with pentafluorobenzyl chloride ($C_6F_5CH_2Cl$). The reaction is summarized in Scheme 1 (Molin *et al.*, 1979). n-Butyllithium was decomposed in the presence of pentafluorobenzyl chloride in boiling hexane. The reaction was performed both in the earth's field ($\sim 1$ G) and at 18 kG. The reaction products were analysed

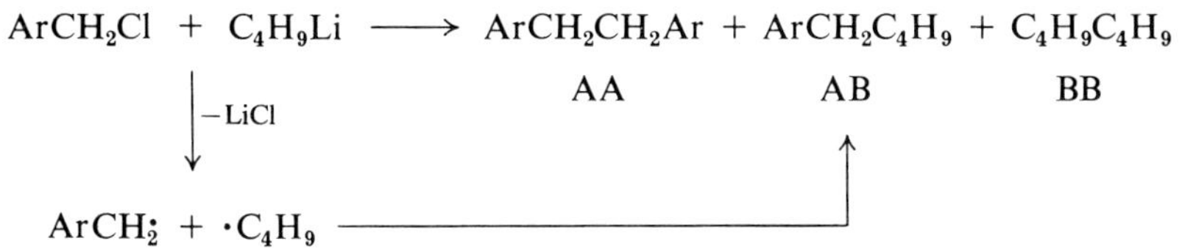

$$\text{ArCH}_2\text{Cl} + \text{C}_4\text{H}_9\text{Li} \longrightarrow \text{ArCH}_2\text{CH}_2\text{Ar} + \text{ArCH}_2\text{C}_4\text{H}_9 + \text{C}_4\text{H}_9\text{C}_4\text{H}_9$$

$$\downarrow -\text{LiCl} \qquad\qquad\qquad \text{AA} \qquad\qquad \text{AB} \qquad\qquad \text{BB}$$

$$\text{ArCH}_2^{\cdot} + \cdot\text{C}_4\text{H}_9 \longrightarrow\uparrow$$

**Scheme 1**

by $^{19}$F nmr spectroscopy. It was found that the amount of the product AB (Scheme 1) relative to product AA, increased by 30% when the reaction was performed in the presence of the applied field. Similarly, an increase in this ratio of 20% was found for the reaction of n-butyllithium and benzyl chloride in the presence of a 25 kG field. Subsequently, more detailed studies (Molin et al., 1979) revealed that the effect was apparent at very low fields and that the ratio reached a constant value above an applied field of about 200 G. It is postulated that both reactions involve the formation of a singlet geminate radical pair (A $\cdot\cdot$ B, benzyl $\cdot\cdot$ butyl) in a primary solvent cage. A competition then is created between cage reaction (recombination to form product AB) and diffusive separation to give free radicals. However, intersystem crossing within the solvent cage, to produce a triplet radical pair, will result in less efficient cage reaction and hence less efficient formation of AB. It is assumed that the free radicals A and B will produce AA, AB and BB in statistical yields (25%, 50%, and 25%, respectively). The application of a magnetic field inhibits the occurrence of $S \rightarrow T_{\pm}$ intersystem crossing. Hence the radical pair retains its singlet character longer and thus more efficient cage reaction results. The observation of the field effects in as weak a field as 200 G is consistent with the Zeeman mechanism as the origin of the magnetic field effect. Since the origin of the magnetic field effect lies in the competition between intersystem crossing and diffusive separation, a change in magnetic field effect should be observed if the rate of diffusion is changed and if the rate of intersystem crossing is held constant. Sagdeev tested this prediction by performing experiments in solvents of differing viscosity. It was found that an increase in the magnetic field effect of 15% could be obtained by increasing the solvent viscosity by 0.6 cP.

The work of Turro et al. (Turro, 1981, 1983; Turro and Kraeutler, 1980) on the photochemistry of aromatic ketones provides many examples of class I magnetic field effects. Photolysis of dibenzyl ketone (DBK) and substituted DBK's has been shown to proceed by the mechanism outlined in Scheme 2 (Engel, 1970). Photolysis initially yields a phenacyl radical and a benzyl radical and leads, after decarbonylation, to two benzyl radicals. Thus, potentially, these systems can exhibit magnetic effects from two radical pairs. Photolysis of an unsymmetrical DBK can lead to the formation of three diphenylethane

## Scheme 2

$$\text{ArCH}_2\text{COCH}_2\text{Ar} \xrightarrow{h\nu} S_1 \longrightarrow T_1 \longrightarrow \text{ArCH}_2\text{CO}\cdot + \cdot\text{CH}_2\text{Ar}$$

Primary radicals

$$\text{ArCH}_2\text{CH}_2\text{Ar} \longleftarrow \text{ArCH}_2^{\cdot} + \text{ArCH}_2^{\cdot} + \text{CO}$$

Secondary radicals

**Scheme 2**

(DPE) products according to (15). The ratio of the products AA : AB : BB

$$\text{PhCH}_2\text{COCH}_2\text{Ar} \rightarrow \underset{\text{AA}}{\text{PhCH}_2\text{CH}_2\text{Ph}} + \underset{\text{AB}}{\text{PhCH}_2\text{CH}_2\text{Ar}} + \underset{\text{BB}}{\text{ArCH}_2\text{CH}_2\text{Ar}} \quad (15)$$

will be 25 : 50 : 25 if there is no cage effect, and 0 : 100 : 0 if cage recombination is 100% efficient. A numerical value for the cage effect for intermediate situations is defined by (16). Experimentally, it is found that the cage effect in homogeneous solution is 0% for photolysis of $p$-xylyl benzyl ketone

$$\% \text{ cage products} = \frac{\text{AB} - (\text{AA} + \text{BB})}{\text{AA} + \text{AB} + \text{BB}} \times 100 \quad (16)$$

(Turro and Cherry, 1978). In this case, because the initial radical pair is in a triplet state (Scheme 2), the formation of cage products requires an intersystem crossing which cannot compete with diffusive separation. Hence a cage effect of zero is observed. Evidently, the observation of cage effects, and thus magnetic field effects, for initially triplet radical pairs requires an environment which allows intersystem crossing to compete with diffusive separation. Certain solution phase systems possess a micro-heterogeneous environment which may severely restrict the diffusional mobility of an organic solute. Specifically, aqueous solutions of ionic detergents, which commonly have a hydrophobic hydrocarbon "tail" and a hydrophilic ionic "head", aggregate above a certain concentration (critical micelle concentration, cmc) to form micelles (Fendler and Fendler, 1975). In Fig. 9 is shown a schematic model of a micelle. Micelles of the most commonly used detergents typically have maximal cross-sections of 20–30 Å, which correspond to $\sim$50–100 detergent molecules. The hydrophobic core provides a volume of hydrophobic space (in an otherwise aqueous environment) that is capable of solubilizing an organic substrate. In effect, micelles can provide a cage environment for organic molecules. A geminate radical pair which is generated by homolytic cleavage of a molecule which is solubilized in a micelle will remain geminate until one or both radicals escape into the bulk aqueous phase. For non-viscous organic solvents at ambient temperature, the residence time of a primary geminate pair in a solvent cage is $\sim 10^{-10}$ to $10^{-11}$s.

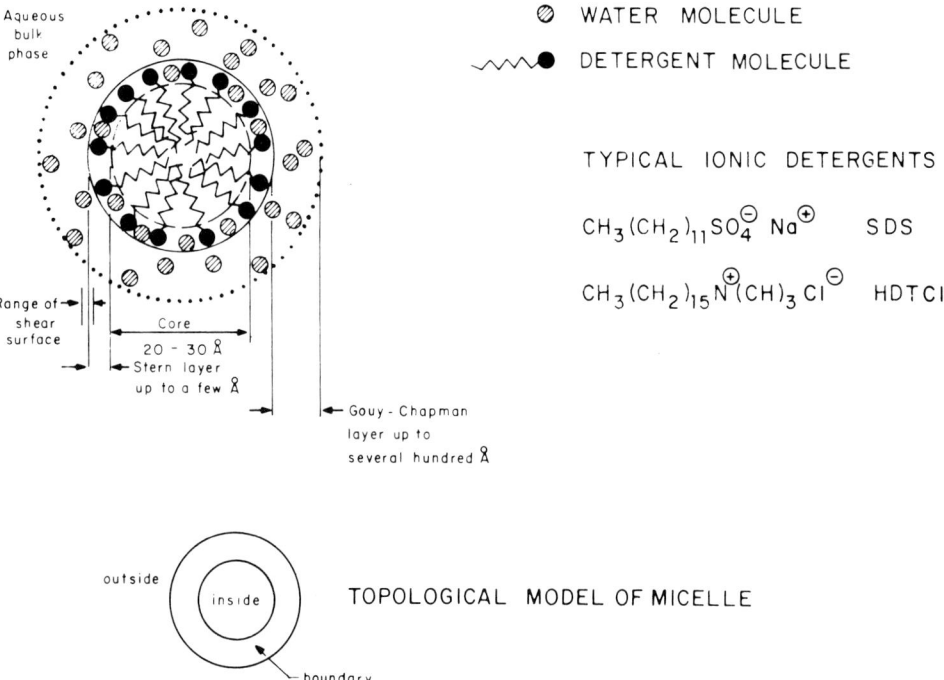

FIG. 9 Schematic model of micelle aggregates formed by addition of hexadecyltrimethylammonium chloride (HDTCl) or sodium dodecylsulphate (SDS) to water

In contrast, for radicals possessing six or more carbons, the residence time of a pair in a micelle cage is $> 10^{-6}$s. Additionally, the "volume" of a solvent cage is roughly the size of the solvated radical pair, whereas the volume of the micelle cage is such that the component radicals of a pair are allowed to separate by distances up to tens of Å. Thus, micellar systems can potentially allow an initially triplet geminate radical pair to separate so that $J$ becomes negligible, but provide a reflecting boundary, at the Stern layer (Fig. 10), to the geminate character of the pair to allow intersystem crossing to compete with diffusive separation. (For reviews on the effects of micelles on reactions of radical pairs, see Turro et al., 1980b; Turro, 1981, 1983.) Experimentally, it is found that photolysis of p-xylyl benzyl ketone in HDTCl micellar solution yields a non-statistical distribution of diphenylethanes. In this case, using (16), a cage value of $\sim 30\%$ may be computed (Turro et al., 1980a).

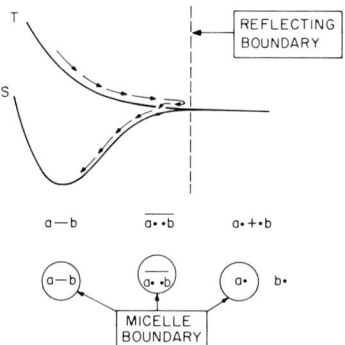

FIG. 10 Diagram showing the effect of a micelle in maintaining the lifetime of an initially triplet radical pair so as to allow more efficient intersystem crossing

Evidently, the micelle environment allows the initially triplet benzyl p-xylyl radical pair to develop sufficient singlet character to allow cage reaction to compete with irreversible diffusive separation. The use of micelles allows an alternative measure of cage reaction which may be used to probe the reactivity of symmetrical ketones. Addition of a sufficiently high concentration of a water soluble selective radical scavenger (e.g. cupric ions) allows the determination of the fraction of cage reaction according to (17), where $[AB]_{US}$ represents the amount of unscavengable coupling product, and $[ACOB]_R$ is

$$\text{Cage} = \frac{[AB]_{US}}{[ACOB]_R} \quad (17)$$

the amount of reacted ketone (Turro et al., 1980a). Only radicals which give coupling products within the micelle cage are not scavenged, and thus the amount of cage reaction is determined directly. Both methods of cage determination have been shown to yield experimentally identical results.

Upon application of an external magnetic field, a change in the % cage products is observed. A plot of the percentage of cage reaction for the micellar photolysis of DBK vs magnetic field is reproduced in Fig. 11. It can be seen that the cage effect drops quickly from 30% at zero G to 20% as fields of a few hundred gauss are imposed upon the sample, and then remains constant from 500 G to 5000 G. It has been proposed that the magnetic field effect on the percentage of cage reaction reflects the behaviour of the secondary (benzyl benzyl) radical pair. The quantum yield for disappearance of DBK (which is equivalent to the quantum yield for DPE production) in homogeneous organic solvent is $\sim 0.7$; however, this value drops to $\sim 0.3$ in

HDTCl micellar solution (Turro et al., 1979b). This decrease must be attributable in part to a more efficient recombination of the primary (benzyl phenacyl) radical pair. In effect, the cage reaction for the primary pair is increased as a result of micellization for the same reason as for the secondary pair, i.e. the radical pair lifetime is increased so that intersystem crossing can compete with the process removing the radical pair. One important difference for the case of this radical pair is that the step which removes the radical pair is loss of carbon monoxide within the micelle, and not exit of one partner from the micellar cage environment. However, this step is magnetic field independent and thus the requisite competition for magnetic field exists. As expected, it is found that the quantum yield for DPE production is magnetic field dependent; the results are illustrated in Fig. 11

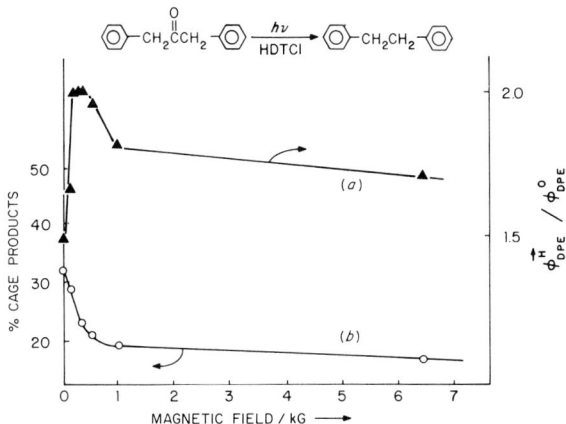

FIG. 11 Plot showing the effect of an external magnetic field of (a) the quantum yield for reaction relative to a standard ketone, and (b) the cage value for DPE production for photolysis of dibenzyl ketone in HDTCl micelles

(Chung, 1982), where the quantum yield relative to a standard DBK at 0 G. is plotted as a function of field. Remarkably for this radical pair an initial increase in relative quantum yield to a maximum at ~200 G is observed, followed by a decrease to an overall higher value. These effects on the two radical pairs are explicable by reference to Fig. 12. For the case of the secondary radical pair ($^3$RP′, Fig. 12), the competition between $k_{ex}$ and $k_{TS}′$ determines the cage value. Since $k_{ex}$ is independent of magnetic field and $k_{TS}′$ is slowed by an external field due to Zeeman splitting of $T_\pm$, a lower cage effect results. For the case of the primary radical pair ($^3$RP, Fig. 12), the competition is between $k_{-CO}$ and $k_{TS}$. Thus, as $k_{TS}′$ is slowed by the external field, recombination to generate the starting material becomes less efficient

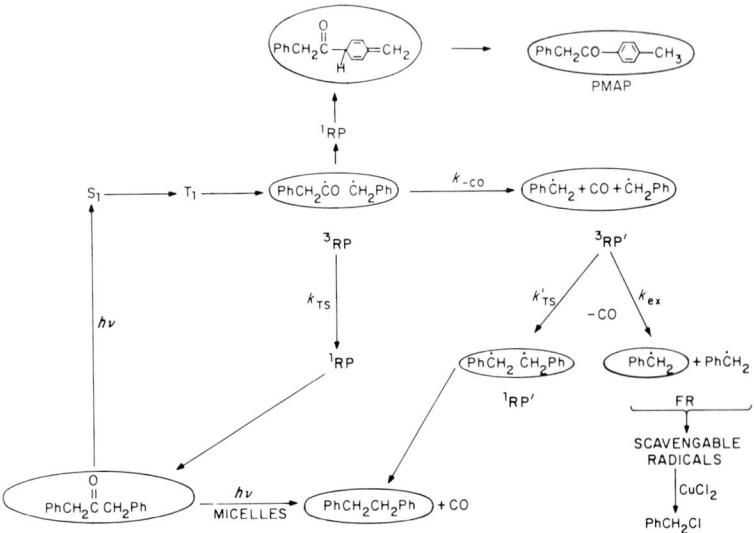

FIG. 12 Reaction scheme for discussion of magnetic field effects on the photochemistry of DBK in micellar solution. The encircled species represent micellized molecules or radicals

and thus, the quantum yield for reaction increases. A magnetic field dependence of the quantum yield for formation of a minor product (PMAP in Fig. 12), an isomer of DBK, is also observed.

An alternative probe for the reactions of the primary radical pairs formed from the photolysis of dibenzyl ketones is 2,4-diphenylpentanone (DPP) (Baretz and Turro, 1983). Irradiation of stereochemically pure ($\pm$) or *meso*-DPP in pentane leads to the products shown in (18) but no observable isomerization in the starting ketone. Irradiation of stereochemically pure DPP in HDTCl micelles results in a different product distribution. In this case the products are isomerized starting ketone (8%), 3% phenylpropionaldehyde [4] is formed, and diphenylbutane [3] formation is reduced to 72%. Thus, the micellar medium again provides the environmental constraint necessary for cage reaction. Photolysis of DPP in micellar media also results in a decrease in quantum yield for reaction of DPP (0.62, 0.50, and 0.33 in pentane, HDTCl and SDS micelles, respectively). Application of an external magnetic field results in a decrease in the isomerization yield for systems in which cage reaction is significant. Isomerization efficiency is determined from plots of

% isomerization *vs* % conversion. The slopes of such plots for DPP under various reaction conditions are reproduced in Table 1. Evidently, in the presence of a 3 kG magnetic field, intersystem crossing in the primary pair is reduced so that the isomerization efficiency is lowered. Interestingly, it has

TABLE 1

Isomerization efficiencies as a function of magnetic field for photolysis of 2,4-diphenylpentanone[a]

| Environment | Field/kG | Efficiency | $\Phi r$ |
|---|---|---|---|
| Pentane | 0 | 0.00 | 0.62 |
| SDS | 0 | 0.15 | 0.33 |
| SDS | 3 | 0.10 | — |
| HDTCl | 0 | 0.11 | 0.50 |
| HDTCl | 3 | 0.07 | — |
| Porous glass | 0 | 0.17 | — |
| Porous glass | 3 | 0.12 | — |

[a]Baretz and Turro, 1983

been shown that micelles are not a unique environment for the observation of these effects. For example, photolysis of DPP in porous glass reveals both cage and magnetic field effects (Baretz and Turro, 1983).

Photolysis of 1,3-diphenylbutan-2-one (DPB) yields, after decarbonylation, three coupling products (AA, AB, BB) which are a direct measure of the extent of cage reaction in the secondary radical pair (19) (Baretz and Turro, 1983). Analysis of the cage effect in a variety of "supercage" environments enables determination of magnetic field effects as a function of these environments (Table 2). The magnetic effects are attributable to the competition

$$\phi\text{-}\underset{O}{\overset{}{\diagdown}}\text{-}\phi \xrightarrow[-CO]{h\nu} \underset{\phi\ \ \phi}{\diagdown\diagup} + \underset{\phi\ \ \phi}{\diagdown\diagup} + \diagdown\diagup \qquad (19)$$

DPB      AA      AB      BB

between intersystem crossing and diffusive separation of the radical pairs in the different environments. Large magnetic effects on the secondary radical pair reactions are possible in certain environments (for example, in porous glass, the magnetic effect was 61%).

Another example for which triplet radical pair reactivity may be modified by the use of micellar environments and a magnetic field is provided by the photolysis of 1,2-diphenyl-2-methylpropanone (DPMP) (Turro and Mattay, 1981). In homogeneous solution the main products are styrene and dicumyl; benzil and benzaldehyde are minor products (Scheme 3). In HDTCl solution

TABLE 2

Magnetic field effects on cage reactions of benzyl: sec-phenethyl radical pairs in micelles and on silicate surfaces[a]

| Environment | % cage products[b] 0 kG | 3 kG | Magnetic field effect (%) |
|---|---|---|---|
| HDTCl | 71 | 52 | −27 |
| SDS | 74 | 50 | −32 |
| Porous glass | 38 | 15 | −61 |
| TLC[c] | 9 | 6 | −32 |
| RPTLC[d] | 15 | 11 | −26 |

[a] Baretz and Turro, 1983
[b] Error limits ± 10%
[c] Photolysis performed on a silica TLC plate in an evacuated cell
[d] Photolysis performed on a reverse phase TLC plate in an evacuated cell

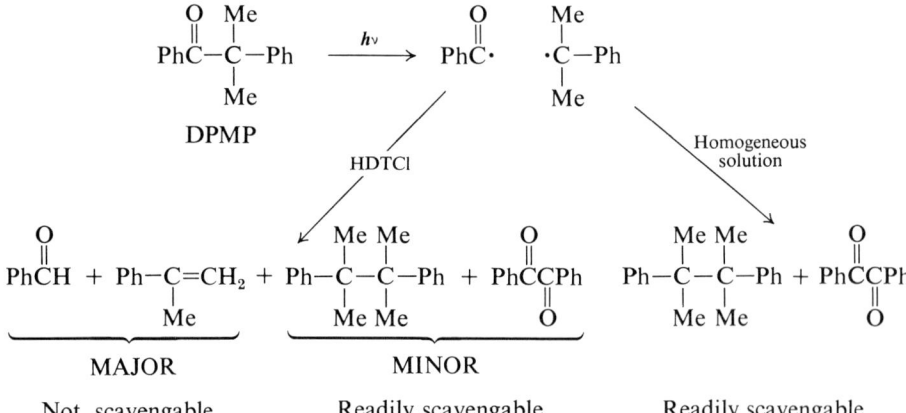

Scheme 3

the major products are styrene and benzaldehyde. As in the case of DBK, the cage effect for photolysis can be evaluated from Cu(II) quenching experiments. In fact, the addition of Cu(II) does not affect the production of styrene and benzaldehyde, and thus the yield of these disproportionation compounds gives the cage effect. The yield of cage products drops from 30% to a value of 20% in a field of 1000 gauss. In this case the initially triplet (benzoyl cumyl) radical pair may undergo intersystem crossing to yield the disproportionation products benzaldehyde and styrene, or escape to yield the coupling products dicumyl and benzil. In the presence of a magnetic field, intersystem crossing from $T_\pm$ is inhibited, and hence the cage effect is lowered.

Sagdeev (Leshina et al., 1980) has reported that the pyrene sensitized *trans-cis* isomerization of stilbene is magnetic field dependent. Irradiation of pyrene in homogeneous solution in the presence of *trans*-stilbene leads to isomerization of the stilbene. The amount of isomerization, measured by nmr methods, was found to decrease by 30% when the reaction was performed in magnetic fields of greater than 100 G. The effect may be explained by reference to Scheme 4. Electron transfer ($k_e$) from the excited state of

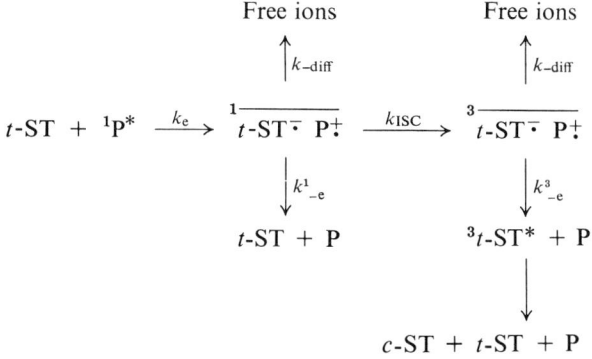

**Scheme 4**

pyrene (P) to *trans*-stilbene (*t*-ST) leads to a singlet radical ion pair. This ion pair may transfer the electron back to form the starting materials ($k'_{-e}$), undergo diffusive separation ($k_{-\text{diff}}$), or undergo intersystem crossing within the solvent cage to form a triplet radical ion pair ($k_{\text{ISC}}$). Back electron transfer within this species leads to the triplet state of stilbene which is the active *trans-cis* isomerization intermediate (Turro, 1978b). Application of a magnetic field will slow down the rate of intersystem crossing within the radical ion pair leading to a decrease in production of triplet-stilbene and hence of isomerization.

One of the first examples of a magnetic field effect on a chemical reaction was reported by Gupta and Hammond (1972). It was found that for photosensitized isomerization of piperylene and stilbene, the application of an external magnetic field changes both the initial quantum yield for the reaction, and the composition of the photostationary states. In addition, the external field effect was dependent upon the nature of the sensitizer. No firm conclusions as to the origin of the effect were drawn; however, in a later article, an explanation was proposed by Atkins (1973). It was suggested that an exciplex between the excited state of the sensitizer and the olefin should be regarded as having the nature of a radical ion pair. Such a radical pair might possess a significant value of $\Delta g$, and hence a magnetic field induced intersystem crossing might occur in a manner analogous to that characteristic of class III reactions (*vide infra*). The suggestion was supported by a quantitative analysis for an assumed value of $\Delta g$; however, as the nature of the exciplex is in fact unknown, the explanation must be regarded as speculative.

Another related example of class I reaction was reported by Turro *et al.* (1980c) who observed an external magnetic field effect upon an emulsion polymerization. Emulsion polymerization (EP) requires an aqueous dispersion medium, dispersed droplets of monomer, a micelle-generating detergent, and an initiator. Dibenzyl ketone (DBK) was used as a photochemical radical initiator for the polymerization of styrene. The DBK is solubilized in the micelle together with monomer styrene molecules from the dispersed droplets; hence polymerization is initiated within the micelle. Experiments were conducted in which a magnetic field was applied at various times after the start of the polymerization, and it was shown that the field effect lay exclusively in the initial stage, i.e. when micelles existed. It was found that the average molecular weight of polymerized styrene could be increased by a factor of 5 if the reaction was performed in the presence of an applied field of > 1000 G (Fig. 13). The effect may be explained with the aid of the highly schematic representation shown in Scheme 5, where the circles represent micelles, and M monomer styrene molecules.

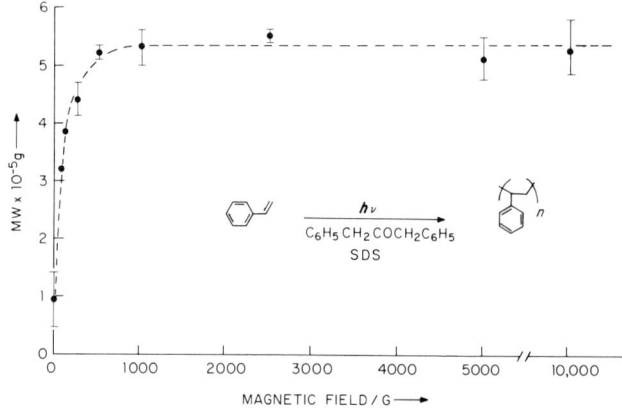

FIG. 13 Plot showing the effect of an applied magnetic field on the average molecular weight of polystyrene, produced in emulsion polymerization of styrene using DBK as the photochemical initiator

The molecular weight of the polystyrene depends upon the rates of termination reactions, which in turn depend upon the availability of terminating free radicals. Termination by one partner in the initial radical pair requires an intersystem crossing. This intersystem crossing is in competition with radical escape into the water phase, and thus the application of an external

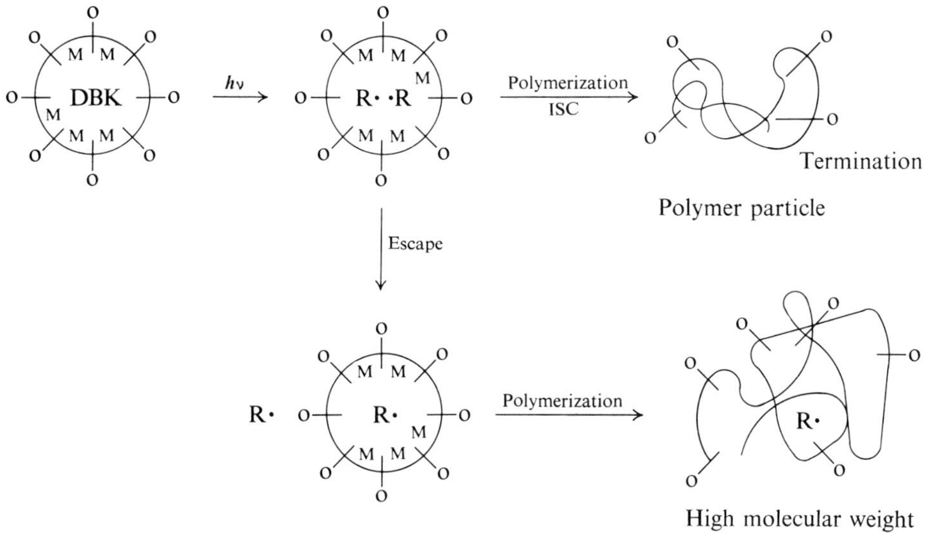

**Scheme 5**

field allows more radicals to escape. Since radicals in the water are less efficient at terminating the polymerization process which initially proceeds in the micelle, higher molecular weights of polymer are achieved in the presence of an external field.

CLASS II EFFECTS: REACTIONS FOR WHICH THE SINGLET-TRIPLET SPLITTING IS NON-ZERO

Examples of class II reactions are less common than those of class I. The requirement of a substantial non-zero $J$ demands that the radical sites remain a few Å from each other, a situation not encountered for radical pairs in homogeneous or micellar solution. Many short chain biradicals possess this property (Turro and Kraeutler, 1982), but as yet no examples of hyperfine induced magnetic field effects on the chemical reactions of biradicals have been reported.

Photolysis of quinoline N-oxides in alcoholic solvents leads to two products: (*i*) rearrangement to the lactam *via* a radical ion pair formed between the quinoline and alcohol, and (*ii*) isomerization to an oxazepin (Bellamy and Streith, 1976). It has been shown that the ion pair is formed between the quinoline and an alcohol molecule which is hydrogen bonded to the N-oxide oxygen atom (Hata, 1976, 1978). The formation of the oxazepin proceeds *via* an oxaziridine intermediate, and is magnetic field independent. However, the yield of the lactam was found to be magnetic field dependent

(Hata et al., 1979). It was found that in ethanol, at fields greater than 6 kG, the yield of lactam dropped from 70% to 55% at 10 kG. At higher fields, the yield increased and at 15 kG had returned to 70% (Fig. 14). Hydrogen bonding in the radical ion pair results in significant exchange overlap, which means that intersystem crossing is energetically unfavourable. For this radical pair it is claimed that the singlet state is higher in energy than the triplet state. Thus, in the presence of a sufficiently high magnetic field, the $T_+$ sublevel is raised in energy, due to the Zeeman interaction, so that at a certain field, $S_0$ and $T_+$ become energetically degenerate. Thus, at this field, intersystem crossing between $S_0$ and $T_+$ is possible due to hyperfine interactions. In this case, resonant interaction is observed at 10 kG, which gives the value of $2J$ for this radical pair. Support for the interpretation came from the following consideration. It was argued that the distance between the radical pair centres and thus the extent of exchange overlap for quinoline N-oxide and a given set of alcohols will depend upon the $pK_a$-value of the alcohol. It was found that the minimum yield of lactam occurred at 7 kG for t-butyl alcohol and at $\sim 18$ kG for methanol. Thus, the magnetic field required for the resonance condition is dependent upon the distance between the radical centres.

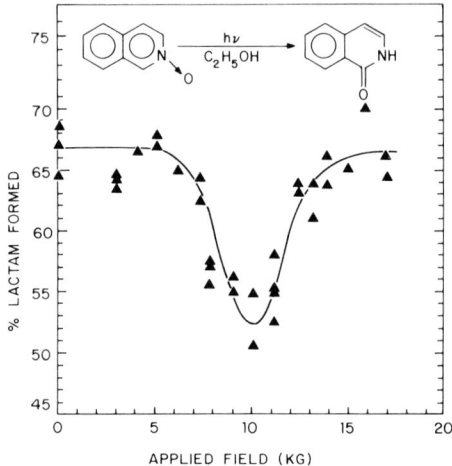

FIG. 14 Plot showing the effect of an external magnetic field on the yield of lactam formed during photolysis of quinoline N-oxide in ethanol. The maximum magnetic field effect occurs when the S-state of the intermediate radical ion pair is degenerate with the $T_+$-state

Similarly, the photosubstitution reaction (20) of 4-methylquinoline-2-carbo-nitrile is affected by an external magnetic field (Hata and Hokawa, 1981). The reaction is again assumed to proceed *via* an initially singlet radical

$$[5] \quad \xrightarrow{h\nu, \text{OH}} \quad [6] \tag{20}$$

ion pair which may (*i*) intersystem cross to give a triplet pair, followed by diffusive separation and ultimate regeneration of starting materials, or (*ii*) undergo a hydrogen-atom transfer from the alcohol to the quinoline nitrogen, followed by formation of the substitution product. In fact, a plot of the yield of the cage product [6] as a function of field reveals a minimum at a value of 10 kG. Again the interpretation is that at this field a resonance between S and $T_+$ occurs, so that hyperfine induced intersystem crossing is enhanced, and thus the yield of cage product decreases.

CLASS III EFFECTS: REACTIONS FOR WHICH THE $\Delta g$-MECHANISM FOR INTERSYSTEM CROSSING IS DOMINANT

The occurrence of $\Delta g$ magnetic field effects on chemical reactions requires either a large $\Delta g$ ($> 10^{-2}$ to $10^{-3}$) or a very large magnetic field ($> 10^5 G$); the former is not often encountered in carbon-based radicals, and the latter is difficult to achieve in the laboratory. Even so, magnetic field effects which have been attributed to the $\Delta g$-mechanism have been reported. They usually accompany Zeeman effects and are observed at higher field strengths.

The pyrene-sensitized photoisomerization of decafluorostilbene is influenced by an external magnetic field in a manner similar to that described earlier for *trans*-stilbene (Leshina *et al.*, 1980). At low fields, intersystem crossing from a singlet radical ion pair to a triplet pair is inhibited; this reduces the efficiency of formation of the free stilbene triplet, and hence lowers the probability of *trans* → *cis* conversion. However, at fields above 10 kG, the yield of *cis*-decafluorostilbene from *trans*-decafluorostilbene increases relative to that at fields in the order of 1 kG, an effect not observed for stilbene itself. The decafluorostilbene-pyrene radical ion pair has a reported $\Delta g$ of $2 \times 10^{-3}$, compared to that for stilbene-pyrene radical ion pair of $2 \times 10^{-4}$. Since the rate of $\Delta g$-promoted $S \to T_0$ intersystem crossing depends directly upon the value of $\Delta g$, a greater effect would be expected for *trans*-decafluorostilbene. Thus, at fields of 10 kG, the rate for the new mechanism of intersystem crossing for this stilbene becomes competitive with the rate of back electron transfer within the singlet ion pair. This effect results in more probable formation of *trans*-decafluorostilbene triplet and hence a higher yield of *cis*-decafluorostilbene.

TABLE 3

Effect of ultrahigh magnetic field on cage reactions of dibenzyl ketones photolyzed in HDTCl micelles[a]

| Radical pair | Field strength/G | % Cage products | $\Delta g$ |
| --- | --- | --- | --- |
| $p$-CH$_3$—PhCH$_2$ $\cdot\cdot$ CH$_2$Ph | 0 | 48 | $3 \times 10^{-5}$ |
| | 14,500 | 25 | |
| | 145,000 | 21 | |
| $p$-ClPhCH$_2$ $\cdot\cdot$ CH$_2$Ph | 0 | 52 | $5.5 \times 10^{-4}$ |
| | 14,500 | 31 | |
| | 145,000 | 37 | |
| $p$-BrPhCH$_2$ $\cdot\cdot$ CH$_2$Ph | 0 | 58 | $2.1 \times 10^{-3}$ |
| | 14,500 | 58 | |
| | 145,000 | 57 | |
| PhCH$_2$ $\cdot\cdot$ CH$_2$Ph | 0 | 32 | 0 |
| | 14,500 | 16 | |
| | 145,000 | 13 | |
| PhCO $\cdot\cdot$ C(CH$_3$)$_2$Ph | 0 | 30 | $2.2 \times 10^{-3}$ |
| | 10,000 | 19 | |
| | 145,000 | 23 | |

[a]Turro *et al.*, 1982a

The use of ultrahigh laboratory magnetic fields has enabled the observation of $\Delta g$ effects in the systems studied by Turro *et al.* (1982a). Dibenzyl ketones which were substituted in the 4-position of a benzene ring were photolyzed in HDTCl micelles and the cage effect on the diphenylethane products were determined as described earlier. The value for the cage effect was determined in the presence of magnetic fields of zero, 14,500, and 145,000 G. The results are summarized in Table 3. For methyl-substituted DBK which has a small $\Delta g$, the effect of the ultrahigh field is similar to that observed for the parent compound, i.e. no change is noted. However, for the case of the chloro-substituted radical, $\Delta g$ is relatively large and an increase in the percentage of cage reaction occurs upon going from high fields to ultrahigh fields. This effect is also observed for photolysis of 1,2-diphenyl-2-methylpropanone (Table 3), where an increase in the cage effect is noted at ultrahigh laboratory magnetic fields. These results were interpreted as evidence for a $\Delta g$-mechanism for intersystem crossing in these two radical pairs. Thus, although intersystem crossing, and hence formation of cage products, is inhibited at low fields due to the Zeeman effect, at high fields the $T_0 \rightarrow S$ crossing is enhanced directly as a result of the high applied field. Of interest are the results of experiments performed using 4-bromo DBK. In this case, the benzyl radical pair possesses the largest value of $\Delta g$, but the percentage of cage reaction is the same at low field and high field as that measured at zero field. These results were interpreted to mean that a new mechanism for intersystem crossing now dominates, i.e. spin-orbit coupling. This result is consistent with the observation that the $^1$H CIDNP spectrum of photoexcited 4-bromo-DBK is very weak compared to those from DBK and 4-chloro-DBK.

The singlet sensitized decomposition of dibenzoyl peroxide in toluene has been shown to be magnetic field dependent (Tanimoto et al., 1976; Sakaguchi et al., 1980). The reaction may be summarized simply as shown in Scheme 6. Chrysene was used as a singlet sensitizer for the reaction; hence the initial radical pair is a benzoxyl $\cdot\cdot$ phenyl pair in a singlet state. Phenyl benzoate is the cage product. Radicals which escape the solvent cage result in a variety of products, many of which involve the solvent toluene. The value of $\Delta g$ for the benzoxyl $\cdot\cdot$ phenyl radical pair is quite large ($\sim 0.01$), and hence a field effect is expected. Experimentally, it is found that the yield of phenyl benzoate (cage product) decreases by 8% in the presence of a magnetic field of up to 43 kG. A corresponding but small (2%) increase in the out of cage products is observed. An analysis of the data according to a theoretical model allows assignment of the mechanism of the effect to a $\Delta g$-phenomenon, although in the later paper (Sakaguchi et al., 1980) more complex effects are claimed.

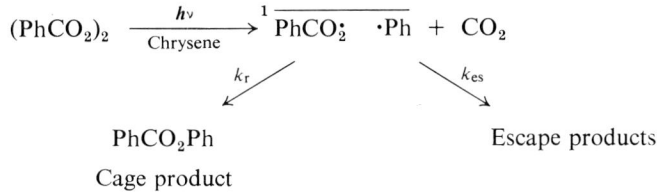

Scheme 6

Thermolysis of certain 1,4-endoperoxides of aromatic compounds produces molecular oxygen and the parent hydrocarbon quantitatively (Rigaudy and Basselier, 1971). From a study of the activation parameters it has been shown that these reactions proceed by two pathways: (i) a concerted mechanism in which singlet oxygen is formed quantitatively, and (ii) a diradical mechanism in which both excited singlet ($^1O_2$) and ground state triplet ($^3O_2$) oxygen are produced (Turro et al., 1979a). Since the spin states of a diradical are entirely analogous to those of a radical pair (Turro and Kraeutler, 1982), it is expected that only the diradical pathways will be influenced by a magnetic field. A simplified diradical mechanism for endoperoxide thermolysis is shown in Scheme 7 (Turro and Chow, 1979). Thermolysis is presumed to lead to an initially singlet diradical ($^1D$), and this species may either fragment to yield singlet oxygen ($k_r^1$), or undergo intersystem crossing ($k_{ISC}$) and fragment ($k_r^3$) to yield triplet oxygen. Since the $g$-values for peroxy radicals are large (typically $\sim 2.01$) (Berndt et al., 1977), these diradicals should have

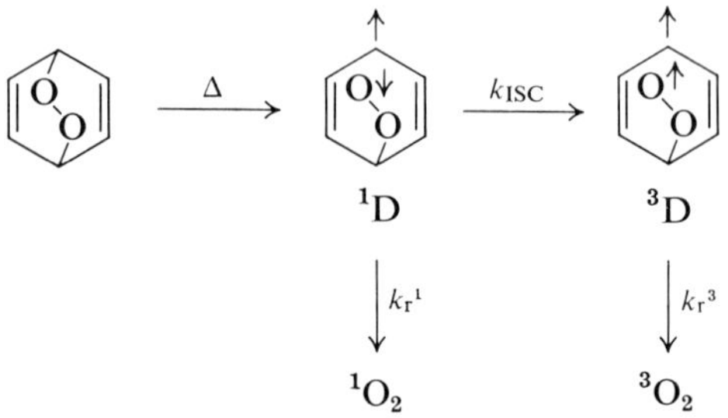

## Scheme 7

large values of $\Delta g$. Hence at sufficiently high magnetic fields, the $\Delta g$-effect will increase $k_{ISC}$ relative to $k_r^1$. A higher yield of $^3O_2$ and a lower yield of $^1O_2$ are therefore expected when endoperoxides, which decompose *via* diradicals, are thermolyzed in a magnetic field. Furthermore, no effect of magnetic field is expected if a concerted mechanism for decomposition is involved. In Fig. 15 are shown plots of $^1O_2$ yield as a function of magnetic field for thermolysis of two endoperoxides (Turro and Chow, 1979). Compound [7] has been shown to decompose by the diradical mechanism and compound [8] by the concerted mechanism. A decrease in $^1O_2$ yield is found for endoperoxide [7] in the range 9,000–15,000 G, whereas for [8] no change is observed. These results support the assignment of a $\Delta g$-mechanism for magnetic field induced intersystem crossing for these diradicals.

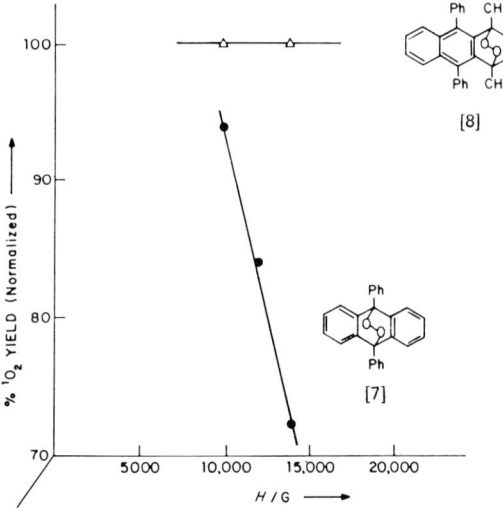

FIG. 15 Magnetic field effects on the yield of singlet oxygen formed upon thermolysis of a 9, 10- and a 1,4-endoperoxide

CLASS IV EFFECTS: MAGNETIC ISOTOPE EFFECTS

Although magnetic isotope effects may be observed in the presence of laboratory fields, for the sake of simplicity and clarity we shall consider first the situation in the absence of any applied field. As for magnetic field effects, essential requirements for the observation of significant magnetic isotope effects (Lawler and Evans, 1971; Buchachenko, 1976) are (*i*) a competition for a spin correlated radical pair between magnetic isotope-induced intersystem crossing (which leads to a capture pathway) and a spin dependent escape pathway, and (*ii*) a set of experimental conditions that allows the magnetic isotopes to exert a significant influence on the rate of intersystem crossing in the spin correlated radical pair. Thus, as for magnetic field effects, magnetic isotope effects are generally very small or negligible in homogeneous solution, but may become significant in microheterogeneous environments such as micelles.

TABLE 4

Some properties of common isotopes

| Nucleus | $\gamma^a$ | $\mu_N^b$ | I (spin) | Natural abundance (%) |
|---|---|---|---|---|
| $^1$H | 27,000 | 2.79 | 1/2 | 100 |
| $^2$H | 4100 | 0.86 | 1 | 0.016 |
| $^3$H | — | 2.98 | 1/2 | 0 |
| $^{12}$C | 0 | 0 | 0 | 99 |
| $^{13}$C | 7000 | 0.70 | 1/2 | 1.11 |
| $^{14}$C | 0 | 0 | 0 | 0 |
| $^{16}$O | 0 | 0 | 0 | 100 |
| $^{17}$O | −3600 | −1.9 | 5/2 | 0.04 |
| $^{18}$O | 0 | 0 | 0 | 0.20 |
| $^{117}$Sn | −9600 | −0.99 | 1/2 | 7.7 |
| $^{119}$Sn | −10,000 | −1.04 | 1/2 | 8.7 |

[a]Magnetogyric ratios in radians s$^{-1}$G$^{-1}$.   [b]Nuclear magnetic moments in erg G$^{-1}$

Table 4 lists the magnetic properties of elements that commonly occur in organic compounds. Consider first the isotopic triad $^1$H, $^2$H, and $^3$H. The magnetic spins and magnetic moments of $^1$H and $^3$H are identical, although their masses differ by a factor of three. Thus, there is no possibility of differentiating radical pairs possessing $^1$H and $^3$H based on their magnetic properties. On the other hand, $^1$H and $^2$H differ both in terms of their spin ($^1$H, spin = 1/2; $^2$H, spin = 1) and their magnetic moments (ratio of magnetic moments for $^1$H/$^2$H = 3). The magnetogyric ratio ($\gamma$), the ratio of the magnetic moment to the spin angular momentum (Carrington and McLauchlan, 1967), is the important quantity. The value of $\gamma$ for $^1$H is six times that of $^2$H. Thus, the hyperfine coupling of $^1$H is six times larger than that of $^2$H. The difference in the hyperfine couplings represents a mechanism for the differentiation of radical pairs possessing $^1$H or $^2$H. Consider next the $^{16}$O, $^{17}$O, and $^{18}$O triad. The even-numbered mass isotopes of oxygen possess no spin and are non-magnetic, while the odd-numbered mass isotope possesses a spin of 1/2 and is magnetic. Thus, it is again conceivable that radical pairs possessing $^{16}$O (or $^{18}$O) may be differentiated from those containing $^{17}$O by magnetic isotope effects. From this discussion it can be seen that the magnetic isotope effect may serve as a novel means of isotopic separation or enrichment, i.e. as a method which employs magnetic properties rather than mass properties as the basis for the separation (Buchachenko et al., 1981; Buchachenko, 1977).

As for magnetic field effects, magnetic isotope effects are expected to be maximal in closed volume reaction environments such as micelles and for initial triplet spin-correlated radical pairs (Turro and Kraeutler, 1980; Turro, 1983; Closs, 1971; Sterna et al., 1980; Tarasov et al., 1981).

The reactivity of a radical pair (RP) within a solvent cage ($\overline{RP}$) depends critically on the spin state of the pair; a singlet RP ($^1\overline{RP}$) within a solvent cage is very reactive toward combination and disproportionation reactions, but triplet RP's in a solvent cage ($^3\overline{RP}$) are completely unreactive toward such reactions. The options (Scheme 8) open to a $^3\overline{RP}$ in a chemically inert solvent are (a) chemical modification of one (or both) of the components of the pair to produce a structurally different triplet pair ($^3\overline{RP'}$), (b) relative diffusion of the pair out of the solvent cage, and (c) intersystem crossing to produce a reactive singlet RP. The general rule concerning the differing reactivities of $^1\overline{RP}$ and $^3\overline{RP}$ is based on the premise that the time of reaction of a caged singlet RP is too short for a change of electronic spin multiplicity.

(a) $^3\overline{RP} \rightarrow {}^3\overline{RP'}$  (b) $^3\overline{RP} \rightarrow a\cdot + \cdot b$  (c) $^3\overline{RP} \rightarrow {}^1\overline{RP}$

(d) $^1\overline{RP} \rightarrow$ combination and/or disproportionation

### Scheme 8

In Scheme 8, path (b) provides an *escape* pathway for pairs that possess non-magnetic spins, and path (c) provides a *capture* for the pairs that possess magnetic spins. A strategy for controlling the reactions of RP's with nuclear spins can now be developed. If the nuclear spins communicate with the electron spin via nuclear-electronic hyperfine-induced intersystem crossing of a RP, then the reactivity of the RP depends on the nuclear-electronic hyperfine coupling. As a general example, consider (Fig. 16) two RP precursors, a*—b and a—b, which differ only in isotope composition (* refers to the occurrence of magnetic nuclei in a fragment of the RP). Suppose

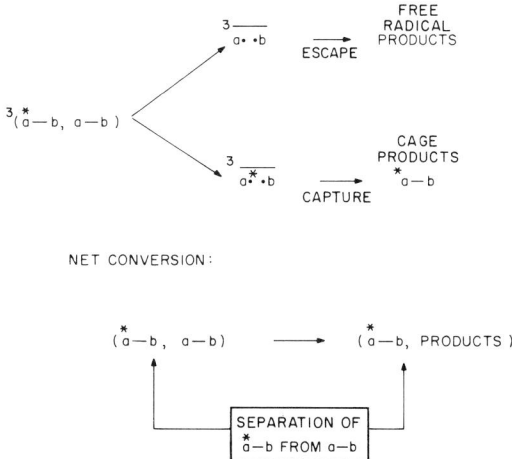

FIG. 16 Diagram which illustrates the concept of the magnetic isotope effect on chemical reactions. Triplet radical pairs which contain a magnetic nucleus (* in the diagram) are more likely to undergo recombination (capture) reactions. Triplet radical pairs which do not contain a magnetic nucleus are more likely to undergo (escape) product forming reactions

now that both precursors are cleaved into a caged triplet RP, $^3\overline{\text{a}^* \cdot\cdot \text{b}}$ and $^3\overline{\text{a} \cdot\cdot \text{b}}$, respectively, Of the two, $^3\overline{\text{a}^* \cdot\cdot \text{b}}$ will convert more rapidly into singlet $^1\overline{\text{a}^* \cdot\cdot \text{b}}$ *because the magnetic nuclei in* a* *provide a mechanism for intersystem crossing which is completely absent in* $^3\overline{\text{a} \cdot\cdot \text{b}}$. If a singlet RP undergoes a cage recombination reaction, then a* — b will be reformed faster than a — b. This means that *as the reaction proceeds, the system will become enriched in the reagent containing magnetic nuclei. In principle, such a system could lead to complete separation of the isotopic materials of* a* — b *and* a — b, *if the escape process occurred for all pairs that did not contain magnetic spins and if the capture process occurred for all pairs containing magnetic spins.*

We may now consider a simple hydrodynamic model which schematically displays the essence of the capture/escape concept (Fig. 17). Suppose a fluid may be pumped into two separate receptacles that possess a switch which prevents the fluids in the two containers from mixing. Both containers possess holes which allow the fluids to be removed. Consider now the analogy of fluid pressure and chemical potential. The fluid on the left is analogous to singlet states producing singlet radical pairs and the fluid on the right is analogous to triplet states producing triplet radical pairs. The switch is analogous to the magnetic effects which allow mixing of the singlet and triplet radical pairs. If one starts from the triplet, and if hyperfine interactions allow

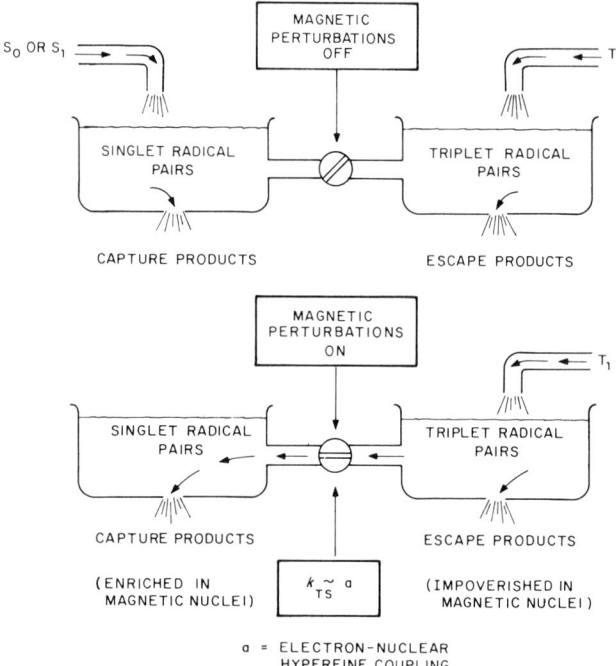

FIG. 17 Schematic illustration of the effect of the perturbations induced by magnetic nuclei on the relative amounts of capture and escape reactions for radical pairs. The number of capture reactions for triplet radical pairs with a magnetic nucleus increases as a consequence of the magnetic perturbations

triplet and singlet to mix, the "flow" of triplet radical pairs into the singlet container is apparent. The rates of mixing, capture, and escape all contrive to determine the efficiency of the processes.

For a high efficiency of isotopic separation, two important criteria must be met: (*i*) the hyperfine interactions in $^3\overline{a \stackrel{*}{.}. b}$ must be large enough to allow for the possibility that intersystem crossing to $^1\overline{a \stackrel{*}{.}. b}$ can be determined mainly by nuclear electron hyperfine interactions, and (*ii*) the formation of free radical products (chemically or by modification of the RP) from $^3\overline{a .. b}$ must occur more efficiently than intersystem crossing to $^1\overline{a .. b}$. In turn, in order for these criteria to be met, certain space/time conditions must exist: (*i*) the caged RP $^3\overline{a \stackrel{*}{.}. b}$ must be able to use the hyperfine interactions fully by being able to have the radical fragments separate in space to a distance such that the singlet-triplet splitting is of the order of or smaller than the hyperfine interaction, and (*ii*) the time of separation must be appropriate for maximal intersystem crossing induced by hyperfine interactions.

Although these conditions are stringent, numerous systems have been discovered for which the efficiency of isotopic separation by the magnetic isotope effect far exceeds that for separation by the mass isotope effect. In the following sections examples are given of magnetic isotope effects involving carbon, hydrogen, oxygen, and tin.

*Magnetic isotope effects involving carbon nuclei*

As a specific example of magnetic isotope effects, let us focus attention on the combination products from photolysis of DBK and PMAP (Fig. 12). These products are formed from the triplet correlated radical pairs ($^3$RP) after intersystem crossing to the singlet correlated radical pair ($^1$RP). If $^{13}$C nuclei in $^3$RP assist in the formation of $^1$RP, then the rate and the efficiency of the $^3$RP $\rightarrow$ $^1$RP process will depend on the occurrence (or non-occurrence) of $^{13}$C nuclei in the radical fragments. Suppose that a given $^3$RP possess a $^{13}$C nucleus which is coupled by the hyperfine interaction to one of the odd electrons. This $^3$RP will enjoy a faster intersystem crossing rate than a $^3$RP that contains only $^{12}$C nuclei, i.e. the $^{13}$C nuclei can promote the $^3$RP $\rightarrow$ $^1$RP process *via* both the spin rephasing ($T_0 \rightarrow S$, Fig. 6a) and spin flipping ($T_+ \rightarrow S$ or $T_- \rightarrow S$, Fig 6b) mechanisms. The recombination process provides a "capture" pathway for $^3$RP containing $^{13}$C nuclei and loss of carbon monoxide provides an "escape" pathway for $^3$RP containing $^{12}$C nuclei (Scheme 9).

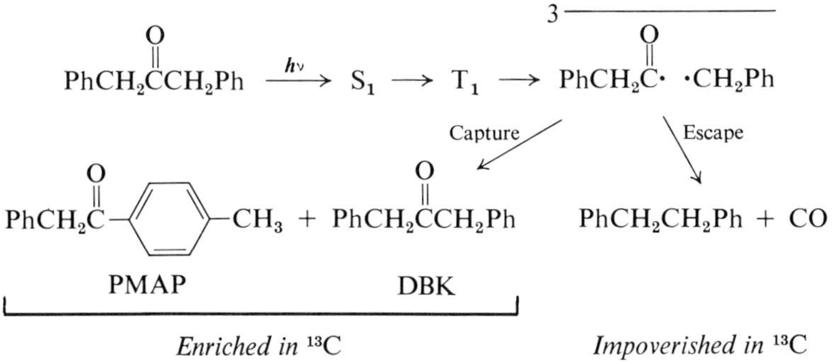

$$\text{PhCH}_2\overset{\overset{\text{O}}{\|}}{\text{C}}\text{CH}_2\text{Ph} \xrightarrow{h\nu} S_1 \rightarrow T_1 \xrightarrow{\quad 3 \quad} \text{PhCH}_2\overset{\overset{\text{O}}{\|}}{\text{C}}\cdot \;\; \cdot\text{CH}_2\text{Ph}$$

Capture ↙         ↘ Escape

PhCH$_2$C(=O)—⟨⟩—CH$_3$ + PhCH$_2$CCH$_2$Ph          PhCH$_2$CH$_2$Ph + CO

PMAP                    DBK

*Enriched in* $^{13}$C                     *Impoverished in* $^{13}$C

**Scheme 9**

The occurrence of a $^{13}$C magnetic isotope effect in the $^3$RP from DBK leads to the following predictions: (*i*) the combination products, DBK and PMAP (Fig. 12), will be enriched in $^{13}$C relative to the starting ketone; (*ii*) the extent of $^{13}$C enrichment at the various carbon atoms of DBK and PMAP will depend on the magnitude of the $^{13}$C hyperfine couplings of the various carbon atoms in $^3$RP; (*iii*) the efficiency of $^{13}$C enrichment in DBK and PMAP will be magnetic field dependent, the efficiency decreasing at high fields

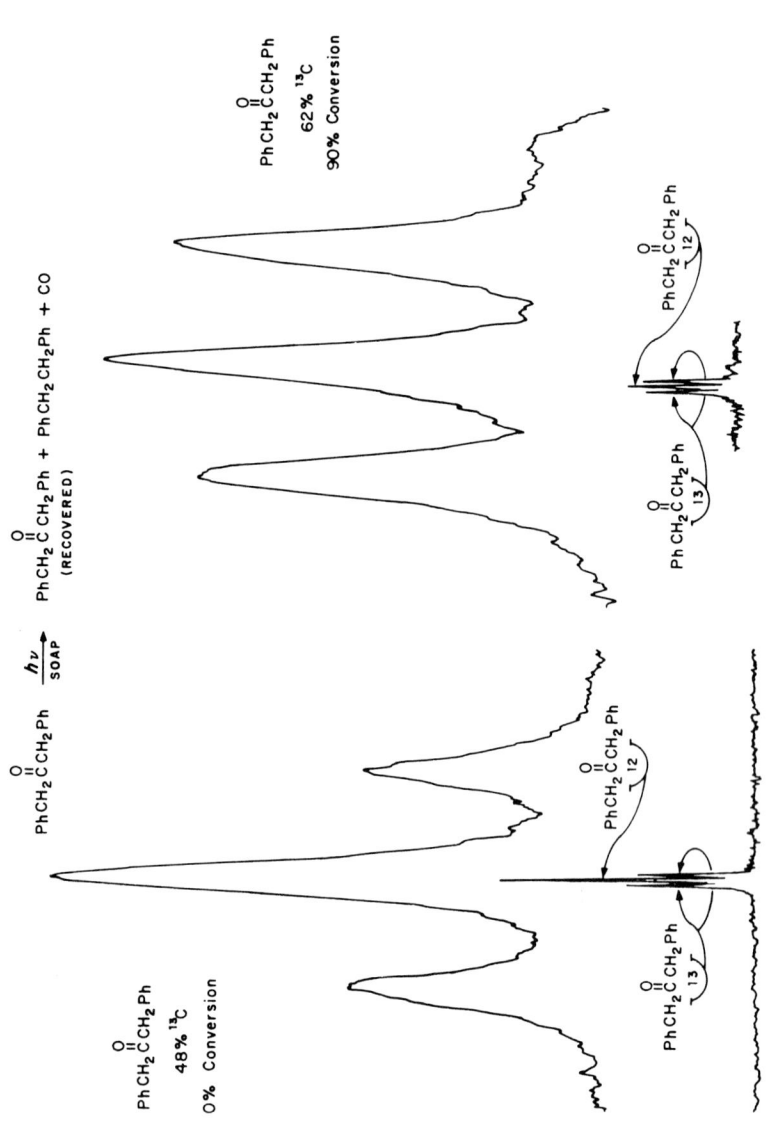

FIG. 18 ¹H nmr spectra of ¹³C-labelled dibenzyl ketone. (*Left*) starting ketone. (*Right*) ketone recovered after 90% photolysis in micellar solution

(when $T_+$ and $T_-$ are "shut off" from intersystem crossing because of Zeeman splitting); (iv) the efficiency of $^{13}C$ enrichment at ultrahigh fields will approach zero and the magnetic isotope effect will disappear (when the $\Delta gH$-mechanism dominates the $^3RP \to {}^1RP$ process which then no longer depends on the occurrence of magnetic nuclei in $^3RP$).

All four of these expectations have been confirmed experimentally (Turro, 1981, 1983; Turro and Kraeutler, 1980). Figure 18 demonstrates the prediction of $^{13}C$ enrichment in DBK recovered from the micellar photolysis of DBK. The $^1H$ nmr spectrum of DBK that has been synthetically enriched to 48% in $^{13}C$ at the carbonyl carbon (natural abundance of $^{13}C$ is 1.1% per carbon atom) is shown at the left of Fig. 18. A very sensitive test for the $^{13}C$ enrichment of the carbonyl of DBK is available from integration of the $^{13}C$ satellites in the $^1H$ nmr spectrum (Turro et al., 1981b). The methylene protons of DBK (Fig. 18) with a $^{12}C$ carbonyl appear as a singlet (at 3.66 ppm, $CHCl_3$, $Me_4Si$ as internal reference); a doublet centred at the same chemical shift with $J_{^{13}C,H} = 6.5$ Hz is caused by $^{13}C$-proton coupling when $^{13}C$ is contained in the carbonyl group. Integration over the singlet and doublet signals allows determination of the $^{13}C$ content of the carbonyl of DBK. Thus, from the satellite method, $62 \pm 4\%$ $^{13}C$ is computed to be in the recovered DBK (after 91% conversion). For the DBK recovered after 91% conversion, a quantitative agreement exists between the mass spectrometrically determined mass increase and the $^{13}C$-enrichment of the carbonyl of DBK determined by $^1H$ nmr spectroscopy. After mass spectroscopy and $^1H$ nmr analysis, these samples of DBK were subjected to $^{13}C$ nmr analysis, which also established qualitatively that the predominant $^{13}C$ enrichment occurs in the carbonyl of DBK (the relative increase of the carbonyl signal corresponds to a $60 \pm 5\%$ $^{13}C$ content). These experiments demonstrate that recovered DBK becomes enriched in $^{13}C$ as a result of photolysis of DBK in micelles and provide strong evidence that there is a magnetic spin isotope effect in the photolysis of DBK in micellar systems. In contrast, the photolysis of DBK in non-viscous homogeneous solvents (e.g. benzene) results in a much less efficient $^{13}C$ enrichment in the recovered DBK (Turro et al., 1982b).

In the photolysis of DBK, if $^{13}C$ nuclear spins control intersystem crossing from $^3RP$ to $^1RP$, starting from natural abundance $^{13}C$ ($\sim 1\%$) at each carbon, the $^{13}C$ enrichment at the various distinguishable carbon atoms in the combination products will be related to the magnitude of the $^{13}C$ hyperfine constant, $a_C$, for each of the corresponding carbon atoms in $^3RP$.

Table 5 lists the measured or estimated values of $a_C$ for the eleven different carbons in $^3RP$ (see Fig. 19 for numbering system) (Turro et al., 1982b). If, as expected for freely diffusing radical pairs, the $^{13}C$ enrichment increases monotonically with $a_C$, it is apparent from Table 5 that the ordering of $^{13}C$ enrichment should be C-1 > C-10 > C-11 > aromatic C. According to

TABLE 5

Relative $^{13}$C isotope ratios for 1-phenyl-4-acetophenone by $^{13}$C nmr spectroscopy[a]

| Carbon[b] | $\delta_C'$ ppm[c] | $S$[d] | $a_C/G$ |
|---|---|---|---|
| 1 | 197.40 | 1.23 ± 0.01 | +124 |
| 2 | 144.12 | (1.00) | −14 |
| 3 | 134.93 | 0.99 | 0 |
| 4 | 134.25 | 0.99 | +11 |
| 5 | 129.59 | 0.99 | 0 |
| 6 | 129.48 | 1.01 | +12 |
| 7 | 128.91 | 0.99 | −9 |
| 8 | 128.79 | 0.99 | 0 |
| 9 | 126.95 | 0.98 | 0 |
| 10 | 45.60 | 1.17 | +51 |
| 11 | 21.84 | 1.06 | +24 |

[a] Turro et al., 1982b
[b] See Fig. 19 for number system
[c] Chemical shifts relative to $\delta_C^{CHCL_3} = 77.27$ ppm. Assignments were made by using selective $^1$H decoupling
[d] Ratio of $^{13}$C nmr intensities for analytical and standard samples divided by the ratio for C–2

the mechanism for photolysis of DBK given in Fig. 12, 1-phenyl-4-methylacetophenone (PMAP) is a combination product which both results after $^3$RP → $^1$RP and which preserves the memory of each distinct carbon atom of $^3$RP in its final structure.

Although mass spectroscopic analysis allows determination of the global $^{13}$C enrichment in PMAP, the measurement of $^{13}$C enrichment at each carbon is best made by a magnetic resonance method. $^{13}$C Satellite intensities in the $^1$H nmr spectrum, for example, were used as a complement to mass spectroscopy for monitoring the $^{13}$C content of specifically labelled DBK and PMAP samples. However, $^{13}$C nmr spectroscopy has the advantage that specific $^{13}$C labelling is not required, because small deviations from natural abundance may be determined directly by comparing peak intensities from the sample of interest with those from a standard. The relative enrichments for PMAP obtained in this way are in Table 5.

The data in Table 5 may be summarized as follows: (*i*) Statistically significant enrichments are obtained for C–1, C–10, and C–11 in the expected order C–1 > C–10 > C–11; (*ii*) The $^{13}$C contents of the other eight carbons varied in this particular sample by <2% from those of the natural abundance sample. The enrichment factor $S$ (*vide infra*), however, seems to depend somewhat non-linearly on $a_C$. For example, C–10 seems to be over-enriched

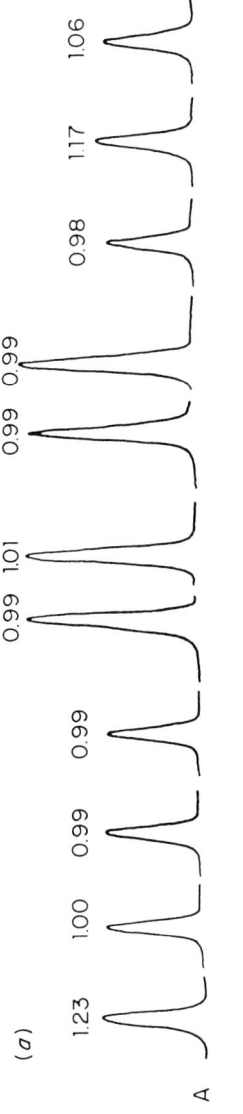

FIG. 19 $^{13}$C nmr analysis of the $^{13}$C-content of 1-phenyl-4-methylacetophenone produced by photolysis of DBK in micellar solution. (a) Integration according to chemical shifts of the $^{13}$C atoms. (b) Structure showing numbering system for the $^{13}$C atoms and C–2 under-enriched relative to the magnitudes of their hyperfine couplings. The data nevertheless are in good qualitative agreement with the expectations for a predominant nuclear spin, rather than mass, isotope effect for the formation of PMAP.

Bernstein (1952, 1957) has shown that for competitive first order isotopic reactions the residual unconverted starting material becomes exponentially enriched in the slower reacting isotope. As a result, if the isotopic rate factor is substantial and if the reaction is run to high conversion, the recovered material may become significantly enriched in the slower reacting isotope. In the case of the photolysis of DBK, if we consider only the $^{12}C/^{13}C$ competitive isotopic reactions, then residual, unconverted DBK will become enriched in $^{13}C$ if molecules containing this isotope proceed to products at a slower rate than molecules that contain $^{12}C$. A parameter $\alpha$ may then be defined as the single stage separation factor, and can be computed from the measurable quantities $S$, the overall separation factor (related to the $^{13}C$ content of product relative to starting material), and $f$, the fractional conversion (Duncan and Cook, 1968). For practical cases the appropriate approximate formula for calculation of $\alpha$ is given by (21). Thus, one can measure the $^{13}C$ content of

$$\log S = \frac{\alpha - 1}{\alpha} [-\log (1 - f)] \qquad (21)$$

the recovered DBK, compare it to the $^{13}C$ content of the initial DBK, compute $S$ and then plot $\log S$ vs $-\log (1 - f)$. The slope of such a plot, if linear, is identified as $(\alpha - 1)/\alpha$. Experimental plots are shown in Fig. 20.

The striking results which are observed are: (i) the value of $\alpha$ for photolysis of DBK in a micellar system is enormously higher than that for photolysis of DBK in a non-viscous homogeneous solvent, and (ii) the value of $\alpha$ for photolysis of DBK in a micellar system is strongly magnetic field dependent (Turro et al., 1981b). The value of $\alpha = 1.03$ for benzene is of the order of the $^{13}C/^{12}C$ mass isotope effects that have been reported. Thus, the value of $\alpha = 1.42$ (global enrichment, normalized to a single carbon atom) represents a magnetic effect far beyond any previously observed for $^{13}C/^{12}C$ mass isotope effects.

The observation that (21) is obeyed experimentally provides a useful parameter, $\alpha$, for quantitative discussion of enrichment efficiency. Consideration of Fig. 21 allows an appreciation of the relationship of $\alpha$ to $^{13}C/^{12}C$ separation efficiency. As $\alpha$ increases, the separation factor $(^{13}C/^{12}C)_f$ increases for any given extent of reaction. For example, for 99% conversion, if $\alpha = 1.05$, 1.5, 3.6, or 20, then $(^{13}C/^{12}C)_f$ will equal 1.4%, 5%, 19%, 37%, and 50%, respectively. Suppose these $\alpha$-values refer to enrichment at a single carbon. Then, starting with natural abundance material for which $(^{13}C/^{12}C)_f = 1/99$,

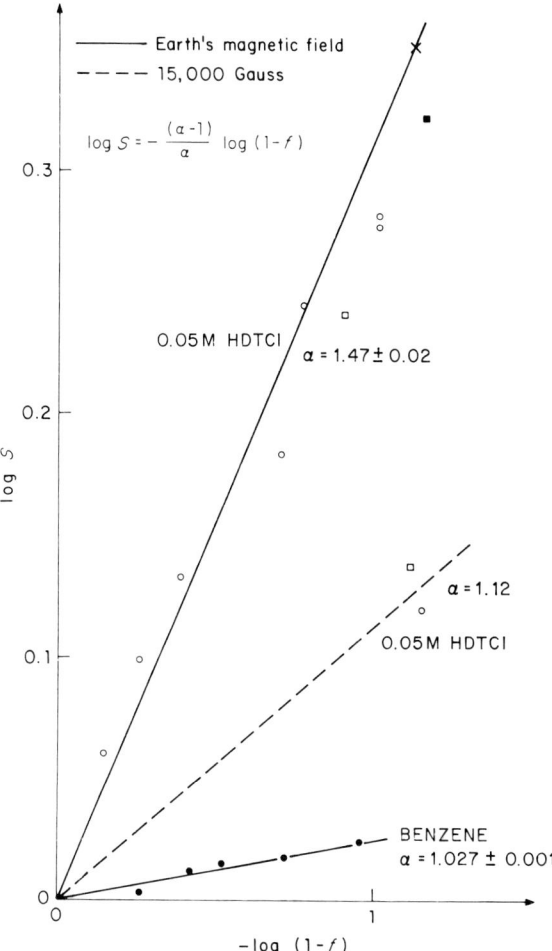

FIG. 20 Typical experimental plots of log $S$ vs $-\log(1-f)$ for the photolysis of DBK. The slope of such plots allows calculation of $\alpha$, the isotopic enrichment factor

for $\alpha = 20$, this carbon will possess $(^{13}C/^{12}C)_f = 1/1$ at 99% conversion. Putting it another way, starting with a mole of natural abundance material, the amount labelled at the pertinent carbon atom will be $\sim 0.01$ mole, and the final residual material after 99% conversion (0.01 mole) will contain $\sim 0.005$ mole of labelled compound.

It was stated above that the efficiency of enrichment should decrease at ultrahigh magnetic fields. Indeed, photolysis of DBK in an applied field of

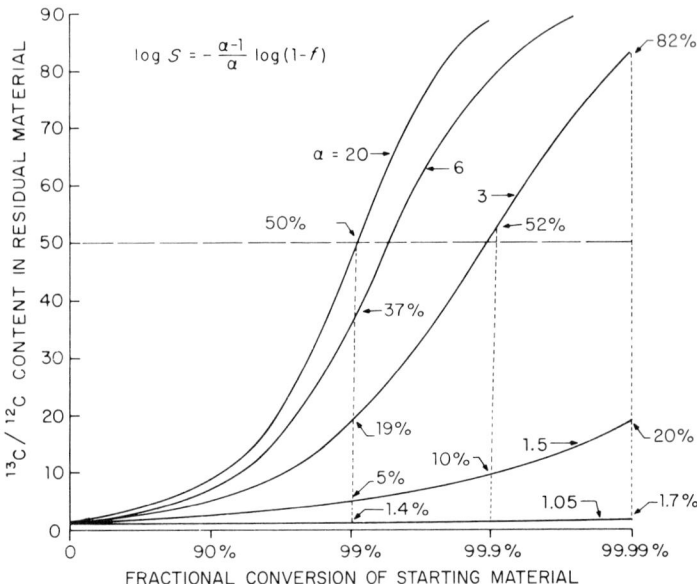

FIG. 21 Relation of experimental $^{13}$C enrichment (calculated for a single carbon atom) as a function of conversion for the photolysis of DBK. The family of curves are generated by using differing values of $\alpha$

150,000 G results in $\alpha = 1.02$, a value comparable to that expected for a $^{12}$C/$^{13}$C *mass* isotope effect.

Several other qualitative predictions of the magnetic isotope effect on $^3$RP in the micellar photolysis of DBK have been confirmed (Turro *et al.*, 1981a,b): (*i*) the quantum yield of disappearance of $^{12}$C-DBK should be *greater* than the quantum yield of disappearance of $^{13}$C-containing DBK because recombination of $^3$RP will be more efficient for $^{13}$C-containing $^3$RP, thereby causing more efficient regeneration of DBK and a lower *net* reaction efficiency; (*ii*) the cage effect for benzyl radical pairs produced by loss of CO from $^3$RP will be higher when $^{13}$C occurs at the methylene carbon of DBK (but not when $^{13}$C occurs at the carbonyl carbon of DBK). Prediction (*i*) is confirmed by the observation that the quantum yields for disappearance of DBK (natural abundance), DBK-$^{13}$CO and DBK-$^{13}$CH$_2$ were found to be 0.30, 0.22, and 0.25, respectively. Prediction (*ii*) is confirmed by the observation that the cage effect of benzyl radical coupling for DBK-$^{13}$CH$_2$ is 43%, while the cage effect for benzyl radical coupling of DBK (natural abundance) is 30%.

As discussed above, the decomposition of dibenzoyl peroxide is subject to magnetic field effects (Scheme 6). A $^{13}$C magnetic isotope effect has been demonstrated (Sagdeev et al., 1977a) in the triplet sensitized decomposition of dibenzoyl peroxide (Scheme 10). Phenyl benzoate, the recombination product, was prepared by irradiation of a $CCl_4$ solution of dibenzoyl peroxide containing acetophenone as a triplet sensitizer until >95% of the starting material had reacted. A low chemical yield (c. 3%) of phenyl benzoate was produced and analyzed for $^{13}C$ content by quantitative $^{13}C$ nmr spectroscopy.

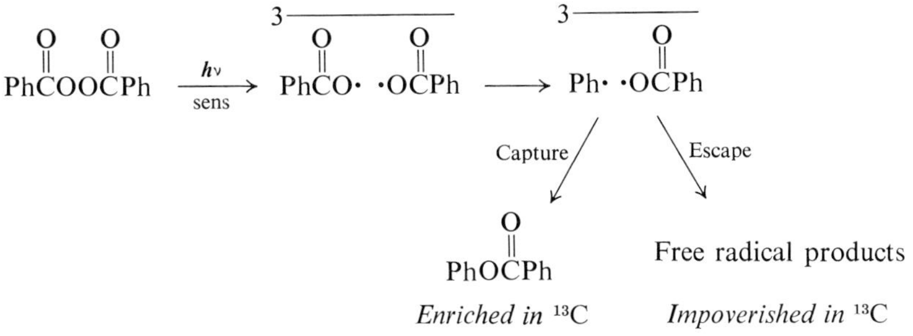

$$PhCOOCPh \xrightarrow[sens]{h\nu} {}^3[PhCO\cdot \ \cdot OCPh] \longrightarrow {}^3[Ph\cdot \ \cdot OCPh]$$

Capture ↙        ↘ Escape

PhOCPh           Free radical products

Enriched in $^{13}C$    Impoverished in $^{13}C$

**Scheme 10**

It was found that the $^{13}C$ content at the *ipso*-position of the phenyl group bound directly to oxygen in phenyl benzoate possessed a $^{13}C$ content which was 23% higher than that found in the other positions of the product molecule. The thermal decomposition of dibenzoyl peroxide also produces phenyl benzoate as a product. However, in this case no $^{13}C$ enrichment was found for any carbon atom in the phenyl benzoate.

These results are readily explained by the hypothesis that triplet sensitization produces a triplet radical pair (Scheme 10) which is capable of the $^{13}C$ capture and $^{12}C$ escape processes needed to produce a magnetic isotope effect, but that the thermolysis produces a singlet radical pair which undergoes cage reaction too rapidly to allow hyperfine interactions to influence the reactivity of the radical pair. The selective $^{13}C$ enrichment at the *ipso*-position of the product is consistent with the large $^{13}C$ hyperfine coupling (>1000 G) of the 1-position of phenyl radicals (Berndt et al., 1977).

*Magnetic isotope effects involving hydrogen nuclei*

Examples of magnetic isotope effects involving deuterium have been observed in the photolysis of DPMP (Scheme 3) (Turro and Mattay, 1981). In this case the substitution of deuterium (D) for hydrogen (H) in the methyl groups of DPMP led to the results shown in Fig. 22. The cage effect for formation of methylstyrene and benzaldehyde is significantly affected by the substitution of D for $^1H$.

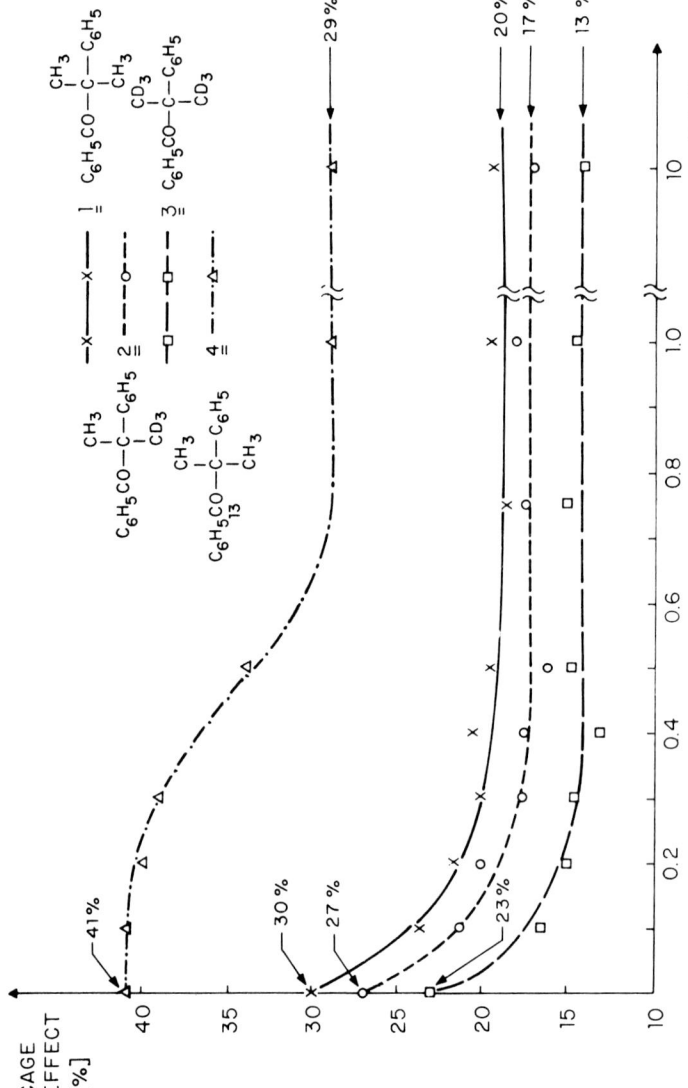

Fig. 22 Percentage cage effect as a function of applied field for variously isotopically substituted DPMP's

In particular it should be noted that substitution of $CD_3$ for $CH_3$ in DPMP reduces the cage effect from 30% to 23%. A check of the kinetic isotope effect on the atom transfer in the disproportionation reaction was made for $PhCOC(CH_3)(CD_3)Ph$. The H/D abstraction ratio was 1 : 1 as is typical for disproportionation reactions. Note also that $^{13}C$ substitution at the carbonyl carbon *increases* the cage effect from 30% to 41%. These results are readily explained in terms of a magnetic isotope effect since the smaller magnetic moment of D compared with H will cause a *slower* intersystem crossing rate in radical pairs possessing $CD_3$ groups relative to $CH_3$ groups, while the magnetic $^{13}C$ nuclei cause a more rapid intersystem crossing in radical pairs than the non-magnetic $^{12}C$ nuclei (Scheme 11). The slower

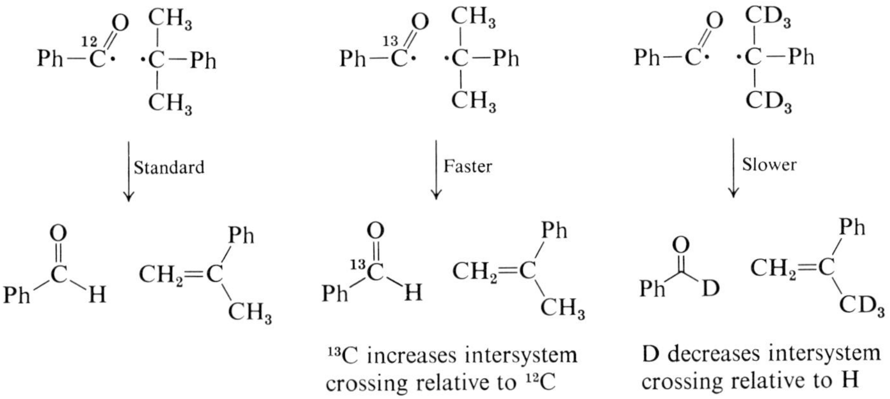

| Standard | Faster | Slower |
|---|---|---|
| $^{13}C$ increases intersystem crossing relative to $^{12}C$ | D decreases intersystem crossing relative to H |

**Scheme 11**

intersystem crossing in the D/H case and the faster intersystem crossing in the $^{13}C/^{12}C$ case lead to a smaller and larger cage effect for the isotopically substituted compounds relative to DPMP. The strong magnetic field effect on the cage effects for the substituted compounds provides convincing evidence that the isotope effect is magnetic in origin.

*Magnetic isotope effects involving oxygen nuclei*
An example of a magnetic isotope effect involving $^{17}O$ has been demonstrated to occur in the thermolysis of endoperoxides of aromatic compounds (Turro and Chow, 1980). As discussed above, the decomposition of certain endoperoxides produces singlet molecular oxygen and the yield of singlet oxygen is magnetic field dependent (Fig. 15). These results were interpreted in terms of a diradical mechanism (Scheme 7) in which intersystem crossing of an initially produced diradical ($^1D$) to a triplet diradical ($^3D$) competes with decomposition of $^1D$ to yield singlet molecular oxygen. This scheme has

the elements needed for a magnetic isotope effect with the $^1D \rightarrow {}^1O_2$ process serving as an escape pathway for $^{16}O$ and $^{18}O$ nuclei (Scheme 12).

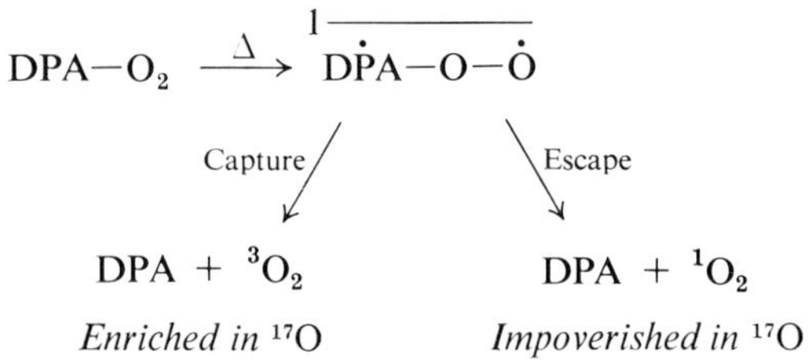

$$DPA-O_2 \xrightarrow{\Delta} \overset{1}{D\dot{P}A-O-\dot{O}}$$

Capture ↙   ↘ Escape

DPA + $^3O_2$     DPA + $^1O_2$

*Enriched in* $^{17}O$     *Impoverished in* $^{17}O$

**Scheme 12**

To test the validity of these ideas, the thermolysis of the endoperoxide of 9,10-diphenylanthracene (DPA-$O_2$) was conducted in the presence of a selective and quantitative $^1O_2$ trap. The reaction of $^1O_2$ with the trap makes escape irreversible. It was found that $^{17}O$-enriched DPA-$O_2$ produced less $^1O_2$ than DPA-$O_2$ containing $^{16}O$ or $^{18}O$. Furthermore, the $^{17}O$ atoms proceed through the capture pathway to $^3D$ which decomposes to yield $^3O_2$ as an irreversibly formed product. Thus, the $^{17}O$ atoms are "captured" in the $^3O_2$ product. Measurement of the isotopic composition of the $^3O_2$ produced in the thermolysis revealed a substantial enrichment in $^{17}O$ content relative to that in DPA-$O_2$ (Table 6). The enrichment was strongly suppressed when the

TABLE 6

Yield of $^1O_2$ formation for thermolysis of DPA-$O_2$ and isotopic composition for thermolysis of $^{17}O$-enriched DPA-$O_2$[a]

| Yield of $^1O_2$[b] | | | Composition of $^{17}O$ in untrappable $O_2$ | Solvent | Magnetic field |
|---|---|---|---|---|---|
| DPA-$^{16}O_2$[c] | DPA-$^{17,18}O_2$[d] | DPA-$^{18}O_2$[e] | | | |
| 0.37 ± 0.02 | 0.34 ± 0.01 | 0.37 ± 0.01 | 0.380 ± 0.005 | CHCl$_3$ | 0.5 kG |
| 0.32 ± 0.01 | 0.31 ± 0.01 | 0.31 ± 0.01 | 0.368 ± 0.002 | CHCl$_3$ | 10 kG |
| | | | 0.369 ± 0.001 | CHCl$_3$ | Control[f] |
| 0.32 ± 0.01 | 0.28 ± 0.02 | 0.32 ± 0.02 | 0.376 ± 0.001 | Dioxan | 0.5 kG |
| 0.27 ± 0.02 | 0.23 ± 0.01 | 0.28 ± 0.02 | 0.376 ± 0.002 | Dioxan | 12 kG |

[a] Turro and Chow, 1980
[b] Yield of $^1O_2$ is related to the consumption of a $^1O_2$ trap
[c] Initial composition is 99.8% $^{16}O$
[d] Initial composition is 60% $^{18}O$; 37% $^{17}O$; (and 3% $^{16}O$)
[e] Initial composition is 92% $^{18}O$; 4% $^{17}O$; (and 4% $^{16}O$)
[f] Thermolysis of DPA-$O_2$ in the absence of a $^1O_2$ trap

endoperoxide thermolyses were performed in the presence of a strong magnetic field.

A radical-pair system for enrichment of $^{17}O$ *via* the magnetic isotope effect (Scheme 13) has been demonstrated (Belyakov *et al.*, 1978). The system takes advantage of interesting chemistry in the liquid-phase chain oxidation

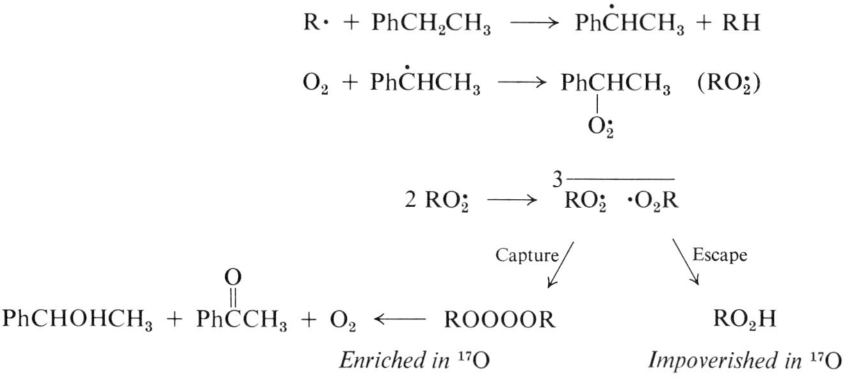

$$R\cdot + PhCH_2CH_3 \longrightarrow Ph\dot{C}HCH_3 + RH$$

$$O_2 + Ph\dot{C}HCH_3 \longrightarrow \underset{\underset{O_2^\cdot}{|}}{PhCHCH_3} \quad (RO_2^\cdot)$$

$$2\ RO_2^\cdot \longrightarrow \overset{3}{\overline{RO_2^\cdot \cdot O_2R}}$$

Capture ↙      ↘ Escape

$$PhCHOHCH_3 + Ph\overset{O}{\overset{\|}{C}}CH_3 + O_2 \longleftarrow ROOOOR \qquad\qquad RO_2H$$

*Enriched in* $^{17}O$      *Impoverished in* $^{17}O$

**Scheme 13**

of organic compounds. These oxidations proceed by the formation of peroxy radicals ($RO_2$) which terminate (at least in part) by radical pair combinations to form unstable tetraoxides ($RO_4R$) (Howard, 1972; Bennett and Howard, 1973). The latter spontaneously decompose to yield molecular oxygen and organic fragments. A magnetic isotope effect can occur at the stage of radical-pair formation. The random encounter of two $RO_2$ radicals is expected to produce three triplet radical pairs for each singlet radical pair (statistics of the spin states). The singlet radical pairs that are formed will probably combine to form $RO_4R$ without any magnetic isotope effect, but the triplet radical pairs will behave as triplet correlated radical pairs and cage reaction will be more efficient for radical pairs containing magnetic isotopes (e.g. $^{17}O$).

Going back to Scheme 13, $RO_4R$ and the molecular oxygen and organic fragments found in $RO_4R$ decomposition will be enriched in $^{17}O$ if a magnetic isotope effect occurs for the triplet $RO_2\cdot\cdot O_2R$ radical pair. Experimentally, the radical-initiated oxidation of ethylbenzene was conducted in a closed volume which allowed for the exchange of $O_2$ between the gas and liquid phases. As the oxidation proceeded, the initial molecular oxygen is consumed and $^{17}O$ enriched molecular oxygen is produced. Thus, as the reaction proceeds, the residual molecular oxygen becomes enriched in $^{17}O$. An enrichment of 13% was found (*c*. 90% consumption of molecular oxygen), whereas the extent of enrichment in $^{18}O$ was negligible (see also Buchachenko *et al.*, 1984).

*Magnetic isotope effects with tin*

Naturally occurring tin is found in numerous isotopic forms, the most abundant of which are non-magnetic ($^{116}$Sn, 14.3%; $^{118}$Sn, 24%; $^{120}$Sn, 33%; $^{124}$Sn, 6%), but several of which are magnetic ($^{115}$Sn, 0.35%; $^{117}$Sn, 7.7%; and $^{119}$Sn, 8.7%) and of spin 1/2.

Applying the general scheme for magnetic isotope enrichment, the system shown in Scheme 14 was investigated (Podoplelov *et al.*, 1979). In this case,

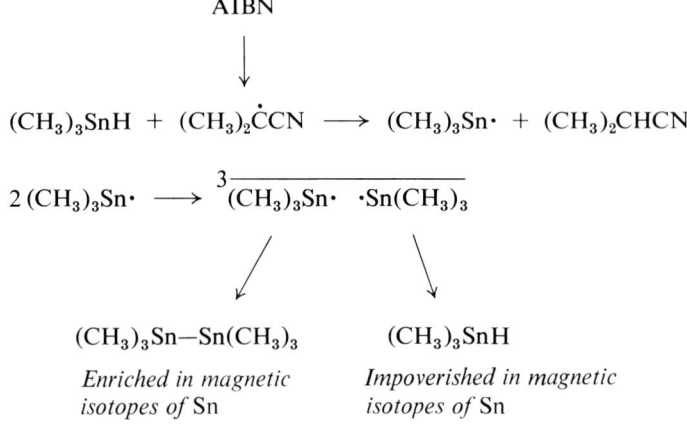

Scheme 14

the electron spins of the radical pairs, which are produced from random free radicals, are determined by the statistics of the singlet and triplet states (1/4 and 3/4, respectively). The singlet pairs are expected to combine without regard to magnetic effects, but the triplet pairs are expected to follow the general rules for the case of initially correlated triplet radical pairs. Thus, the radical pair combination product is enriched in magnetic tin isotopes (16.0% initial; 17.8% final), whereas the free radical product was impoverished in magnetic tin isotopes (16.0% initial; 13.5% final). The enrichment (impoverishment) was found to depend on the concentration of reagents.

Other hetero-organic radicals (containing silicon, lead, etc.) also possess high values of hyperfine coupling constants (Harris and Mann, 1978). Thus, radical reactions involving hetero-organic compounds are candidates for separation of magnetic heteroatomic isotopes from their non-magnetic counterparts.

## 3 Conclusion

The crucial requirements for the observation of magnetic effects on chemical reactions may be summarized as follows:

1 An intermediate in which exchange interactions are sufficiently small as to allow intersystem crossing and thus production of a different spin state for the intermediate.

2 A reaction of the intermediate which is in competition with the magnetic field influenced intersystem crossing process, which enables a sorting of the reaction products of the different spin states of the intermediate to take place.

3 A proper balance of the rates of the intersystem crossing, electron exchange and the competition processes.

As we have seen, the competition process is almost exclusively the diffusive separation of the intermediate radical (ion) pair. Diffusive separation, together with the intersystem crossing process, controls the yield of cage products for radical pair reactions. For systems in which hyperfine interactions are dominant, the application of an external magnetic field either increases (for initially singlet radical pairs) or decreases (initially triplet radical pairs) the amount of cage reaction. For systems with a large value of $\Delta g$, the opposite effect may be observed.

The observations of external magnetic field effects on chemical reactions have been confined to changes in the distributions of reaction products, compared to those obtained for reactions performed in the earth's magnetic field. At first sight this may not appear to be a dramatic change in the reactivity of the system. However, it is important to realize that the product distribution reflects the dynamic behaviour of the intermediate radical pairs. If the changes in the dynamic properties of the radical pairs can be harnessed, then more dramatic effects may be observed in secondary processes. For example, the work of Turro on the photoinitiated emulsion polymerization of styrene has demonstrated that changes in the product polymer weight by a factor of five can be achieved by using a magnetic field which produces changes in the energy of the participating radicals of only $c.\ 10^{-4}$ kcal mol$^{-1}$.

In summary, a large number of well-documented examples of the magnetic field effect on chemical reactions exist. These effects have been observed in many different kinds of chemical systems, and are readily interpreted in terms of the well established principles of chemically induced magnetic polarization.

## Acknowledgements

The authors thank the National Science Foundation and the Department of Energy for their generous support of this work.

## References

Atkins, P. W. (1970). "Molecular Quantum Mechanics". Oxford University Press, London

Atkins, P. W. (1973). *Chem. Phys. Lett.* **18**, 355
Atkins, P. W. (1976). *Chem. Brit.* 214
Atkins, P. W. and Lambert, T. P. (1975). *Ann. Rep. Prog. Chem.* **A72**, 67
Baretz, B. H. and Turro, N. J. (1983). *J. Am. Chem. Soc.* **105**, 1309
Bellamy, F. and Streith, J. (1976). *Heterocycles* **4**, 1391
Belyakov, V. A., Maltsev, V. I., Galimov, E. M. and Buchachenko, A. L. (1978). *Proc. Russ. Acad. Sci.* **243**, 561
Bennett, J. E. and Howard, J. A. (1973). *J. Am. Chem. Soc.* **95**, 4008
Berndt, A., Fischer, H. and Paul, H. (1977). "Magnetic Properties of Free Radicals". Landolt-Börnstein, New Series, Vol. 9b, Springer Verlag, Berlin
Bernstein, R. B. (1952). *J. Phys. Chem.* **56**, 893
Bernstein, R. B., (1957). *Science* **126**, 119
Buchachenko, A. L. (1976). *Russ. Chem. Rev.* **45**, 761
Buchachenko, A. L. (1977). *Russ. J. Phys. Chem.* **51**, 1445
Buchachenko, A. L. and Zhidomirov, F. M. (1971). *Russ. Chem. Rev.* **40**, 801
Buchachenko, A. L., Tarasov, V. F. and Maltsev, V. I. (1981). *Russ. J. Phys. Chem.* **55**, 936
Buchachenko, A. L., Fedorov, A. V., Yasina, L. L. and Galimov, E. M. (1984). *Chem. Phys. Lett.* **103**, 405
Carrington, A. and McLauchlan, K. A. (1967). "Introduction to Magnetic Resonance". Harper and Row, New York
Chung, C.-J. (1982). Ph.D. Dissertation, Columbia University, USA
Closs, G. L. (1971). *Proc. XXIIIrd Congr. Pure Appl. Chem.* **4**, 19
Duncan, J. F. and Cook, G. B. (1968). "Isotopes in Chemistry". Clarendon Press, Oxford, U.K.
Engel, P. S. (1970). *J. Am. Chem. Soc.* **92**, 6074
Fendler, E. J. and Fendler, J. H. (1975). "Catalysis in Micellar and Macromolecular Systems". Academic Press, London and New York
Frith, P. G. and McLauchlan, K. A. (1975). *Ann. Rep. Prog. Chem.* 378
Gupta, A. and Hammond, G. S. (1972). *J. Chem. Phys.* **57**, 1789
Harris, R. H. and Mann, B. E. (1978). "NMR and the Periodic Table". Academic Press, London and New York
Hata, N. (1976) *Chem. Lett.* 547
Hata, N. (1978). *Chem. Lett.* 1359
Hata, N. and Hokawa, M. (1981). *Chem. Lett.* 507
Hata, N., Ono, Y. and Nakagawa, F. (1979). *Chem. Lett.* 603
Howard, J. A. (1972). *Adv. Free Radical Chem.* **4**, 49
Kaptein, R. (1975). *Adv. Free Radical Chem.* **5**, 381
Kaptein, R., (1977). In "Chemically Induced Magnetic Polarization" (L. T. Muus, ed.). Reidel, Dordrecht, p. 1
Lawler, R. G. (1972). *Ace. Chem. Res.* **5**, 25
Lawler, R. G. and Evans, G. T. (1971). *Ind. Chim. Belg.* **36**, 1087
Leshina, T. V., Salikhov, K. M., Sagdeev, R. Z., Belyaeva, S. G., Maryasova, V. I., Purtov, P. A. and Molin, Y. N. (1980). *Chem. Phys. Lett.* **18**, 355
Molin, Y. N., Sagdeev, R. Z. and Salikhov, K. M. (1979). In "Soviet Scientific Reviews. Chemistry", Vol. 1. O.P.A., Amsterdam, Ch. 1
Podoplelov, A. V., Leshina, T. V., Sagdeev, R. Z., Molin, Y. N. and Goldanskii, V. I. (1979). *JETP Lett.* **29**, 380
Rigaudy, J. and Basselier, J. J. (1971). *Pure Appl. Chem.* **1**, 383
Sagdeev, R. Z., Molin, Y. N., Salikhov, K. M., Leshina, T. V., Kamha, M. A. and Shein, S. M. (1973a). *Org. Mag. Res.* **5**, 599

Sagdeev, R. Z., Molin, Y. N., Salikhov, K. M., Leshina, T. V., Kamha, M. A. and Shein, S. M. (1973b). *Org. Mag. Res.* **5**, 603

Sagdeev, R. Z., Leshina, T. V., Kamkha, M., Belchenko, O. I., Molin, Y. N. and Rezvukhin, A. L. (1977a). *Chem. Phys. Lett.* **48**, 89

Sagdeev, R. Z., Salikhov, K. M. and Molin, Y. N. (1977b). *Russ. Chem. Rev.* **46**, 297

Sakaguchi, Y., Hisahara, H. and Nagakura, S. (1980). *Bull Chem. Soc. Jpn.* **53**, 39

Sterna, L., Ronis, D., Wolfe, S. and Pines, A. (1980). *J. Chem. Phys.* **73**, 5493

Tanimoto, Y., Hayashi, H., Nagakura, S. and Tokumaru, K. (1976). *Chem. Phys. Lett.* **41**, 267

Tarasov, V. F., Buchachenko, A. L. and Maltsev, V. I. (1981). *Russ. J. Phys. Chem.* **55**, 1095.

Turro, N. J. (1978a). "Modern Molecular Photochemistry". Benjamin/Cummings, Menlo Park, p. 275

Turro, N. J. (1978b). *Ibid.*, p. 473

Turro, N. J. (1981). *Pure Appl. Chem.* **53**, 259

Turro, N. J. (1983). *Proc. Natl. Acad. Sc. (USA)* **80**, 609

Turro, N. J. and Cherry, W. R. (1978). *J. Am. Chem. Soc.* **100**, 7431

Turro, N. J. and Chow, M.-F. (1979). *J. Am. Chem. Soc.* **101**, 3701

Turro, N. J. and Chow, M.-F. (1980). *J. Am. Chem. Soc.* **102**, 1190

Turro, N. J. and Kraeutler, B. (1980). *Acc. Chem. Res.* **13**, 369

Turro, N. J. and Kraeutler, B. (1982). *In* "Diradicals". (W. T. Borden, ed.). Wiley, New York

Turro, N. J. and Mattay, J. (1981). *J. Am. Chem. Soc.* **103**, 4200

Turro, N. J., Chow, M.-F. and Rigaudy, J. (1979a). *J. Am. Chem. Soc.* **101**, 1300

Turro, N. J., Kraeutler, B. and Anderson, D. R. (1979b). *J. Am. Chem. Soc.* **101**, 7435

Turro, N. J., Chow, M.-F., Chung, C.-J. and Tung, C.-H. (1980a). *J. Am. Chem. Soc.* **102**, 7391

Turro, N. J., Chow, M.-F., Chung, C.-J., Weed, G. C. and Kraeutler, B. (1980b). *J. Am. Chem. Soc.* **102**, 4843

Turro, N. J., Gratzel, M. and Braun, A. M. (1980c). *Angew. Chem. Int. Ed. Engl.* **19**, 675

Turro, N. J., Anderson, D. R., Chow. M.-F., Chung, C.-J. and Kraeutler, B. (1981a). *J. Am. Chem. Soc.* **103**, 3892

Turro, N. J., Chow, M.-F., Chung, C.-J. and Kraeutler, B. (1981b). *J. Am. Chem. Soc.* **103**, 3886

Turro, N. J., Chung, C.-J., Jones, G. and Becker, W. G. (1982a). *J. Phys. Chem.* **86**, 3677

Turro, N. J., Chung, C.-J., Lawler, R. G. and Smith, W. J. (1982b). *Tetrahedron Lett.* 3223

# Kinetics and Mechanisms of Reactions of Organic Cation Radicals in Solution

OLE HAMMERICH[1] and VERNON D. PARKER[2]

[1] *University of Copenhagen, The H.C. Ørsted Institute, Copenhagen, Denmark*
[2] *Norwegian Institute of Technology, University of Trondheim, Trondheim, Norway*

1 Introduction  56
2 Dimerization and cyclization reactions of cation radicals  56
    Dimerization of aryl ether cation radicals  57
    Aromatic hydrocarbon cation radicals  60
    Dimer forming reactions of aromatic amine cation radicals  60
    Additive dimerization of arylolefin cation radicals  62
    Cyclization reactions of cation radicals  63
3 Mechanisms of the reactions of cation radicals with nucleophiles  68
    Comments on equilibria and reaction kinetics  70
    Reactions occurring at the dication stage  72
    Reactions occurring at the cation radical stage  80
    Reaction pathways of the cation radicals of anthracenes in the presence of weak nucleophiles  87
    Theoretical relationships for the estimation of cation radical reactivity toward nucleophiles  90
4 Electron-transfer reactions initiated by cation radicals  94
    Cation radicals as oxidants  95
    Redox catalysis  100
    Catalysis by cation radicals  106
5 Fragmentation reactions of cation radicals  123
    Acidity of cation radicals  124
    Products of phenolic cation radical reactions  132
    Synthetic and mechanistic aspects of alkylbenzene oxidations  134
    Cleavage of carbon—carbon bonds  141
    Cleavage of bonds between carbon and heteroatoms  143
    Cleavage of sulfur—sulfur and selenium—selenium bonds  147
    Cleavage of nitrogen—nitrogen bonds  148

6 Cation radicals as intermediates in conventional organic reactions 151
    The nitramine rearrangement 152
    Electrophilic aromatic substitution 155
    Aromatic nitration 160
    The abnormal Wittig reaction 170
    Addition of radicals to diazonium ions and protonated heteroaromatic
        bases 172
    Reactions of aliphatic ammoniumyl radicals 174
7 Concluding remarks 180
References 180

## 1 Introduction

The formation, properties and reactions of cation radicals in solution have been reviewed earlier in this series (Bard et al., 1976). At that time, very much of the work dealing with cation radical chemistry had been carried out by specialists in technique-oriented disciplines, i.e. esr spectroscopy, radiochemistry and organic electrochemistry. Since then interest in ion radical chemistry has become very much more widespread. This is due, at least in part, to the increasing recognition that electron transfer can play important roles in diverse homogeneous reactions of organic compounds. Electron-transfer reactions in organic chemistry are being studied intensively and have been reviewed recently (Eberson, 1982). The chemistry of excited complexes as well is dominated by the formation and reactions of ion radical pairs (Davidson, 1983).

In this chapter we will concentrate on the developments in cation radical chemistry that have taken place since 1976. The coverage of the literature will not be exhaustive. For example, practically all anodic reactions of organic compounds proceed *via* the formation of the corresponding cation radical but only those papers dealing specifically with the reaction pathways of the intermediates will be discussed. Papers dealing primarily with the spectroscopic properties of cation radicals will not be reviewed. The electrochemical methods for the formation and investigation of cation radical reactions have been covered in detail in other reviews (Bard et al., 1976; Parker, 1983a) and will not be discussed here.

A number of reviews dealing with specific aspects of cation radical chemistry have appeared (D. H. Evans and Nelsen, 1978; Asmus, 1979; Musker, 1980; Hammerich and Parker, 1981a; Nelsen, 1981; Bock and Kaim, 1982; Minisci et al., 1983).

## 2 Dimerization and cyclization reactions of cation radicals

Dimerization and cyclization reactions are among the most important reactions of cation radicals employed in organic synthesis. The cation radicals may be generated either electrochemically or with chemical oxidants such as VOF$_3$ (Kupchan et al., 1978a,b, 1975; Kupchan and Kim, 1975; Elliot, 1977, 1979), thallium trifluoroacetate (McKillop et al., 1977; Taylor et al., 1977), or manganic tris(acetylacetonate) (Ronlán and Parker, 1974). One of the primary advantages of the electrochemical method is that high concentrations of the ion radicals can be generated close to the electrode surface. A further advantage of the electrochemical method is that the reactions of the reactive intermediates can be monitored during the generation (Parker, 1983a). The preparative aspects of these reactions will not be discussed here. For details the reader is referred to a number of reviews dealing with the electrochemical oxidation of aromatic hydrocarbons (Nyberg, 1978), aromatic amines (Nelson, 1974), olefins (Schäfer, 1981) and phenol derivatives (Parker et al., 1978).

DIMERIZATION OF ARYL ETHER CATION RADICALS

The dimerization of cation radicals of simple benzene derivatives is usually accompanied by the formation of the corresponding biphenyl derivative. This by virtue of the extended conjugation is more easily oxidized, and frequently only intractable tars are found as final products. On the other hand, if the reaction medium is carefully selected so that the biphenyl cation radicals are stable, useful synthetic procedures can be developed. This was found to be the case when anisoles and substituted anisoles were subjected to anodic oxidation in CH$_2$Cl$_2$/TFA (2 : 1) (Ronlán et al., 1973a). Cyclic voltammetry is uniquely suited to diagnose the reaction as is evident from the voltammograms shown in Fig. 1. The oxidation peaks (O$_1$) for the three anisole derivatives correspond to the following sequence of reactions: (a) one-electron oxidation of the substrate to the cation radical is followed by (b) dimerization and loss of two protons from the dimer dication yielding the corresponding biphenyl and finally (c) two consecutive one-electron oxidations of the biphenyl to the corresponding dication. The reduction of the dications are observed at peak R$_3$ and the cation radical is reversibly reduced at R$_2$. While the cyclic voltammograms give a clear picture of the overall reactions involved, little information as to the mechanisms of the reactions is obtained under the conditions used in the measurements shown in Fig. 1.

Two mechanisms were considered to be probable for the dimerization of cation radicals of anisoles and related compounds. The two mechanisms are commonly referred to as radical-radical dimerization (RRD) and radical-substrate coupling (RSC). The general features of these two mechanisms have recently been discussed in some detail (Parker, 1983a) and they need not be repeated here. A derivative cyclic voltammetry (DCV) kinetic investigation of the reactions of 4-methoxybiphenyl (MBH) cation radical, chosen as a

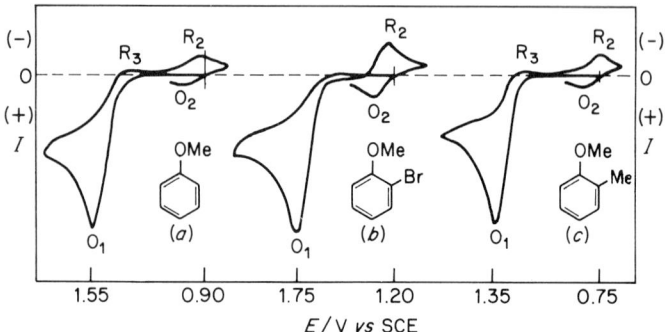

FIG. 1 Cyclic voltammetry for the anodic coupling of anisoles in CH$_2$Cl$_2$/TFA (2:1) containing n-Bu$_4$NBF$_4$ (0.2 M). (a) Anisole, (b) 2-bromoanisole, (c) 2-methylanisole. Voltage sweep rate = 156 mV s$^{-1}$. (Reprinted with permission from Ronlán et al., 1973a)

model substance because of its reduced reactivity as compared to anisole cation radical, revealed a complex rate law with reaction orders changing with the initial substrate concentration. Competing mechanisms were proposed with RRD (1) dominating at high substrate concentrations and RSC (2–3) taking over at low concentrations (Aalstad et al., 1981a). Calculations were carried out on the data reported and the conclusions of the study [as well

$$2\ \text{MBH}^{+\cdot} \xrightarrow{k_1} (\text{MBH})^+ - (\text{MBH})^+, \tag{1}$$

$$\text{MBH}^{+\cdot} + \text{MBH} \underset{k_{-2}}{\overset{k_2}{\rightleftharpoons}} (\text{MBH})^+ - (\text{MBH})\cdot \tag{2}$$

$$(\text{MBH})^+ - (\text{MBH})\cdot\ +\ \text{MBH}^{+\cdot} \xrightarrow{k_3} (\text{MBH})^+ - (\text{MBH})^+\ +\ \text{MBH} \tag{3}$$

as the method of electrode mechanism analysis used (Parker, 1981b)] were criticized (Amatore and Savéant, 1983). The calculations showed that the DCV data (Aalstad et al., 1981a) could be explained by a single mechanism (2–3) with step (2) rate-determining at high and step (3) at low substrate concentrations. This conclusion was supported by an alternative method of data analysis and a comparison of the results from the two analysis methods is given in Table 1 (Parker, 1984).

An interesting aspect of the kinetic study (Aalstad et al., 1981a) is the magnitude of the apparent activation energies ($E_a$) observed. At low substrate concentration where electron-transfer reaction (3) is apparently rate-limiting the Arrhenius plot was only approximately linear with an apparent $E_a$ of 0. At higher substrate concentrations (8 mM) the linearity was excellent and

TABLE 1

A comparison of rate and equilibrium constants for the dimerization of 4-methoxybiphenyl cation radical

| Quantity | "Reaction Order Approach"[a] | Theoretical data[b] |
|---|---|---|
| $k_2$ | $4.14 \times 10^4$ M$^{-1}$s$^{-1}$ | $4.07 \times 10^4$ M$^{-1}$s$^{-1}$ |
| $k_{-2}$ | $(2.98 \times 10^6$ s$^{-1})^c$ | |
| $k_3$ | $(10^{10}$ M$^{-1}$s$^{-1})^c$ | |
| $K_2$ | $(1.39 \times 10^{-2}$ M$^{-1})^c$ | |
| $k_3 K_2$ | $1.39 \times 10^8$ M$^{-2}$s$^{-1}$ | $1.65 \times 10^8$ M$^{-2}$s$^{-1}$ |

[a] Parker, 1984
[b] Amatore and Savéant, 1983
[c] Estimated values based on the assumption of diffusion control for reaction (3)

$E_a$ was observed to be 10.6 kcal mol$^{-1}$. The former result is easily understood if the pre-equilibrium (2) has a negative $\Delta H°$ which is reasonable. The relatively high activation energy for forward reaction (2) is significant with regard to ion radical dimerization in general. There should be no steric restraints in the transition state for reaction (2). The approach giving the maximum overlap would be expected to be the parallel plane arrangement illustrated in Fig. 2. Since only one of the MBH moieties is charged, there is no charge repulsion term to contribute to $E_a$. One factor which is common to this and other ion radical dimer forming reactions is that the delocalization energy for both species is significantly reduced in the transition state. This seems likely to be the dominating factor here. The original interpretation of the mechanism (Aalstad et al., 1981a) led to the assumption that at least part of $E_a$ is due to charge repulsion in the transition state. Since the current view of the mechanism is that reaction (2) is rate-determining at high substrate concentration and no evidence for the occurrence of (1) has been found, the activation energy for (1) has to be even greater than 11 kcal mol$^{-1}$ (Parker et al., 1983). Activation energies for the dimerization of anion radicals of activated olefins have also been observed to be greater than 11 kcal mol$^{-1}$ (Parker, 1983b).

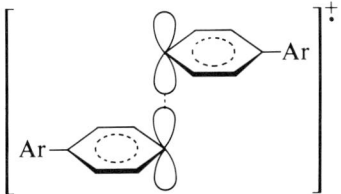

FIG. 2 Possible transition state for the coupling of 4-methoxybiphenyl cation radical with substrate

AROMATIC HYDROCARBON CATION RADICALS

The oxidation of aromatic hydrocarbons and alkylaromatic compounds in media of low nucleophilicity results in the formation of dimers. The reactions of the cation radicals of alkylbenzenes and of naphthalene have been studied in some detail (Nyberg, 1970, 1971a–c, 1973; Nyberg and Trojanek, 1975; Eberson et al., 1973). No kinetic studies have been reported for these reactions and mechanistic considerations have relied upon the effect of reaction conditions, i.e. substrate concentrations and current densities, on the product distribution. Both the RRD and the RSC mechanisms were considered to be likely possibilities. The RSC mechanism was deemed most likely on the basis that the yield of biphenyls is dependent upon the substrate concentration (Nyberg, 1971a) but not on the current density and that co-electrolysis of naphthalene and alkylbenzenes under conditions where only naphthalene is oxidized results in good yields of 1-arylnaphthalenes (Nyberg, 1971c, 1973). The RSC mechanism is also implicated in the reactions of anthracene cation radical in acetonitrile containing water (Parker, 1970). In this case the trimer [3], in which the central ring is most probably derived from unoxidized

substrate, becomes the major product at high water concentrations. The distribution of [1], [2] and [3] as a function of water concentration showed that the cation radical reactions depend strongly upon the nature of the reaction medium.

DIMER FORMING REACTIONS OF AROMATIC AMINE CATION RADICALS

The oxidation of aromatic amines and the reaction pathways of the resulting cation radicals were studied intensively in the 1960s (Adams, 1966, 1969).

The reaction pathways of these intermediates are very diversified and dependent upon the reaction conditions. Dimerization is one of the more important pathways but in most cases the details of the mechanisms have not been elucidated. This work has been well reviewed (Nelson, 1974).

The cation radicals of triphenylamines form dimers in high yield when generated in acetonitrile (Seo et al., 1966). The products are tetraphenylbenzidines, [4] in the case of the parent compound. The mechanism of the reaction was concluded to most likely be RRD although RSC was not ruled out (Nelson and Feldberg, 1969). The reaction was observed to be general for a wide variety of substituted triphenylamines of structure [5] (Creason et al., 1972) and a linear Hammett relationship was observed with ρ equal to 2 when log $(k_x/k_o)$ was plotted vs $\sigma^+$.

The rates of dimerization of cation radicals of substituted carbazoles [6] were compared to the corresponding triphenylamine cation radicals [5]

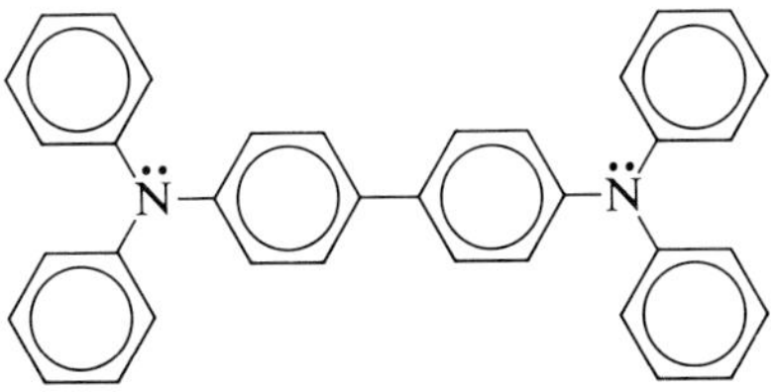

TPB
[4]

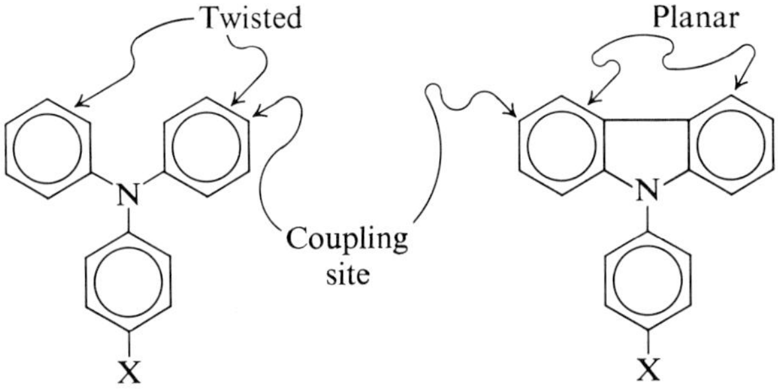

[5]        [6]

(Ambrose et al., 1975). As indicated, the structures of the parent compounds are very similar and coupling takes place at sites *para* to the nitrogen atom in both cases. The most significant structural feature in which the two differ is that the rings of the carbazole system are planar while those of the triphenylamine are twisted out of plane due to the interactions between hydrogen atoms *ortho* to nitrogen. In the three comparisons made, the rate constants observed for the carbazole cation radicals were $10^4$–$10^6$ greater than those for

## ADDITIVE DIMERIZATION OF ARYLOLEFIN CATION RADICALS

Additive dimerization of the cation radicals of arylolefins and related substances in the presence of nucleophiles, exemplified by reaction (4), has been developed by Schäfer and coworkers into a valuable synthetic method

$$2 \text{ ArCH=CH}_2 + 2 \text{ CH}_3\text{OH} \xrightarrow{-2e^-} \underset{\underset{\text{OCH}_3}{|}}{\text{ArCH}}-\text{CH}_2-\text{CH}_2-\underset{\underset{\text{OCH}_3}{|}}{\text{CHAr}} + 2 \text{ H}^+ \quad (4)$$

(Schäfer, 1981). In general, the substrates of synthetic interest form cation radicals that are highly reactive and mechanism studies have not been carried out in these cases.

Three initial reaction steps are plausible and correspond to nucleophilic attack on the cation radical by the nucleophile Nu (5), the RSC step (6) and RRD (7). Reaction (5) is very highly unlikely for a number of reasons. Two

$$(\text{ArCH=CH}_2)^{+\cdot} + \text{Nu} \rightarrow \underset{\underset{\text{Nu}}{|\ +}}{\text{Ar}-\text{CH}}-\text{CH}_2^\cdot \quad (5)$$

$$(\text{ArCH=CH}_2)^{+\cdot} + \text{ArCH=CH}_2 \rightarrow \text{Ar}-\overset{\cdot}{\text{CH}}-\text{CH}_2-\text{CH}_2\overset{+}{\text{CH}}-\text{Ar} \quad (6)$$

$$2 (\text{ArCH=CH}_2)^{+\cdot} \rightarrow \text{Ar}-\overset{+}{\text{CH}}-\text{CH}_2-\text{CH}_2-\overset{+}{\text{CH}}-\text{Ar} \quad (7)$$

obvious reasons are that the less stable of the two possible radicals is formed and that if this radical were formed in low concentration it would not be expected to undergo selectively second order dimerization or reaction with substrate without interference from hydrogen atom abstraction reactions. As in all the other cases discussed, it is impossible to predict on the basis of available data whether (6) or (7) would be the more favourable.

The reactions of 4,4'-dimethoxystilbene cation radical in acetonitrile were studied in order to gain more information about the mechanism. The first steps in this reaction involve the formation of dimeric species similar to those in (6) and (7). The initial mechanism studies (Steckhan, 1978; Burgbacker and Schäfer, 1979) resulted in the proposal that reaction involves rate determining RRD (7). A more recent study (Aalstad et al., 1981b) indicated that the reaction is much more complex than the earlier studies indicated and that RSC (6) is probably the initial step. This data and discussion has been recently reviewed and that review (Parker, 1983a) can be consulted for more details of the mechanistic arguments.

CYCLIZATION REACTIONS OF CATION RADICALS

Compounds containing more than one aromatic group, usually separated by an aliphatic side chain, form cation radicals capable of undergoing intramolecular coupling analogous to the RRD and RSC mechanisms. These cyclization reactions have been studied in a number of cases and can be useful synthetic methods.

*Tetraphenylethylene cation radical and dication cyclization*

Several alternative pathways were considered to be likely for the cyclization of tetraphenylethylene (TPE) cation radical in acetonitrile (Stuart and Ohnesorge, 1971) including disproportionation to the dication which undergoes cyclization. The reaction is very rapid and difficult to study in acetonitrile. The reaction (8) was observed to be quantitative in $CH_2Cl_2/TFA$ and

$$2\ TPE^{+\cdot} \rightarrow DPP + TPE + 2\ H^+ \tag{8}$$

$TPE^{+\cdot}$ is sufficiently long-lived in that medium for kinetic studies (Svanholm *et al.*, 1974). The product of the reaction, 9,10-diphenyl phenanthrene (DPP) is about 300 mV more difficult to oxidize than TPE which greatly simplifies mechanism studies. The course of the reaction is readily shown by the cyclic voltammograms in Fig. 3. The single sweep voltammogram (Fig. 3a) includes the oxidation of TPE to $TPE^{+\cdot}$ ($O_1$), the formation of $TPE^{2+}$ ($O_2$), the oxidation of DPP ($O_3$), the reduction of $DPP^{+\cdot}$ ($R_3$) and the reduction of $TPE^{+\cdot}$ ($R_1$). The steady state voltammogram after several cycles (Fig. 3b) accentuates the build-up of DPP in the diffusion layer and Fig. 3c is the voltammogram for the oxidation of DPP.

The kinetic study revealed rate law (9). The rate was observed to be strongly

$$-d[TPE^{+\cdot}]/dt = k_{obs}[TPE^{+\cdot}]^2/[TPE] \tag{9}$$

affected by both [TFA] and [$Bu_4NBF_4$]. The disproportionation mechanism

$$2\ TPE^{+\cdot} \underset{}{\overset{K_{10}}{\rightleftharpoons}} TPE^{2+} + TPE \tag{10}$$

$$TPE^{2+} \overset{k_{11}}{\rightarrow} DPP + 2\ H^+ \tag{11}$$

(10)–(11) was found to be consistent with the data and the salt effect was attributed to the highly polarized nature of transition state [7]. The Arrhenius activation energy was observed to be 14.5 kcal mole$^{-1}$ with $\Delta S^{\ddagger}$ near zero.

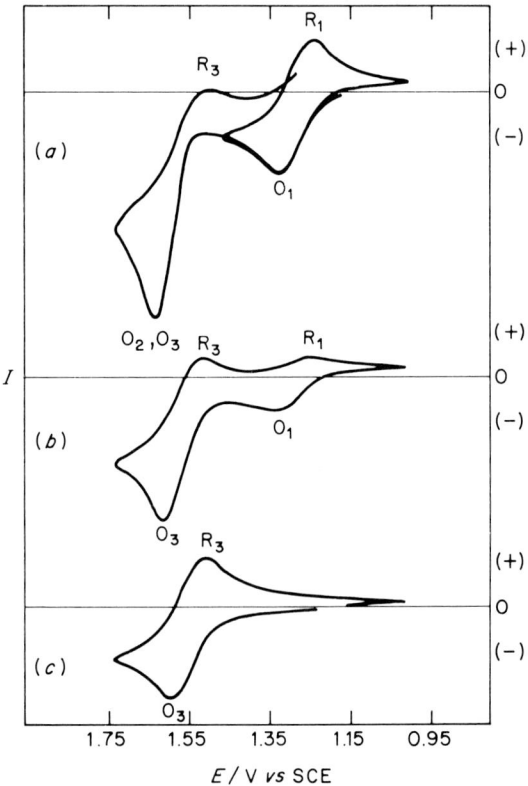

FIG. 3 Cyclic voltammograms in TFA/dichloromethane (1:3) containing Bu$_4$NBF$_4$ (0.2 M): (a) single sweep voltammogram of TPE (1 mM), (b) steady state voltammogram of TPE (1 mM), (c) voltammogram of DDP (1 mM). (Reprinted with permission from Svanholm et al., 1974)

The large activation energy was attributed to the localization of charge in [7]. The mechanistic assignment was supported by rotating disk electrode measurements on the kinetics of the cylization of $TPE^{2+}$. The value of $k_{11}$ was observed to be in the range $30$–$100$ $s^{-1}$ under the reaction conditions, in excellent agreement with that implied from the kinetic studies on the reaction of $TPE^{\pm}$ (Svanholm et al., 1974).

A more precise method was recently developed to study the cyclization of $TPE^{2+}$ (Aalstad et al., 1982a) inspired by the need for activation parameter data to compare with ion radical dimerization data. The method involves derivative linear sweep voltammetry and is illustrated in Fig. 4. The first two peaks, $P_1$ and $P_2$, correspond to $O_1$ and $O_2, O_3$ in Fig. 3a. $P_3$ is due to oxidation products of DPP and the ratio of this peak height to the other two gives a direct measure of $k_{11}$ by comparison with theoretical calculations. Under the conditions of the experiments $k_{11}$ was $3.2 \times 10^3 s^{-1}$ at 298 K, $E_a$ was equal to 11.4 kcal mol$^{-1}$ and $\Delta S^{\ddagger}$ was $-6$ cal K$^{-1}$ mol$^{-1}$. This study verified the general conclusions of the earlier study and provides more reliable kinetic and activation parameters for reaction (11).

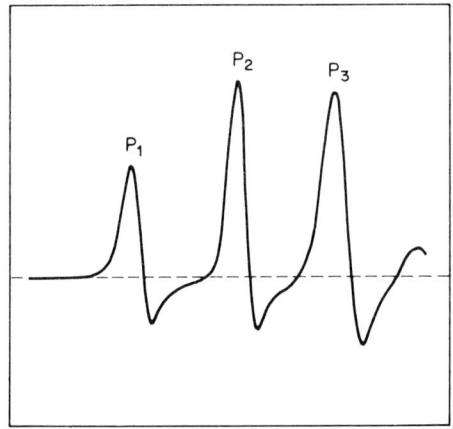

FIG. 4  Derivative linear sweep voltammogram for the oxidation of tetraphenylethylene in dichloromethane/trifluoroacetic acid/trifluoroacetic acid anhydride (8.5 : 1.0 : 0.5) containing Bu$_4$NBF$_4$ (0.1 M) at $-33°C$. Voltage sweep rate $= 10$ V s$^{-1}$. (Reprinted with permission from Aalstad et al., 1982a)

*Cation radicals derived from 1,2-diarylethanes and related compounds*

A great deal of interest has been shown in these reactions, much of it stemming from the analogy with naturally occurring transformation of laudanosine [8] to O-methylflavinatine [9] and related transformations of natural products. A compilation of the literature references on these reactions can be found in a recent publication (Aalstad et al., 1982b).

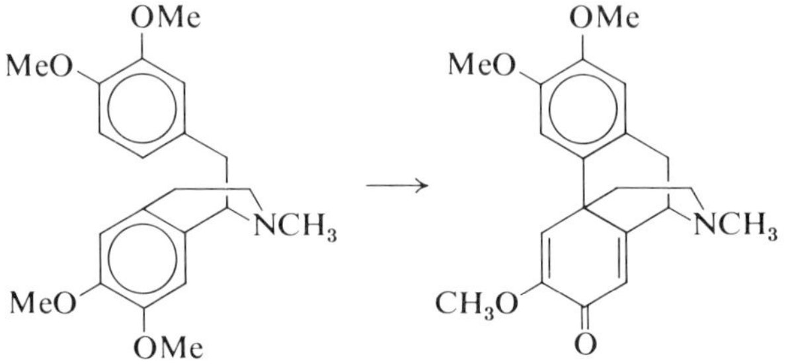

(12)

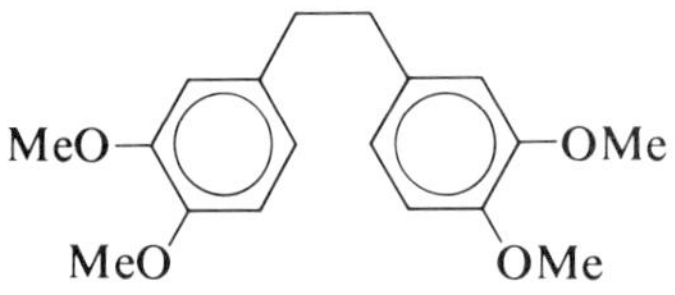

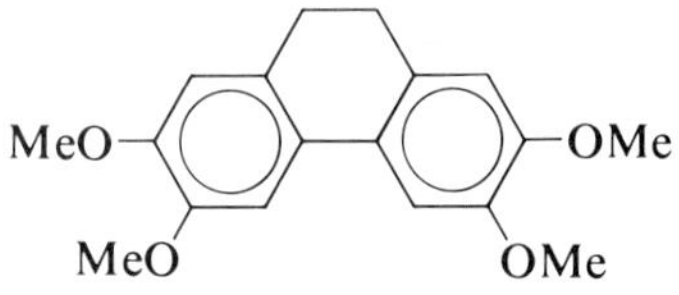

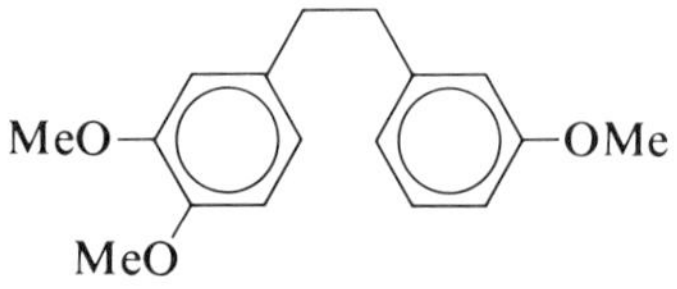

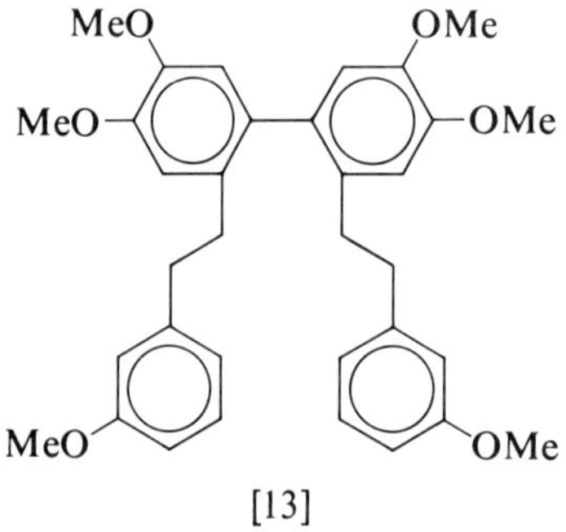

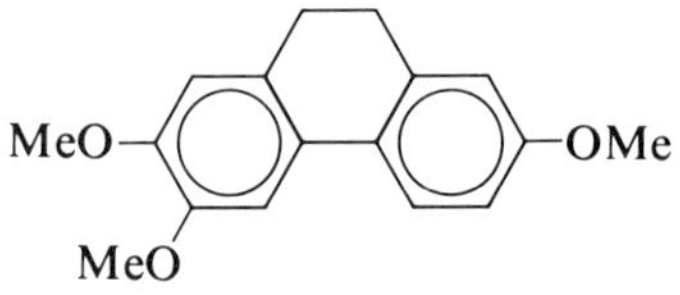

The first mechanism study of the reaction was carried out on the reactions of the cation radicals of 3,3′,4,4′-tetramethoxy bibenzyl [10] and 3,3′,4-trimethoxybibenzyl [12] in acetonitrile. The product in the former case is [11] while either [13] or [14] were observed to be formed during oxidation of [12], depending upon the oxidation potential. It was shown both by cyclic voltammetry and controlled potential electrolysis that [13] arises from dimerization of the cation radical while [14] is formed from the dication diradical of [12]. From this it was inferred that the dication diradical of [10] is an intermediate in the formation of [11] (Ronlán et al., 1973b). The mechanism studies were extended to other unsymmetrically substituted diarylalkanes in which the length of the aliphatic carbon chain was varied (Parker and Ronlán, 1975). It was observed that the dication diradical mechanism is not general and cation radical cyclization was observed to occur as well. An intermediate was observed by cyclic voltammetry during the cyclization of 1-(3,4-dimethoxyphenyl)-3-(4-methoxyphenyl)propane (Parker et al., 1974) and it was concluded that the intermediate was a cation radical in which the charge and odd electron were in different aromatic $\pi$-systems as in [15].

[15]    [16]

Voltammetric kinetic studies were carried out on the cyclization of the cation radicals of [8] and [16] (Kerr et al., 1979). Two different rate laws were proposed (13)–(14) corresponding to mechanisms (15)–(17) and (18)–(19), respectively. The difference in mechanism was attributed to the fact

$$-d[8\overset{+}{\cdot}]/dt = k_{17}K_{15}K_{16}[(8-H^+)\overset{+}{\cdot}]^2/[H^+] \quad (13)$$

$$-d[16\overset{+}{\cdot}]/dt = k_{19}K_{18}[16\overset{+}{\cdot}]^2/[16] \quad (14)$$

$$(8-H)\overset{+}{\cdot} \underset{}{\overset{K_{15}}{\rightleftharpoons}} (C-H)\overset{+}{\cdot} \quad |\text{(C indicates cycle)} \quad (15)$$

$$(C-H)\overset{+}{\cdot} \underset{}{\overset{K_{16}}{\rightleftharpoons}} C\overset{+}{\cdot} + H^+ \quad (16)$$

$$\overset{+}{C}\cdot + (8-H)\overset{+}{\cdot} \overset{k_{17}}{\rightarrow} \text{Products} \quad (17)$$

$$2[16\overset{+}{\cdot}] \underset{}{\overset{K_{18}}{\rightleftharpoons}} [16^{2+}] + [16] \quad (18)$$

$$[16^{2+}] \overset{k_{19}}{\rightarrow} \text{Products} \quad (19)$$

that the side chain nitrogen in [8] is protonated under the reaction conditions.

A linear sweep voltammetry kinetic study was carried out on the cyclization of [10$_+^+$] in acetonitrile and in acetonitrile containing TFA (Aalstad et al., 1982b). This study resulted in rate law (20). It was pointed out that the disproportionation mechanism and a mechanism involving reversible

$$-d[10^{+}]/dt = k_{obs}[10^{+}]^2 \qquad (20)$$

cyclization to intermediates such as [15] as in (21)–(23) cannot be distinguished by the form of the rate law and that rate law (20) is consistent with

$$\overset{\frown}{\text{Ar–H}\quad\text{Ar–H}^{+\cdot}} \;\rightleftarrows\; \overset{\frown}{\underset{\underset{H}{|}}{\text{Ar}^+}\text{—}\underset{\underset{H}{|}}{\text{Ar}\cdot}} \qquad (21)$$

$$\overset{\frown}{\underset{\underset{H}{|}}{\text{Ar}^+}\text{—}\underset{\underset{H}{|}}{\text{Ar}\cdot}} + \overset{\frown}{\text{Ar–H}\quad\text{Ar–H}^{+\cdot}} \;\rightleftarrows\; \overset{\frown}{\underset{\underset{H}{|}}{\text{Ar}^+}\text{—}\underset{\underset{H}{|}}{\text{Ar}^+}} + \overset{\frown}{\text{ArH}\quad\text{ArH}} \qquad (22)$$

$$\overset{\frown}{\underset{\underset{H}{|}}{\text{Ar}^+}\text{—}\underset{\underset{H}{|}}{\text{Ar}^+}} \;\longrightarrow\; \text{products} \qquad (23)$$

either rate-determining disproportionation, analogous to forward reaction (18), or rate-determining (22).

Since the cyclization takes place after rate-determining electron transfer in the disproportionation case, i.e. under conditions where rate law (20) applies, and before the rate determining step in mechanism (21)–(23), the mechanisms can be differentiated by the presence or absence of a secondary deuterium kinetic isotope effect (Aalstad et al., 1982b). For the reactions of [10$^{+\cdot}$] and [10–d$_6^{+\cdot}$], where the latter has deuterium only on the ring positions, $k_H/k_D$ was observed to range from 0.7 to 0.9. This is consistent with the change in hybridization of the carbons undergoing bond formation from sp$^2$ to sp$^3$ (do Amaral et al., 1972). In fact $k_H/k_D$ was observed to be in the same range for the dimerization reactions of 4-methoxybiphenyl cation radical (Aalstad et al., 1981a).

## 3 Mechanisms of the reactions of cation radicals with nucleophiles

The cation radicals of most organic compounds react so rapidly that it is not possible to prepare solutions of the intermediates for kinetics and mechanism studies. There are exceptions and compounds having structures [17] and [18] where X and Y are O, S or N—R are transformed upon oxidation

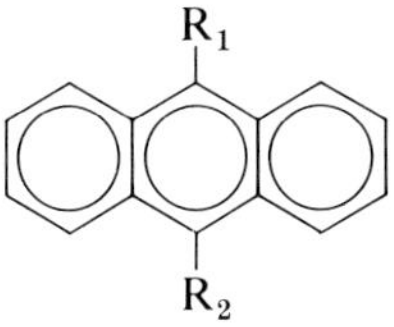

ORGANIC CATION RADICALS

[17] 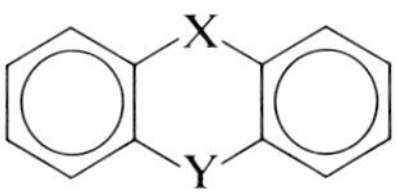     [18]

in solvents such as acetonitrile to reasonably stable cation radicals. The kinetics of the reactions of the cation radicals with nucleophiles can be studied by monitoring the decay in concentration after the addition of nucleophiles by conventional techniques. A number of such studies have been reported.

*Disproportionation*

$$2\,\mathrm{A}^{+\cdot} \underset{k_{-24}}{\overset{k_{24}}{\rightleftharpoons}} \mathrm{A}^{2+} + \mathrm{A} \tag{24}$$

$$\mathrm{A}^{2+} + \mathrm{X} \overset{k_{25}}{\rightarrow} \text{Products} \tag{25}$$

$$-d[\mathrm{A}^{+\cdot}]/dt = 2k_{24}k_{25}\,[\mathrm{A}^{+\cdot}]^2[\mathrm{X}]/(k_{-24}[\mathrm{A}] + k_{25}[\mathrm{X}]) \tag{26}$$

*Complexation*

$$\mathrm{A}^{+\cdot} + \mathrm{X} \underset{k_{-27}}{\overset{k_{27}}{\rightleftharpoons}} (\mathrm{A/X})^{+\cdot} \tag{27}$$

$$(\mathrm{A/X})^{+\cdot} + \mathrm{A}^{+\cdot} \underset{k_{-28}}{\overset{k_{28}}{\rightleftharpoons}} (\mathrm{A/X})^{2+} + \mathrm{A} \tag{28}$$

$$(\mathrm{A/X})^{2+} \overset{k_{29}}{\rightarrow} \text{Products} \tag{29}$$

$$-d[\mathrm{A}^{+\cdot}]/dt = 2\,k_{28}k_{29}K_{27}[\mathrm{A}^{+\cdot}]^2[\mathrm{X}]/(k_{-28}[\mathrm{A}] + k_{29}) \tag{30}$$

*Half-regeneration*

$$\mathrm{A}^{+\cdot} + \mathrm{X} \underset{k_{-31}}{\overset{k_{31}}{\rightleftharpoons}} \mathrm{A}\cdot-\mathrm{X}^+ \tag{31}$$

$$\mathrm{A}\cdot-\mathrm{X}^+ + \mathrm{A}^{+\cdot} \underset{k_{-32}}{\overset{k_{32}}{\rightleftharpoons}} \mathrm{A}^+-\mathrm{X}^+ + \mathrm{A} \tag{32}$$

$$\mathrm{A}^+-\mathrm{X}^+ \overset{k_{33}}{\rightarrow} \text{Products} \tag{33}$$

$$-d[\mathrm{A}^{+\cdot}]/dt = 2\,k_{32}k_{33}K_{31}[\mathrm{A}^{+\cdot}]^2[\mathrm{X}]/(k_{-32}[\mathrm{A}] + k_{33}) \tag{34}$$

Scheme 1

By virtue of the fact that the initial reaction between a cation radical and a nucleophile produces an unstable radical intermediate, the reactions are inevitably complex. Further electron-transfer reactions accompanied by proton transfer or reaction with nucleophiles are necessary in order to reach a stable product. These multistep reactions can give rise to complicated rate laws and the observed mechanism can be highly dependent upon the reaction conditions, especially on the magnitudes of the concentrations of the reactants. Three mechanisms either have been or are believed to describe the reaction pathways of cation radicals with nucleophiles. The mechanisms and the corresponding rate laws are shown in Scheme 1.

The mechanisms are designated Disproportionation, Complexation and Half-regeneration. The most important difference between the Disproportionation mechanism and the other two is that the dication, rather than the cation radical, is the species reacting with the nucleophile. The primary difference between the Complexation and the Half-regeneration mechanisms is that in the former the initial interaction results in the formation of a $\pi$-complex and covalent bonding does not occur until the dication stage while in the latter the bond formation takes place in the first step. Thus, the Complexation mechanism can be considered as a compromise between Disproportionation where the dication is the electrophile and Half-regeneration where only ions with a single charge are involved.

COMMENTS ON EQUILIBRIA AND REACTION KINETICS

The rate laws given in Scheme 1 were derived making use of both the steady state and the equilibrium approximations. These approximations are necessary in order to obtain tractable relationships. However, a degree of caution must be exercised in using the approximations, especially that relating to equilibrium. It has recently been pointed out that it is very important to consider the position of equilibrium and that using the approximation in cases where the equilibrium constant is large is not a valid procedure and can lead to rate laws that are not consistent with the mechanisms being considered (Hammerich and Parker, 1982a).

Disproportionation to doubly charged ions (DI) and neutral parent compounds is always a distinct possibility during ion radical (IR) reactions. The DI will invariably be more reactive toward nucleophiles (dications) or electrophiles (dianions) than the corresponding IR. The likelihood that such reactions will follow the disproportionation pathway depends upon three factors; (a) $k_{DI}/k_{IR}$, the relative rate constants for the reactions of the two species, (b) the magnitude of $K_{disp}$, the disproportionation equilibrium constant and (c) the magnitude of $k_{disp}$, the rate constant for disproportionation. For cation radical $A^{+\cdot}$, $K_{disp}$ for reaction (37) can be calculated from the difference in standard potentials for electrode reactions (35) and (36) according to relationship (38) which reduces to (39) at 298 K.

$$A^{\cdot +} + e^- \underset{}{\overset{E°_1}{\rightleftharpoons}} A \qquad (35)$$

$$A^{\cdot +} \underset{}{\overset{-E°_2}{\rightleftharpoons}} A^{2+} + e^- \qquad (36)$$

$$2A^{\cdot +} \underset{}{\overset{K_{disp}}{\rightleftharpoons}} A^{2+} + A \qquad (37)$$

$$-\Delta G° = RT \ln K_{disp} = F(E°_1 - E°_2) \qquad (38)$$

$$\log K_{disp} = (E°_1 - E°_2)/0.0591 \qquad (39)$$

In general, values of $K_{disp}$ for cation radicals of aromatic compounds in solvents such as acetonitrile are very small. Since the dications are very reactive, a difficulty arises in the determination of $K_{disp}$. It is necessary to measure the reversible potential for reaction (36) and this requires that $A^{2+}$ be observable under the conditions of the measurement. The measurements can be made in solvents containing strong mineral acids (Hammerich and Parker, 1972) or even in neutral solvents if nucleophilic impurities are effectively removed (Hammerich and Parker, 1973). The values of $-(E°_1 - E°_2)$ for the oxidations of substrates of structures [1] and [2] in acetonitrile were observed to be of the order of 500 mV or greater and in general, $K_{disp} < 10^{-9}$.

Once reliable values of $K_{disp}$ are available, the feasibility of the disproportionation mechanism can be tested by comparing the observed rate constants ($k_{obs}$) with the maximum possible values predicted for (37) and (40) (Parker, 1972). Reactions such as the pyridination of 9,10-diphenylanthracene (DPA)

$$A^{2+} + \text{Nucleophile} \rightarrow \text{Products} \qquad (40)$$

cation radical and the hydroxylation of thianthrene [18; X = Y = S] cation radical had been observed in early studies (Manning et al., 1969; Murata and Shine, 1969; Shine and Murata, 1969) to be first order in the nucleophiles. This requires that if the disproportionation mechanism applies, forward reaction (37) must be at least 10 times more rapid than (40). Thus, the observed pseudo second order rate constant ($k_{obs} = 2k_{40}K_{37}$) can be no greater than $10^{-1}k_f$ where $k_f$ refers to forward reaction (37). The maximum possible value of $k_f$ can be estimated from (41) where $(k_b)_{diff}$ refers to a diffusion

$$(k_f)_{max} = (k_b)_{diff} K_{disp} \qquad (41)$$

controlled rate constant for back reaction (37). For example, if $K_{disp}$ is $10^{-9}$ and $(k_b)_{diff}$ is $10^{10}$ M$^{-1}$s$^{-1}$, $(k_f)_{max}$ is 10 M$^{-1}$s$^{-1}$ and the maximum value of $k_{obs}$ consistent with the disproportionation mechanism is 1 M$^{-1}$s$^{-1}$.

In order for the Complexation mechanism to be favorable relative to Disproportionation, the formation of the complexed cation radical must be accompanied by a significant lowering of the oxidation potential. A rough estimate of the relative rate constants for electron transfer can be made by assuming that the reverse reaction is diffusion controlled which would give rise to a decade change in the forward rate constant for a 60 mV change in the oxidation potential. If the ratio of the complexed to the uncomplexed cation radical concentrations is $10^{-2}$ an oxidation potential shift of about 180 mV would be required for the rate of the Complexation reaction to be 10 times that of Disproportionation.

There are no pertinent data to use in making predictions of the effect of complex formation on the oxidation potentials of the cation radicals. However, the extreme of complex formation could be considered to be covalently bonded adduct formation. The nucleophile adducts [20] of 9,10-diarylanthracene cation radicals [19] where Nu is $CF_3CO_2^-$, $CH_3O^-$ or $HO^-$ are oxidized at potentials about 1.0 to 1.2 V less positive than the corresponding

[19]   [20]

cation radicals (Hammerich and Parker, 1974a). The oxidation potentials of the cation radical-nucleophile complexes would be expected to be intermediate between those of [19] and [20]. The ease of oxidation of the complexes would depend upon the identity of the nucleophile as well as the strength of the interactions in the complex.

Since the favorable effect of lowering the oxidation potential brought about by complex formation is counteracted by a low equilibrium concentration of the complex, the Complexation mechanism passing through the dication is not very favorable. If the initially formed complex has another favorable reaction pathway this will be expected to compete. In that case, the Complexation mechanism and the Half-regeneration mechanism cannot be distinguished by kinetic measurements.

REACTIONS OCCURRING AT THE DICATION STAGE

The initial studies of the reactions of the thianthrene (Th) cation radical with water (Murata and Shine, 1969; Shine and Murata, 1969) as well as with anisole (Silber and Shine, 1971) and phenol (Kim et al., 1974) resulted in the proposal of the Disproportionation mechanism for all of these reactions. The reaction of DPA$^{+\cdot}$ with pyridine was also proposed to take place by the Disproportionation mechanism (Marcoux, 1971). Later, the Half-regeneration mechanism was found to be more consistent with the results of a voltammetric study of the hydroxylation of Th$^{+\cdot}$ (Parker and Eberson, 1970) and arguments were presented which render the Disproportionation mechanism highly unlikely for the pyridination of DPA$^{+\cdot}$ (Parker, 1972). All of these reactions were later studied in more detail and the results of the most recent studies are discussed in the following paragraphs.

*Reactions of thianthrene cation radical with arene nucleophiles*

The value of $K_{disp}$ for Th$^{+\cdot}$ in acetonitrile was observed to be equal to $2.3 \times 10^{-9}$ (Hammerich and Parker, 1973) which results in a value for $(k_{obs})_{max}$, estimated in the manner described in a previous section, of 2.3 M$^{-1}$s$^{-1}$ for the reactions of Th$^{+\cdot}$ going through the Disproportionation mechanism with rates directly dependent upon the nucleophile concentration. Values as high as 0.2 M$^{-1}$s$^{-1}$ for the reaction with water (Murata and Shine, 1969) and 3 M$^{-1}$s$^{-1}$ for the reaction with anisole (Silber and Shine, 1971) were reported. Thus, it is conceivable on the basis of this data that both of these reactions follow the Disproportionation mechanism. However, the voltammetric evidence (Parker and Eberson, 1970) is most consistent with the cation radical as the species undergoing nucleophilic attack.

A detailed kinetic analysis of the reaction between Th$^{+\cdot}$ and anisole (AnH) was carried out in in CH$_2$Cl$_2$/TFA/TFAn (97/2/1), where TFA is trifluoroacetic acid and TFAn is the anhydride (Svanholm et al., 1975). The stoichiometry of the reaction is that shown in (42). Under conditions where [AnH] and [Th] are large relative to [Th$^{+\cdot}$] these quantities can be considered to be constant and useful relationships involving $1/k_{app}$, the apparent second order rate constants can be derived for the three mechanisms. This analysis serves

$$2 \; \text{[Th}^{+\cdot}\text{]} + \text{[AnH]} \longrightarrow \text{[Th-An}^{+}\text{]} + \text{[Th]} + H^+ \tag{42}$$

to differentiate between the Disproportionation mechanism and the other two mechanisms. The distinguishing feature is the intercepts of plots of $1/k_{app}$ vs [Th]. Disproportionation requires that the intercept be a constant independent of [AnH] while the intercepts for the Complexation and Half-regeneration mechanisms are predicted to be directly proportional to $[AnH]^{-1}$. The kinetic study (Svanholm et al., 1975) resulted in a linear relationship (43), where $a$ and $b$ are constants, for the reactions of Th$^{+\cdot}$ in the presence of excess

$$1/k_{app} = a[Th]/[AnH] + b/[AnH] \quad (43)$$

Th and AnH. This analysis ruled out the Disproportionation mechanism.

Since the kinetics do not distinguish between the Complexation and the Half-regeneration mechanisms it was necessary to rely on other considerations to assign a mechanism. If the reaction involved the formation of the covalently bonded intermediate [21], proton loss would be expected to intervene before involvement of the second Th$^{+\cdot}$ moiety. Further evidence regarding

(44)

this point was obtained from a comparable study of the reaction of Th$^{+\cdot}$ with phenol (Svanholm and Parker, 1976a) a reaction with general features very similar to (42). If [23] were a common intermediate in all cases and undergoes competing electron transfer (45) and proton loss (46), the nature of R (H or CH$_3$) in [23] would be expected to have a profound effect on the partitioning between reactions (45) and (46). The rate of reaction (45) should be more or less independent of R while (46) would be expected to be several orders of magnitude faster when R is H. This follows from the fact that proton transfer (46) can be from oxygen when R is H but must be from carbon in the other case. Under nearly identical conditions in $CH_2Cl_2$ containing TFA, reactions involving both AnH and phenol were observed to go nearly exclusively by pathways second order in $Th^{+\cdot}$ when $[Th^{+\cdot}]$ ranged from $10^{-4}$ to $10^{-3}$ M. At lower $[Th^{+\cdot}]$ the reaction became first order in $Th^{+\cdot}$ when the other reactant was AnH but not in the phenol case. This appears to be inconsistent with competing reactions (45) and (46) since (46) is far more favorable when R is H and would be expected to compete at higher $[Th^{+\cdot}]$. Furthermore, the deprotonation of dication [24] especially when R is H, would be expected to be so rapid that reverse reaction (45) could not compete effectively which would cause deviations from the observed kinetics.

## Reactions of anthracene cation radicals

The cation radicals of anthracene and 9-phenylanthracene (PA) were observed to undergo arylation reactions in the presence of a number of benzene derivatives including anisole (Svanholm and Parker, 1976b). The reaction between $PA^{+\cdot}$ and AnH was studied in the most detail and the kinetics were observed to be similar to those of the reactions of $Th^{+\cdot}$. The complexation mechanism was proposed for these reactions as well.

The cation radical of dibenzo-*p*-dioxin [18; X = Y = O], $DBO^{+\cdot}$, does not undergo nucleophilic attack by AnH. However, in the presence of $DBO^{+\cdot}$ at the same concentration, the apparent rate constant for the reaction of $PA^{+\cdot}$ with AnH was about 150 times greater than when $PA^{+\cdot}$, was the only cation radical present. This rate enhancement was attributed to electron-transfer catalysis (47) and, along with a similar effect observed during the anisylation

$$(PA/AnH)^{+\cdot} + DBO^{+\cdot} \rightleftharpoons (PA/AnH)^{2+} + DBO \qquad (47)$$

of $Th^{+\cdot}$, is an early example of electron-transfer catalysis which has been studied in detail in recent years.

The pyridination of $DPA^{+\cdot}$ can be considered to belong to the same class of reactions as the arylation of cation radicals with benzene derivatives. The most important difference between the nucleophilic action of pyridine and that of the other arene nucleophiles is that bonding takes place to nitrogen in the pyridine case and C—C bonding is involved with the other nucleophiles. The mechanism of the reaction was first studied by rotating disk electrode voltammetry and $DPA^{+\cdot}$ was concluded to be the species undergoing attack by pyridine (Manning et al., 1969). After the proposal of the Disproportionation mechanism (Marcoux, 1971) was deemed unlikely (Parker, 1972), kinetic studies were carried out (Svanholm and Parker, 1973; Blount, 1973; J. F. Evans and Blount, 1978) and all of them resulted in the conclusion that the reaction in acetonitrile follows rate law (48). The final product of the

$$-d[DPA^{+\cdot}]/dt = 2\, k_{49}[DPA^{+\cdot}][\text{pyridine}] \qquad (48)$$

reaction was characterized as the bis-pyridinium adduct [26] (Shang and Blount, 1974). Thus, it appeared that the overall reaction can be described by steps (49)–(51).

More recent studies have resulted in conflicting results. While only rate law (48) was found at [DPA$^{\ddot{+}}$] of the order of $10^{-4}$ M, (52) was proposed for data obtained at lower concentrations (J. F. Evans and Blount, 1979).

$$-[\text{DPA}^{\ddot{+}}]/dt = 2 k_{50}K_{49}[\text{DPA}^{\ddot{+}}]^2[\text{pyridine}] \quad (52)$$

Linear sweep voltammetry studies (Ahlberg and Parker, 1980; Parker, 1980) resulted in rate law (53). Difficulty is encountered in attempting to relate (53)

$$-[\text{DPA}^{\ddot{+}}]/dt = k_{\text{obs}}[\text{DPA}^{\ddot{+}}]^2[\text{pyridine}]/[\text{DPA}] \quad (53)$$

to mechanism (49)–(51). The observed reaction orders, 2 in DPA$^{\ddot{+}}$ and $-1$ in DPA, are consistent with reactions (49) and (50) being in equilibrium and followed by irreversible product-forming reactions. However, if the product-forming reaction is (51) the reaction order in pyridine would be 2.

On the other hand, the Complexation mechanism (54)–(56) provides a rationale for all of the rate data. Under conditions where the reaction is first

$$\text{DPA}^{\ddot{+}} + \text{C}_5\text{H}_5\text{N} \underset{k_{54}}{\overset{k_{54}}{\rightleftharpoons}} (\text{DPA}/\text{C}_5\text{H}_5\text{N})^{\ddot{+}} \quad (54)$$

$$(\text{DPA}/\text{C}_5\text{H}_5\text{N})^{\ddot{+}} + \text{DPA}^{\ddot{+}} \underset{k_{55}}{\overset{k_{55}}{\rightleftharpoons}} (\text{DPA}/\text{C}_5\text{H}_5\text{N})^{2+} + \text{DPA} \quad (55)$$

$$(\text{DPA}/\text{C}_5\text{H}_5\text{N})^{2+} \overset{k_{56}}{\rightarrow} \text{DPA}^+\text{—}\overset{+}{\text{N}}\text{C}_5\text{H}_5 \quad (56)$$

order in DPA$^{\ddot{+}}$, forward reaction (54) could be rate-determining and rate law (48) can be modified by replacing $k_{50}$ by $k_{54}$. When rate law (52) is applicable, equilibrium (54) could be followed by rate-determining electron transfer (55) and $k_{\text{obs}} = k_{55} K_{54}$. The final case arises when both (54) and (55) are at equilibrium and internal reorganization of the dication-pyridine complex (56) becomes rate-determining, resulting in rate law (53) with $k_{\text{obs}} = 2 k_{56} K_{54} K_{55}$. At the present time, mechanism (54)–(56) is the only one proposed which can account for all of the kinetic data.

It appears that in cases where the initially formed complex between a cation and a nucleophile does not have an energetically favorable rearrangement pathway available which could result in covalent bond formation, further electron transfer takes place and the reaction passes through the dication stage. This version of the Complexation mechanism has only been observed with arene nucleophiles. If the cation radical-nucleophile complex does have another favorable reaction pathway, this will be expected to compete. In the case of the reaction between Th$^{\ddot{+}}$ and AnH the complex apparently dissociates (57) and oxidation products of AnH are formed when

$$(\text{Th}/\text{AnH})^{\ddot{+}} \overset{k_{57}}{\rightarrow} \text{AnH}^{\ddot{+}} + \text{Th} \quad (57)$$

[Th$^{\dot{+}}$] is too low for reaction (58) to compete (Svanholm et al., 1975). This

$$(Th/AnH)^{\dot{+}} + Th^{\dot{+}} \xrightarrow{k_{58}} (Th/AnH)^{2+} + Th \tag{58}$$

reaction was not observed when the nucleophile was phenol but in $CH_2Cl_2$ in the absence of TFA the reaction is first order in $Th^{\dot{+}}$ and can be formulated as in (59) where $—C_6H_5O$ is a cyclohexadienonyl group (Svanholm and

$$(Th/C_6H_5OH)^{\dot{+}} \xrightarrow{k_{59}} Th\cdot—C_6H_5O + H^+ \tag{59}$$

Parker, 1976a). In general, if the nucleophile has an easily expelled proton or is negatively charged, rearrangement of the initially formed π-complex to the covalently bonded adduct would be expected to be more favorable than further oxidation to the dication.

### Reactions of nucleophiles with dications

Until now there have been only a few studies of the reactions of dications derived from oxidation of the corresponding cation radicals with nucleophiles. Under conditions where $DPA^{\dot{+}}$ is stable, i.e. in $CH_2Cl_2/TFA/TFAn$ (45/5/1), $DPA^{2+}$ reacts rapidly with $CF_3CO_2^-$ to form the adduct $DPA^+—OCOCF_3$ which is the oxidation product of [20]. The preparation of a solution of $DPA^+—OCOCF_3$ during two-electron oxidation of DPA is illustrated in Fig. 5 (Hammerich and Parker, 1974a). The cyclic voltammogram for the oxidation of DPA (Fig. 5a) shows peaks for the formation of $DPA^{\dot{+}}$ ($O_1$), oxidation of $DPA^{\dot{+}}(O_2)$, reduction of $DPA^{\dot{+}}(R_1)$ and reduction of $DPA^+—OCOCF_3(R_3)$. After one-electron oxidation, the cyclic voltammogram shows that the predominant species in solution is $DPA^{\dot{+}}$ (Fig. 5b) and after one-electron oxidation of $DPA^{\dot{+}}$ a stable solution of $DPA^+—OCOCF_3$ is obtained (Fig. 5c). Experiments of this nature were used to establish the ease of oxidation of the nucleophile adducts [20].

No reaction can be detected between $Th^{\dot{+}}$ and AnH, in low concentration, in acetonitrile under cyclic voltammetric conditions. On the other hand, $Th^{2+}$ reacts rapidly with AnH under these conditions and the kinetics of the reaction were studied by a voltammetric method (Svanholm et al., 1975). The kinetic method was that developed earlier (Jensen and Parker, 1973) and involved the observation of a pre-peak before the main oxidation peak of $Th^{\dot{+}}$ when [AnH] was limiting, i.e. lower than [$Th^{\dot{+}}$]. Determination of the rate constant for reaction (60) is illustrated by the voltammograms shown

$$Th^{2+} + AnH \xrightarrow{k_{60}} Th^+—C_6H_5OCH_3 + H^+ \tag{60}$$

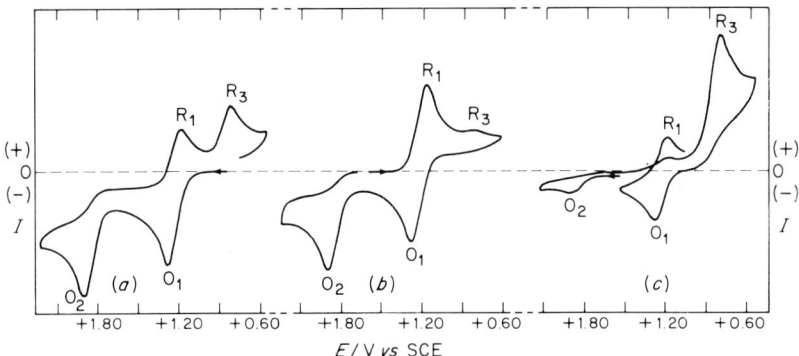

FIG. 5 Cyclic voltammograms of (a) DPA (1.0 mM), (b) DPA$^{+\cdot}$, (1.0 mM), (c) (DPA—OOCF$_3$)$^+$ (1.0 mM) in CH$_2$Cl$_2$/TFAn/TFA (45 : 5 : 1) containing n-Bu$_4$NBF$_4$ (0.2 M); voltage sweep rate, 86 mV s$^{-1}$. (Reprinted with permission from Hammerich and Parker, 1974a)

in Fig. 6. In Fig. 6a, O$_1$ corresponds to the formation of Th$^{+\cdot}$ and O$_2$ to the oxidation of Th$^{+\cdot}$ to Th$^{2+}$. Figures 6b–e show the effect of the addition of increments of AnH until [AnH] = [Th] = 1 mM (Fig. 6e). Measurement of the potential difference between the pre-peak and the main oxidation peak allowed $k_{60}$ to be evaluated. This resulted in a value of $10^7$ M$^{-1}$s$^{-1}$ at ambient temperature.

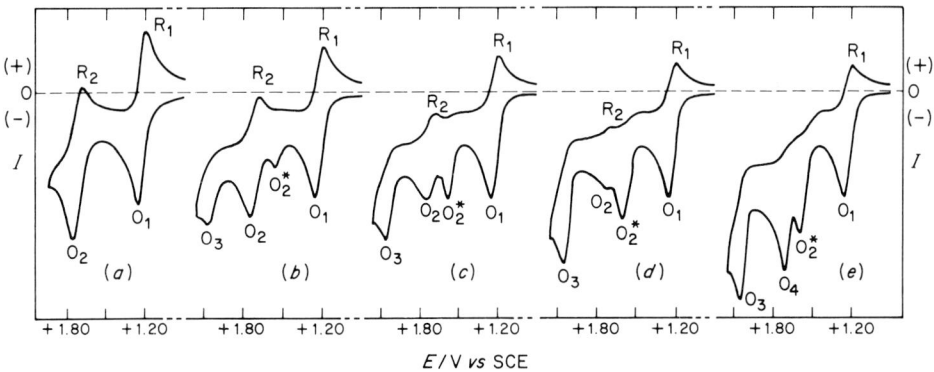

FIG. 6 Cyclic voltammograms for the oxidation of Th (1.0 mM) in acetonitrile containing n-Bu$_4$NBF$_4$ (0.2 M) in the presence of neutral alumina and anisole, [AnH] = 0 (a), 0.2 mM (b), 0.4 mM (c), 0.6 mM (d), and 1.0 mM (e). Voltage sweep rate = 86 mV s$^{-1}$. (Reprinted with permission from Svanholm et al., 1975)

REACTIONS OCCURRING AT THE CATION RADICAL STAGE

A number of these reactions, as limiting cases for the Complexation mechanism, have already been mentioned. In the case of nucleophiles which are negatively charged or those that have easily expelled protons, as is the case with $H_2O$, bonding usually occurs at the cation radical stage. Under these conditions the Complexation and Half-regeneration mechanisms are kinetically indistinguishable. Kinetic studies of the reactions of [17$\overset{+}{\cdot}$] and [18$\overset{+}{\cdot}$] with a variety of nucleophiles have been reported (Svanholm and Parker, 1973; Blount, 1973; Svanholm et al., 1975; Svanholm and Parker, 1976a–c, 1980; J. F. Evans and Blount, 1976a,b, 1977a,b, 1978, 1979; Cheng et al., 1978a,b; Neptune and McCreery, 1978; Sackett and McCreery, 1979; Mayausky and McCreery, 1983; Hammerich and Parker, 1982a,b,d, 1983; Parker and Hammerich, 1982).

*The reactions of thianthrene cation radical with water*

The mechanism studies of this reaction have attracted the most attention (Hanson, 1980; Shine, 1981). After the initial contradictory conclusions (Murata and Shine, 1969; Shine and Murata, 1969; Parker and Eberson, 1970) the reaction was not studied for several years. A detailed study (J. F. Evans and Blount, 1977a) indicated that the hydroxylation of $Th\overset{+}{\cdot}$ in acetonitrile is second order in $Th\overset{+}{\cdot}$, third order in $H_2O$ and inhibited by acid. These observations led to the proposal of mechanism (61)–(63) and rate law (64).

$$Th\overset{+}{\cdot} + H_2O \underset{}{\overset{K_{61}}{\rightleftharpoons}} Th\cdot\text{—}\overset{+}{O}H_2 \qquad (61)$$

$$Th\cdot\text{—}\overset{+}{O}H_2 + H_2O \underset{}{\overset{K_{62}}{\rightleftharpoons}} Th\cdot\text{—}OH + H_3O^+ \qquad (62)$$

$$Th\cdot\text{—}OH + Th\cdot\text{—}\overset{+}{O}H_2 \overset{k_{63}}{\rightarrow} \text{Products} \qquad (63)$$

$$-d[Th\overset{+}{\cdot}]/dt = 2k_{63}K_{61}{}^2 K_{62}[Th\overset{+}{\cdot}]^2[H_2O]^3/[H_3O^+] \qquad (64)$$

However, it has recently been pointed out that rate law (64) is inconsistent with mechanism (61)–(63) (Hammerich and Parker, 1982d) and that the analysis (J. F. Evans and Blount, 1977a) involved the improper use of the equilibrium approximation on reaction (61). The mechanism predicts reaction orders in $H_2O$ either equal to 1 or 2 depending upon the magnitude of $K_{61}$. If $K_{61}$ is large the predominant cation radical species will be $Th\cdot\text{—}\overset{+}{O}H_2$, a condition necessary for reaction (63) and a first order dependence upon $[H_2O]$ will result from reaction (62). On the other hand, if $K_{61}$ is small, the stronger oxidant $Th\overset{+}{\cdot}$ will participate in reaction (63) and a second order dependence on $[H_2O]$ is expected to arise from reactions (61) and (62).

The mechanism of the reaction between $Th^{+\cdot}$ and $H_2O$ in acetonitrile was reinvestigated under a wide variety of conditions (Hammerich and Parker, 1982d). The order in $H_2O$ was observed to be a complex function of the reaction conditions and ranged from 1 to 4.5. The reactions were studied in neutral acetonitrile and in the same solvent in the presence of acidic and basic buffers. The buffers $LH^+OCOCF_3^-$ were prepared from 2,6-lutidine (L) and $CF_3CO_2H$ in the presence of excess TFA (acidic) or L (basic). Both $H_2O$ and $D_2O$ were used as nucleophiles under all of the reaction conditions.

Even in the presence of very low concentrations of the base (L), the reaction order in $H_2O$ changed from the high variable value observed under neutral or acidic conditions to 1. This suggested that the complex water reaction order is simply a consequence of acid-base equilibria. In general terms, where the nature of base B is not specified, the mechanism of the hydroxylation reaction was formulated as in reactions (65)–(68). The Complexation mechanism versions of the first two steps are given in parentheses.

$$Th^{+\cdot} + H_2O \underset{}{\overset{K_{65}}{\rightleftharpoons}} Th\cdot\text{—}\overset{+}{O}H_2 \ (Th^{+\cdot} + H_2O \overset{K_{65}}{\rightleftharpoons} (Th/H_2O)^{+\cdot}) \quad (65)$$

$$Th\cdot\text{—}\overset{+}{O}H_2 + B \overset{K_{66}}{\rightleftharpoons} Th\cdot\text{—}OH + BH^+ \ ((TH/H_2O)^{+\cdot} + B \overset{K_{66}}{\rightleftharpoons} Th\cdot\text{—}OH + BH^+) \quad (66)$$

$$Th\cdot\text{—}OH + Th^{+\cdot} \underset{k_{-67}}{\overset{k_{67}}{\rightleftharpoons}} Th^+\text{—}OH + Th \quad (67)$$

$$Th^+\text{—}OH + B \overset{k_{68}}{\rightarrow} ThO + BH^+ \quad (68)$$

When B is $H_2O$, for every mole of ThO produced, two moles of $H_3O^+$ are formed which can show their inhibiting influence *via* reaction (66). Under conditions where electron-transfer reaction (67) is rate-determining (Evans and Blount, 1977a) a reaction order of 2 in water is predicted in the absence of other complications. The higher order in water can be accounted for by equilibrium (69) assuming that $H_3O^+$ is a more effective participant in reaction (66) than is $H_3O^+(H_2O)_n$.

$$H_3O^+ + nH_2O \overset{K_{69}}{\rightleftharpoons} H_3O^+(H_2O)_n \quad (69)$$

Mechanism (65)/(68) accounts for the fact that the reaction order in $H_2O$ is 1 in the presence of L. In this case, $B/BH^+$ is $L/LH^+$ and the only participation of $H_2O$ is as the nucleophile in reaction (65). Equilibrium (69) is insignificant under these conditions due to the very low $[H_3O^+]$.

The deuterium kinetic isotope effects ($k_H/k_D$) for the reactions of $Th^{+\cdot}$ with $H_2O$ and $D_2O$ were correlated (Parker and Hammerich, 1982) with the composite reaction order in cation radical and substrate, $R_{A/B}$ (Parker, 1981b).

Three rate laws consistent with mechanism (65)–(66) can be of importance depending upon the relative magnitudes of $k_{67}$, $k_{-67}$ and $k_{68}$. The rate laws (70)–(72) must account for the kinetic isotope effects as well as the observed

$$-d[Th\overset{+}{\cdot}]/dt = 2\, k_{68} K_{65} K_{66} K_{67} [Th\overset{+}{\cdot}]^2 [H_2O][B]^2/[BH^+][Th] \tag{70}$$

$$-d[Th\overset{+}{\cdot}]/dt = 2\, k_{67} K_{65} K_{66} [Th\overset{+}{\cdot}]^2 [H_2O][B]/[BH^+] \tag{71}$$

$$-d[Th\overset{+}{\cdot}]/dt = 2\, k_{61} k_{68} K_{65} K_{66} [Th\overset{+}{\cdot}]^2 [H_2O][B]^2/[BH^+]\,(k_{68}[B] + k_{-67}[Th]) \tag{72}$$

reaction orders. In neutral or acidic acetonitrile B is most probably $H_2O$ and the large $k_H/k_D$ observed (9–12 at 291–3 K) suggests that proton transfers (66) and (68) may both contribute to the rate and rate law (72) provides for the non-integral values of $R_{A/B}$ (1.3–1.6). In acidic buffer $R_{A/B}$ was observed to be equal to 1.83 and $k_H/k_D$ was 1.08 + 0.8. Under these conditions rate law (71) holds approximately and the low $k_H/k_D$ could be a consequence of the fact that proton transfer takes place in an equilibrium step (66). In basic buffer $R_{A/B}$ was observed to be about 1.7 and $k_H/k_D$ was about 2 which suggests that the rates of back reaction (67) and reaction (68) are of comparable magnitude and that rate law (72) is approximated under these conditions.

The reactions of $Th\overset{+}{\cdot}$ with water in $CH_2Cl_2/TFA/TFAn$ (97/2/1) give rise to kinetics consistent with the complexation mechanism (Hammerich and Parker, 1982a). In this case, a first order dependence of the rate on $[H_2O]$ provides strong evidence that bond formation takes place at the dication stage. Water was certainly the strongest base in the system and the alternative Half-regeneration scheme predicts a second order dependence on $[H_2O]$ by virtue of reactions (65) and (66). This suggests that the first step in the reaction, in either solvent system, involves the formation of the $\pi$-complex (65). In acetonitrile, but not in acidic $CH_2Cl_2$, reaction (66) results in bond formation at the cation radical stage. In the less polar and acidic solvent (66) is apparently not significant.

*The reaction of thianthrene cation radical with pyridine and $CF_3CO_2^-$*

In wet acetonitrile the reaction of $Th\overset{+}{\cdot}$ with pyridine is very rapid and irreversible. The reaction is illustrated by the voltammograms shown in Fig. 7. The reversible couple in Fig. 7a is for the formation and reduction of Th . When pyridine is added (Fig. 7b) the oxidation peak corresponds to a two-electron process and is irreversible. The addition of TFAn to the same solution resulted in a quasi-reversible two-electron process (Hammerich and Parker, 1982d). Since TFAn reacts rapidly with residual water to form

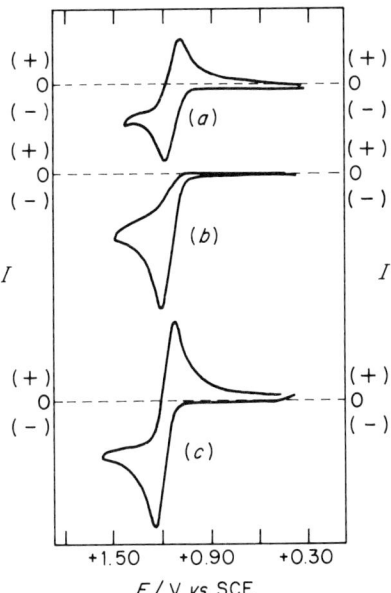

FIG. 7 Cyclic voltammograms of thianthrene in acetonitrile containing Bu$_4$NBF$_4$ (0.2 M). (a) In the presence of neutral Al$_2$O$_3$, (b) Solution a plus pyridine (50 mM) and (c) Solution b plus TFAn (6%). Sweep rate = 100 mV s$^{-1}$, θ = 22°C. (Reprinted with permission from Hammerich and Parker, 1982d)

TFA which in this case results in the formation of $CF_3CO_2^-$ $\overset{+}{H}NC_5H_5$, the reversible process is most probably (73). In fact, similar voltammetry was observed in acetonitrile containing $CF_3CO_2^-$ and TFAn in the absence of

$$2 \overset{+}{Th^{\cdot}} + CF_3CO_2^- \rightleftharpoons Th^+\text{—OCOCF}_3 + Th \qquad (73)$$

pyridine. In any event the reactions are too rapid to involve dication intermediates.

*Reactions of cation radicals of chlorpromazine and related compounds with water and anionic nucleophiles*

The reaction of chlorpromazine [27] cation radical with water in aqueous buffers results in the formation of the oxide [28], and the sulfurane [29] has been postulated to be an intermediate (Cheng et al., 1978a, 1978b; Sackett and McCreery, 1979). The mechanism of the reaction was described as (74)–(76) for reactions in phosphate and citrate buffers. This mechanism was criticized (Hammerich and Parker, 1982b); it was pointed out that no reasonable structure exists for [29$\bar{\cdot}$] and a more conventional mechanism can

[27]          [28]          [29]

account for the kinetic data. Although mechanism (74)–(76) was meant to apply only when the nucleophile is phosphate or citrate and not carboxylates

$$[27\overset{+}{\cdot}] + RCO_2^- + H_2O \overset{K_{74}}{\rightleftharpoons} [29\bar{\cdot}] + H^+ \tag{74}$$

$$[29\bar{\cdot}] + [27\overset{+}{\cdot}] \underset{k_{-75}}{\overset{k_{75}}{\rightleftharpoons}} [29] + [27] \tag{75}$$

$$[29] \overset{k_{76}}{\rightarrow} [28] + RCO_2H \tag{76}$$

in general (McCreery and Mayausky, 1982), it was pointed out (Hammerich and Parker, 1983) that the mechanism is unacceptable regardless of the nature of nucleophile. Besides the problem with the structure of [29$\bar{\cdot}$], the homogeneous electron transfer (75) cannot be treated as an equilibrium with a backward rate constant, $k_{-75}$ of comparable magnitude to $k_{76}$. Under the experimental conditions the standard potential for the redox couple [27]/[27$\overset{+}{\cdot}$] was observed to be close to 0.6 V vs the saturated calomel electrode. $E°$ for the reduction of sulfuranes like [29] are not available but may be estimated to be close to the peak potential for reduction of [30], i.e. about −1.9 V (Liao et al., 1974). From these electrode potentials, $K_{75}$ can be estimated to be of the order of $10^{42}$. This then predicts the maximum possible value of $k_{-75}$ to be of the order of $10^{-32} M^{-1} s^{-1}$ assuming $k_{75}$ to be about $10^{10} M^{-1} s^{-1}$. This value of $k_{-75}$ is so small that electron-transfer reaction (75) must be treated kinetically as an irreversible process and this results in a rate law inconsistent with that observed experimentally.

[30]

The formation of [28] in aqueous buffers has recently been discussed in terms of the more likely version of the Half-regeneration mechanism (Mayausky and McCreery, 1983) for monobasic anions, $B^-$, as in (77)–(79).

$$[27\overset{+}{\cdot}] + B^- \overset{K_{77}}{\rightleftharpoons} [27\cdot{-}B] \tag{77}$$

$$[27\cdot{-}B] + [27\overset{+}{\cdot}] \overset{K_{78}}{\rightleftharpoons} [27^+{-}B] + [27] \tag{78}$$

$$[27^+{-}B] + H_2O \overset{k_{79}}{\rightarrow} [28] + 2H^+ + B \tag{79}$$

An interesting aspect of this study is that a very big rate enhancement was found when the corresponding dibasic nucleophiles, $HPO_4^{2-}$ vs $H_2PO_4^-$, were used, providing that the anionic centres were spatially situated so that both could interact to give [27=B], which is a tetravalent S derivative analogous to [29]. This, along with the fact that the reaction order in $B^-$ was in all cases found to be 1, led to the proposal of mechanism (80)–(83) for these reactions.

$$[27\overset{+}{\cdot}] + B^{2-} \overset{K_{80}}{\rightleftharpoons} [27\cdot{-}B^-] \tag{80}$$

$$[27\cdot{-}B^-] + [27\overset{+}{\cdot}] \overset{K_{81}}{\rightleftharpoons} [27\cdot{-}B{-}27\cdot] \tag{81}$$

$$[27\cdot{-}B{-}27\cdot] \underset{k_{-82}}{\overset{k_{82}}{\rightleftharpoons}} [27{=}B] + [27] \tag{82}$$

$$[27{=}B] + H_2O \overset{k_{83}}{\rightarrow} [28] + 2H^+ + B^{2-} \tag{83}$$

Although this is consistent with the observed rate law, there are other factors which must be considered in assessing the probability of the mechanism. The equilibrium approximation was invoked for reactions (80) and (81). Thus, in order for mechanism (80)–(83) to be consistent with the observed rate law, both (80) and (81) must be shifted to the left. The equilibrium constant for reaction (84) can be estimated from related systems (Hammerich and Parker, 1974a) to be of the order of $10^7$, i.e. $[27\cdot{-}B^-]$ is expected to be oxidized at a potential over 400 mV less positive than [27]. This implies that $k_{84}$ will be a diffusion-controlled second order rate constant. Reaction (85),

$$[27\cdot{-}B^-] + [27\overset{+}{\cdot}] \overset{K_{84}}{\rightleftharpoons} [27^+{-}B^-] + [27] \tag{84}$$

$$[27^+{-}B^-] \overset{k_{85}}{\rightarrow} [27{=}B] \tag{85}$$

which is simply bond formation between oppositely charged centers held in close proximity to one another, could be exceedingly rapid. The rate law for the two mechanisms differ only in the meaning of the rate constants. It seems almost inconceivable that (81) could compete effectively with (84) and it could certainly not exclude the latter altogether.

The feasibility of reactions (84)–(85) in these systems is implied from the results on the reactions of the cation radical of N-methylphenothiazine [31] in aqueous acetonitrile in the presence of acetate buffers (Hammerich and

[structure of N-methylphenothiazine]

[31]

Parker, 1983). The reaction order in AcO$^-$ was observed to be of the order of 1.5 and $R_{A/B}$ was close to the same value. These kinetics are consistent with mechanism (86)–(88). Reaction (86) is written as a single step but

$$[31^{\stackrel{+}{\cdot}}] + AcO^- \underset{}{\overset{K_{86}}{\rightleftharpoons}} [31\cdot\text{---}OAc] \qquad (86)$$

$$[31\cdot\text{---}OAc] + [31^{\stackrel{+}{\cdot}}] \underset{k_{-87}}{\overset{k_{87}}{\rightleftharpoons}} [31^+\text{---}OAc] + [31] \qquad (87)$$

$$[31^+\text{---}OAc] + AcO^- \overset{k_{88}}{\rightarrow} [27^+\text{---}O^-] + Ac_2O \qquad (88)$$

probably involves Complexation in a pre-equilibrium step. It was impossible to determine whether or not water is kinetically involved in the reaction since $k_H/k_D$ was very close to unity and the effect of water on the activity of acetate ion complicates the situation. Thus, reaction (88) could conceivably result in the tetravalent S compound which undergoes hydrolysis in analogy to reaction (83).

*The reactions of 10-phenylphenothiazine in pyridine*

No fewer than five Disproportionation mechanisms (Evans and Blount, 1977b) were considered for the decomposition of 10-phenylphenothiazine cation radical in pyridine. The kinetic data were analysed by the method used earlier for the reaction between Th$^{\stackrel{+}{\cdot}}$ and AnH (Svanholm et al., 1975). It was concluded that disproportionation is not involved and the Half-regeneration mechanism was postulated. The Disproportionation mechanism can also be ruled out without carrying out detailed kinetic studies. The value of $K_{disp}$ was found to be $2.2 \times 10^{-12}$. This leads to $(k_{obs})_{max}$ of the order of $10^{-2} M^{-1} s^{-1}$ (Parker, 1972) which is $10^5$ times lower than the observed apparent second order rate constants, a fact that clearly eliminates the necessity to consider disproportionation mechanisms.

REACTION PATHWAYS OF CATION RADICALS OF ANTHRACENES IN THE PRESENCE OF WEAK NUCLEOPHILES

When nucleophilic impurities are effectively removed from solvents such as acetonitrile or dichloromethane either by deactivation with strong acids or by adsorption on neutral alumina, the solutions of the cation radicals of 9,10-diarylanthracenes are stable (Hammerich and Parker, 1974a). On the other hand, oxidation of anthracene in nucleophile-free acetonitrile results in the formation of the corresponding nitrilium ion according to reaction (89) (Hammerich and Parker, 1974b).

$$\text{anthracene} \xrightarrow[-2e^-, -H^*]{CH_3CN/TFAn} \text{anthracene-N}^+\equiv\text{C-CH}_3 \qquad (89)$$

The reactions of a series of 9-substituted anthracene cation radicals were studied in acetonitrile with the objective of determining the effect of the substituents on the rate of reaction (89) (Hammerich and Parker, 1982e). However, the preparative studies of the reactions showed that the products and the reaction mechanisms are strongly dependent upon the nature of the 9-substituent as shown in (91)–(94).

The dimerization of $PA^{\ddot{+}}$ which is involved in (94) gave kinetic data consistent with rate law (90). The apparent activation energy ($E_a$) for the

$$-d[PA^{\ddot{+}}]/dt = k_{app}[PA^{\ddot{+}}]^2 \qquad (90)$$

reaction was observed to be close to 0. Since related cation radical dimerizations have been observed to have appreciable values of $E_a$ (Aalstad et al., 1981a), $k_{app}$ in rate law (90) cannot apply to the simple irreversible dimerization mechanism. A reversible formation of a π-complex involving two $PA^{\ddot{+}}$ was postulated to be the primary reaction in analogy to related cases (Hammerich and Parker, 1981b; Parker and Bethell, 1981).

Under some conditions, kinetic data for the acetamidation of the anthracene cation radical ($AN^{\overset{+}{\cdot}}$) shown in (91) was observed to conform to rate law (95) which suggests mechanism (96)–(98) and $k_{obs} = 2 k_{98} K_{96} K_{97}$. Reaction

(98) corresponds to the loss of a proton from the 9-position giving rise to the formation of the nitrilium ion in (89). Acetamidation was also observed to be the major reaction pathway during the oxidation of 9-methylanthracene.

$$-d[AN\overset{+}{\cdot}]/dt = k_{obs}[AN\overset{+}{\cdot}]^2/[AN] \tag{95}$$

$$AN\overset{+}{\cdot} + CH_3CN \underset{}{\overset{K_{96}}{\rightleftharpoons}} AN\cdot\text{—}\overset{+}{N}\text{≡}CCH_3 \tag{96}$$

$$AN\cdot\text{—}\overset{+}{N}\text{≡}CCH_3 + AN\overset{+}{\cdot} \underset{}{\overset{K_{97}}{\rightleftharpoons}} AN^+\text{—}\overset{+}{N}\text{≡}CCH_3 + AN \tag{97}$$

$$AN^+\text{—}\overset{+}{N}\text{≡}CCH_3 \overset{k_{98}}{\rightarrow} \text{Products} \tag{98}$$

The rate of the reaction of $AN\overset{+}{\cdot}$ with $CH_3CN$ was observed to be increased upon decreasing the temperature. This was attributed to the inverse temperature dependence of equilibrium (96).

The cation radical of 9-nitroanthracene was observed to be long-lived in nucleophile-free acetonitrile but in the presence of TFA or $CF_3CO_2^-$, rapid trifluoroacetoxylation was observed to take place quantitatively by cyclic voltammetry but upon work-up anthraquinone was the only product detected as shown in (92). In the presence of $CF_3CO_2^-$, a kinetic study revealed a second order rate law (99) which corresponds to reaction (100) with an activation energy of about 7.5 kcal mol$^{-1}$.

$$-d[AN\text{—}NO_2\overset{+}{\cdot}]/dt = 2\,k_{100}\,[AN\text{—}NO_2\overset{+}{\cdot}][CF_3CO_2^-] \tag{99}$$

$$AN\text{—}NO_2\overset{+}{\cdot} + CF_3CO_2^- \overset{k_{100}}{\rightarrow} \text{Adduct} \tag{100}$$

Thus, under essentially the same conditions, in acetonitrile containing TFA, 9-substituted anthracene cation radicals undergo three distinctly different reactions; dimerization, acetamidation, or trifluoroacetoxylation. Dimerization predominates when the 9-substituent is phenyl or methoxy. Both of these substituents strongly stabilize the charges in the 10,10'-positions of the initially formed dimer dication. Formation of the dimer dication is highly unfavorable with electron-withdrawing nitro groups in the 10,10'-positions. The reason suggested for the failure of 9-nitroanthracene cation radical to undergo reaction with acetonitrile is that the electron transfer analogous to (97) is unfavorable in this case, once again because of the electron-withdrawing properties of the nitro group. The neutral adduct formed in reaction (100) is more easily oxidized and the reaction can readily go to completion. The reactions of anthracene and 9-methylanthracene cation radicals with the solvent are favored relative to reaction with TFA which is probably less nucleophilic and present in lower concentrations.

## THEORETICAL RELATIONSHIPS FOR THE ESTIMATION OF CATION RADICAL REACTIVITY TOWARD NUCLEOPHILES

The reactivity of cation radical-nucleophile pairs have been discussed in terms of orbital symmetry relationships (Eberson *et al.*, 1978c) and thermochemical calculations based upon comparisons with corresponding aromatic compound-radical reactions (Eberson and Nyberg, 1978). The methods and some of the results are summarized below.

### Application of the Dewar–Zimmerman rules

It was pointed out that the largest single body of consistent experimental work on the reactivity of cation radicals in homogeneous solution by Shine and coworkers (reviewed in Bard *et al.*, 1976) presents a rather perplexing reactivity pattern toward nucleophiles with respect to main reaction types (Eberson, 1975; Eberson and Nyberg, 1976). The reactions considered are electron-transfer oxidation of the nucleophile (101) and neutral adduct formation (102). The application of the Dewar–Zimmerman rules was made

$$ArH^{\overset{+}{\cdot}} + Nu^- \rightarrow ArH + Nu\cdot \quad (ET) \quad (101)$$

$$ArH^{\overset{+}{\cdot}} + Nu^- \rightarrow ArH\cdot\text{---}Nu \quad (N) \quad (102)$$

in an attempt to gain some insight into the reasons for the reactivity pattern.

The initial assumption was that the approach to the transition states for reactions (101) and (102) involves attack by the nucleophile perpendicular to the $\pi$-system at the midpoint between two centers. The transition state for the adduct formation was assumed to involve strong interactions while electron transfer was assumed to occur at greater distances. It was therefore supposed that if unfavorable interactions between the orbitals involved ensue in the early stages of the reaction that electron transfer will be favoured.

Both suprafacial and antarafacial interactions were assumed to be possible in all cases. The orbital representations shown in Fig. 8 illustrate the two types of interactions of a nucleophile p orbital with the $\pi$-system of the benzene cation radical. The first step in the analysis of the interactions was to add an electron to complete the half-vacant orbital. Then the transition state of Fig. 8a corresponds to a Hückel system containing 8 electrons, whereas that of Fig. 8b depicts a Möbius system with 8 electrons. It was concluded that these transition states are antiaromatic and aromatic, respectively, and that the suprafacial case would not lead to any reaction. With the N process "forbidden", ET was then concluded to be the only feasible reaction to take place in that system. The orbital representation for halide ions which have both s and p orbitals to be considered is somewhat more complicated with both favorable and unfavorable interactions but it was concluded that the $p_z$ interaction (illustrated in Fig. 8) predominates and the suprafacial interaction is unfavorable.

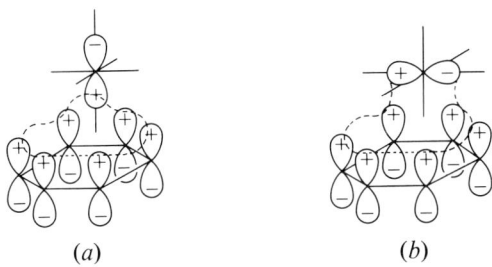

FIG. 8 Orbital representation of the transition state between benzene radical cation and a nucleophile p orbital interacting (a) suprafacially and (b) antarafacially. (Reprinted with permission from Eberson et al., 1978c)

The overall conclusion based on the orbital representations was that a nucleophile interacting suprafacially with a cation radical derived from a $4n + 2$ (Hückel) system corresponds to an antiaromatic transition state and should be less favored than the competing electron-transfer reaction. Since most of the mechanisms of the reactions which were treated by this approach are now known to be much more complex than the simple irreversible attack by the nucleophile (102) the details of the conclusions will not be pursued here. It was also concluded later (Eberson and Nyberg, 1978) that the standard redox potentials of the species undergoing reaction are of greater importance than orbital symmetry in determining the reactivity of cation radical-nucleophile pairs.

*Application of thermochemical calculations*

This approach (Eberson and Nyberg, 1978) is based upon the analogy between the cation radical/nucleophile and the corresponding neutral molecule/neutral radical reactions exemplified by (103) and (104). It was concluded that (a) the initial states are different, (b) the final states are identical, and

$$\mathrm{Ar}^{\overset{+}{\cdot}} + \mathrm{Nu}^- \rightleftharpoons (\mathrm{Ar}\ldots:\ldots\mathrm{Nu})^{\ddagger} \to \mathrm{Ar}\cdot-\mathrm{Nu} \qquad (103)$$

$$\mathrm{Ar} + \mathrm{Nu}\cdot \rightleftharpoons (\mathrm{Ar}\ldots:\ldots\mathrm{Nu})^{\ddagger} \to \mathrm{Ar}\cdot-\mathrm{Nu} \qquad (104)$$

(c) the transition states in all likelihood must be identical, assuming identical external reaction conditions in the two cases. When the assumption of identical transition states was made, it was then possible to construct the hypothetical energy diagram shown in Fig. 9. The free energy of activation for the reaction of interest ($\Delta G_1^{\ddagger}$), i.e. (103), was obtained as the sum of the free energy of activation for the Ar/Nu reaction ($\Delta G_2^{\ddagger}$) and the standard free energy difference between the initial states ($\Delta G_x^{\circ}$) which is assumed to be $>0$ in Fig. 9. $\Delta G_x^{\circ}$ can be estimated from thermochemical data and $\Delta G_2^{\ddagger}$ values were available from kinetic studies of aromatic free radical substitution reactions.

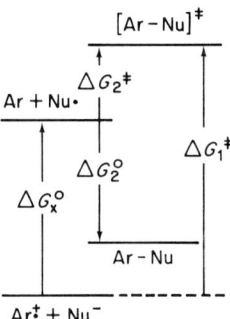

FIG. 9 Energy diagram showing the relationships between $\Delta G_x^{\circ}$, $\Delta G_1^{\ddagger}$, $\Delta G_2^{\ddagger}$ and $\Delta G_2^{\circ}$ for the case that $\Delta G_x^{\circ} \gg 0$. (Reprinted with permission from Eberson and Nyberg, 1978)

Since the reactants in (103) and (104) only differ in oxidation state, providing the external conditions are the same, $\Delta G_x^{\circ}$ is simply related to the difference in standard potentials for reactions (105) and (106), i.e. $\Delta G_x^{\circ} = -F(E_{106}^{\circ} - E_{105}^{\circ})$. The results were presented as plots of $\Delta G_x^{\circ}$ as a function of

$$\text{Ar}^{+\cdot} + e^- \underset{}{\overset{E_{105}^{\circ}}{\rightleftharpoons}} \text{Ar} \tag{105}$$

$$\text{Nu}\cdot + e^- \underset{}{\overset{E_{106}^{\circ}}{\rightleftharpoons}} \text{Nu}^- \tag{106}$$

$E_{105}^{\circ}$ which are a series of parallel straight lines, one for each nucleophile. This representation is illustrated in Fig. 10.

The data in Table 2 are the results of the treatment for the reaction of benzene cation radical with a series of pertinent nucleophiles. The most interesting features of these data are that both $\Delta G_x^{\circ}$ and $\Delta G_1^{\ddagger}$ are strongly negative for all nucleophiles other than fluoride ion. This is in qualitative agreement with the known very high reactivity of the cation radical. The data indicate that the reactions should be diffusion controlled but do not allow an unequivocal choice between whether the nucleophilic reaction takes place or if electron transfer should predominate. In this case there were no pertinent kinetic data with which to compare.

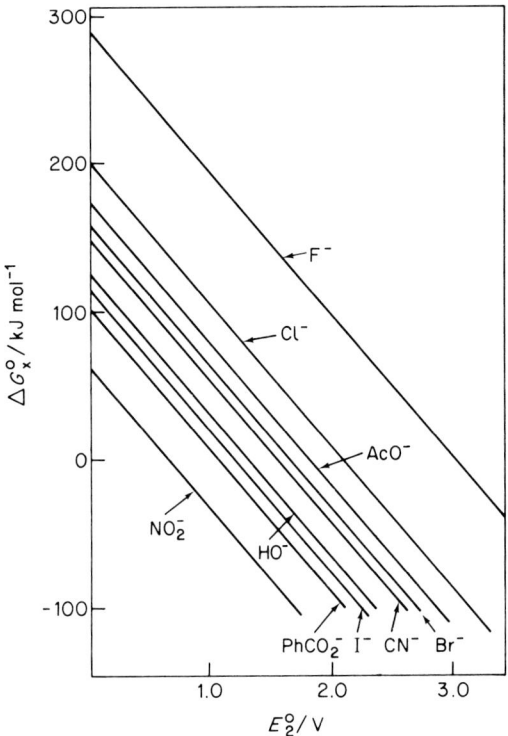

FIG. 10 Relationship between $\Delta G_x^\circ$ and $E_2^\circ$ for different nucleophiles. (Reprinted with permission from Eberson and Nyberg, 1978)

Rate data have been published for the reactions of perylene cation radical with a number of nucleophiles (T. R. Evans and Hurysz, 1977). Thus, these reactions would appear to be a good test case for the method. However, a difficulty arose in treating this cation radical since the pertinent $\Delta G_2^\ddagger$ are not available. Two different situations had to be treated, i.e. that when $\Delta G_2^\ddagger$ was either less than or greater than that for benzene. Analysis of the available data allowed some conclusions to be made. When $\Delta G_x^\circ < 0$, electron transfer was concluded to be diffusion controlled. When $\Delta G_x^\circ > 0$, electron transfer is slower and bond formation between $Ar^{\overset{+}{\cdot}}$ and $Nu^-$ could take place directly (unless symmetry restrictions operate). In general the conclusions based on the thermochemical method were in qualitative agreement with the results from kinetic and products studies in most cases.

TABLE 2

$\Delta G_x^\circ$, $\Delta G_2^\circ$, $\Delta G_2^\ddagger$, and "$\Delta G_1^\ddagger$" values for the benzene system ($E_2^\circ = 3.00$ V). All values are in kJ mol$^{-1}$. (Reprinted with permission from Eberson and Nyberg, 1978)

| Nu$^-$ | $\Delta G_x^\circ$ | $\Delta G_2^{\circ a}$ | $\Delta G_2^\ddagger$ | "$\Delta G_1^\ddagger$" |
|---|---|---|---|---|
| F$^-$ | 3 | $-150$ | 0 | 3 |
| Cl$^-$ | $-85$ | $-24$ | 25 | $-60$ |
| Br$^-$ | $-128$ | 37 | 37 | $-91$ |
| I$^-$ | $-173$ | 99 | 99 | $-74$ |
| CN$^-$ | $-138$ | $-192$ | 0 | $-138$ |
| HO$^-$ | $-160$ | $-86$ | 0 | $-160$ |
| AcO$^-$ | $-113$ | 12 | 28 | $-85\,(-50^b)$ |
| PhCOO$^-$ | $-185$ | 11 | 63 | $-122$ |
| NO$_2^-$ | $-225$ | 94 | 94 | $-131$ |

$^a$ Gas phase values; since all participating species in (104) are neutral, no serious error is likely to be introduced by using the same values for acetonitrile
$^b$ Assuming $\Delta G_2^\ddagger$ has the same value as for PhCOO·

*Comments on the theoretical methods*

At the time that both of these methods were developed (Eberson, 1975; Blum et al., 1977; Eberson et al., 1978c; Eberson and Nyberg, 1978) the reactions of cation radicals with nucleophiles were generally treated as elementary reactions such as (102). However, in more recent years it has become increasingly evident that the reactions are much more complex and usually involve coupled equilibria. The apparent low reactivity of cation radicals in some cases is not necessarily due to a high activation energy for the initial reaction. In the opinion of the authors, it is more likely that the initial interaction with a nucleophile is indeed very rapid but the equilibrium constant for the reaction is relatively small. Both of the methods described above presume an irreversible elementary reaction.

The assumption of identical transition states for reactions (103) and (104) could only be valid when the transition state for reaction (103) is very far along the reaction coordinate. Both Ar$^{\cdot+}$ and Nu$^-$ are very much more strongly solvated than is either Ar or Nu·, and in order for the transition states to be identical these solvation shells will have to be lost before arriving there. This limitation should be kept in mind.

## 4 Electron-transfer reactions initiated by cation radicals

It is obvious that electron transfer is an inherently important reaction of cation radicals. In this section emphasis will be placed on reactions in which the role of the cation radical is that of an oxidizing agent and may be used in either stoichiometric amounts or as a catalyst for reactions during which there is no net change in oxidation state. The latter type of reaction has been demonstrated to be of both synthetic and theoretical importance. In the earlier review (Bard et al., 1976) a significant amount of space was devoted to the chemistry of cation radicals related to the bipyridinium system. This topic has recently been extensively reviewed (Hünig, 1980). The progress in the field since then has mainly been of relevance to the possible aspects of these redox systems in energy conversion and the photoreduction of water to hydrogen, a topic which is beyond the scope of this chapter. The chemistry of this class of compounds has therefore been deliberately omitted. References in this area may be found in several recent papers (Okura et al., 1982; Darwent et al., 1982; Amouyal et al., 1982; Deronzier and Esposito, 1983; Sullivan et al., 1982).

CATION RADICALS AS OXIDANTS

The primary event of many, but not all, organic oxidation reactions is the transfer of one electron from substrate to the oxidant. Mechanistic information concerning these reactions can be obtained conveniently by electrochemical methods or by more conventional techniques together with the use of so-called one-electron transfer agents. Included in this group are ions such as Ce(IV), Co(III), Mn(III) and Fe(III), salts of inorganic ions like $NO^+PF_6^-$ and Fremy's salt, $K_2(NO(SO_3)_2)$ as well as some stable organic cation radicals. Among these, cation radicals play a special role since the electron-transfer reactions of these species are generally very fast in contrast to the slow electron transfer often encountered when organic molecules react with the inorganic oxidants.

Although a large number of organic cation radical salts, which have been isolated and characterized, may serve as one-electron oxidants the cation radicals of *p*-substituted triphenylamines and in particular the *p*-bromo derivative have most often been used. The popularity of these species as oxidants is without doubt connected to the fact that they are easily prepared from readily available starting materials (Walter, 1955; Bell et al., 1969b; Barton et al., 1975) and show exceptional thermal and photochemical stability. Particularly, the hexachloroantimonate salts of the cation radicals are of interest in this sense since they may be stored almost indefinitely. In solution, monitoring the decay of the intense blue to bluish-green colors can provide a convenient way to study the kinetics of the reactions. The triarylamines resulting from the electron-transfer reactions are in most cases inert and have no influence on the course of the reactions. The basicity of the amines is low due to the presence of the aryl groups and they are not susceptible to reactions with electrophilic reagents. A further advantage from the preparative point of view is that the triarylamines are only slightly soluble in many organic solvents and often precipitate during the reactions. In other cases products may simply be distilled from the amine.

The oxidation potentials of triarylamines may be adjusted within a broad range by suitable substitution as shown in Table 3 (Reynolds et al., 1974; Schmidt and Steckhan, 1980). The effect of variation of the *p*-substituent is as expected; electron donating substituents tend to lower the oxidation potential while electron withdrawing substituents have the opposite effect. In other words, electron withdrawing substituents provide cation radicals that are stronger oxidants than those with electron donating substituents.

TABLE 3

Reversible oxidation potentials for triarylamines[a]

| Compound | R | $X^1$ | $X^2$ | $X^3$ | $Y^b$ | $E$/V vs SCE |
|---|---|---|---|---|---|---|
| [32] | $CH_3O$ | | | | | 0.52 |
| [33] | $CH_3$ | | | | | 0.75 |
| [34] | F | | | | | 0.95 |
| [35] | Cl | | | | | 1.04 |
| [36] | Br | | | | | 1.06[c] |
| [37] | Br | Br | | | | 1.18[c] |
| [38] | Br | Br | Br | | | 1.32[c] |
| [39] | Br | Br | Br | Br | | 1.50[c] |
| [40] | Br | Br | Br | Br | Br | 1.72[c] |

[a] Reynolds et al., 1974; Schmidt and Steckhan, 1980
[b] $X^1 = X^2 = X^3 = Y = H$ unless otherwise stated
[c] Converted from V vs NHE by subtraction of 0.24 V

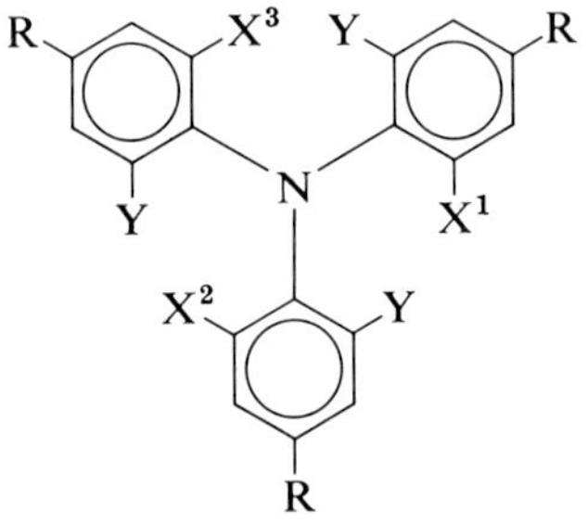

[32—40]

The effect of the successive introduction of bromine atoms in the *o*-positions of tris(*p*-bromophenyl)amine has been discussed in terms of both electronic and steric effects (Schmidt and Steckhan, 1980). A comparison of the potentials for the oxidation of compounds [36] to [39] shows that substitution of one hydrogen atom by bromine causes the oxidation potential of the triarylamine to increase from 120 to 180 mV (Table 3) while the effect of the additional three bromine atoms in [40] amounts to only 220 mV or approximately 70 mV per bromine atom. In addition to the electronic effect of exchanging the o-hydrogens with bromine, a steric effect is also observed. The aryl groups are significantly twisted about the carbon—nitrogen bond resulting in less effective conjugation of the π-system with the nitrogen lone pair. This effect was considered to have reached its maximum after the introduction of one bromine atom in the o-position of each ring and the three additional bromine atoms in [40] are then expected to affect the oxidation potential mainly through the electronic effect.

Advantage has been taken of the small, but significant, differences in the redox potentials of compounds [36], [37], [38] and [39] in the application of the corresponding cation radicals in the oxidative cleavage of benzyl- and p-methoxybenzyl ethers which are often used as alcohol protecting groups. It was demonstrated (Schmidt and Steckhan, 1978a) that p-methoxybenzyl ethers could be effectively cleaved to the alcohols by oxidation with the cation radical salt of [36] (107). Yields between 75 and 95% were observed depending upon the nature of R and on whether the reaction was carried out using

$$\text{ROCH}_2\text{An} \xrightarrow[\text{2) H}_2\text{O}]{\text{1) 2 [36}\overset{+}{\cdot}\text{]}} \text{ROH} + \text{AnCHO} \qquad (107)$$

stoichiometric amounts of the cation radical salt or by indirect electrolysis using only catalytic amounts of [36]. It appeared that the cation radical of [36] was not a sufficiently strong oxidant to cleave the benzyl ethers, but a very clean reaction was observed when the cation radicals of [38] and especially [39] were used (Schmidt and Steckhan, 1979a,b). The yields were between 90 and 95% except in one case where R contained a secondary alcohol function which was apparently also attacked by the cation radical.

The difference in behavior of p-methoxybenzyl ethers and unsubstituted benzyl ethers was used in an elegant selective bromination of the two alcohol groups in the diol [41] to give either [42] or [43] (Schmidt and Steckhan, 1979b). The results are summarized in Scheme 2. Treatment of [41] with sodium in boiling dioxan followed by the addition of p-methoxybenzyl chloride afforded [44] in 83% yield after which the secondary alcohol group was converted to the corresponding benzyl ether by reaction with NaH and benzyl chloride in DMF. The p-methoxybenzyl ether linkage in [45] was then selectively cleaved by oxidation with [36$\overset{+}{\cdot}$] to give the mono-protected alcohol [46]. Conversion to the bromide [47] could be carried out by reaction with

Ph$_3$PBr$_2$ in DMF and [42] was then finally obtained by oxidative cleavage with [39$^{\ddot{+}}$]. The overall yield of [42] in the five step synthesis was 50%. If alternatively, [44] was treated directly with the brominating agent, [48] was obtained from which the isomeric monobromo alcohol could be obtained by cleavage with [36$^{\ddot{+}}$] in an overall yield of 62%.

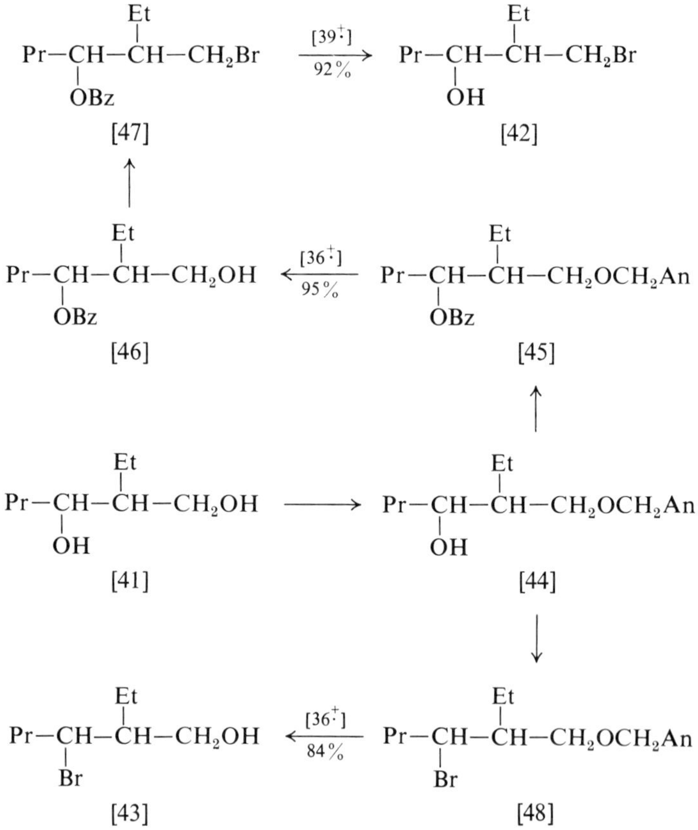

$$\begin{array}{ccc}
\text{Et} & & \text{Et} \\
| & & | \\
\text{Pr}-\text{CH}-\text{CH}-\text{CH}_2\text{Br} & \xrightarrow[92\%]{[39^{\ddot{+}}]} & \text{Pr}-\text{CH}-\text{CH}-\text{CH}_2\text{Br} \\
| & & | \\
\text{OBz} & & \text{OH} \\
{[47]} & & {[42]} \\
\uparrow & & \\
\text{Et} & & \text{Et} \\
| & & | \\
\text{Pr}-\text{CH}-\text{CH}-\text{CH}_2\text{OH} & \xleftarrow[95\%]{[36^{\ddot{+}}]} & \text{Pr}-\text{CH}-\text{CH}-\text{CH}_2\text{OCH}_2\text{An} \\
| & & | \\
\text{OBz} & & \text{OBz} \\
{[46]} & & {[45]} \\
& & \uparrow \\
\text{Et} & & \text{Et} \\
| & & | \\
\text{Pr}-\text{CH}-\text{CH}-\text{CH}_2\text{OH} & \longrightarrow & \text{Pr}-\text{CH}-\text{CH}-\text{CH}_2\text{OCH}_2\text{An} \\
| & & | \\
\text{OH} & & \text{OH} \\
{[41]} & & {[44]} \\
& & \downarrow \\
\text{Et} & & \text{Et} \\
| & & | \\
\text{Pr}-\text{CH}-\text{CH}-\text{CH}_2\text{OH} & \xleftarrow[84\%]{[36^{\ddot{+}}]} & \text{Pr}-\text{CH}-\text{CH}-\text{CH}_2\text{OCH}_2\text{An} \\
| & & | \\
\text{Br} & & \text{Br} \\
{[43]} & & {[48]} \\
\end{array}$$

**Scheme 2**

*p*-Methoxybenzyl and unsubstituted benzyl ethers may also be oxidatively cleaved by anodic oxidation, but even with careful control of the anode potential it is not possible to oxidize *p*-methoxybenzyl ether selectively in the presence of an unsubstituted benzyl ether. The selective cleavage offered by the appropriate choice of triarylamine cation radical may have important applications in carbohydrate chemistry. A related reaction, the cleavage of benzyl esters by Ar$_3$N$^{\ddot{+}}$, has been reported recently (Dapperheld and Steckhan, 1982).

The electron-transfer induced decomposition of diazoacetophenone provides an illustration of the possibility to monitor the redox reaction between a triarylamine cation radical and an organic substrate by the decay in absorption due to the cation radical (C. R. Jones, 1981). Under pseudo first order conditions it was shown that the rate of disappearance of the cation radical was dependent upon the oxidation potential of the corresponding amine in the expected manner. The more powerful the oxidant, the faster the electron-transfer reaction was observed to be and an approximate linear relationship was found between $E°$ and $\ln k$. Irreversible loss of nitrogen, which was suggested to take place either in or after the rate-determining step, would result in formation of the carbene cation radical complex [49] which in turn might dissociate or undergo back-electron transfer to [50] and $Ar_3N^{+\cdot}$. The competition between these two pathways was deemed to favour dissociation, since attempts to trap the carbene with cyclohexene or methanol and subsequent isolation of [51] or [52] were unsuccessful. In this respect the decomposition of diazoacetophenone induced by $Ar_3N^{+\cdot}$ differs from that initiated by, e.g. tetraphenylethylene in which case products indicating the intermediate presence of [50] have been isolated in good yields (Ho et al., 1974).

Oxidation of a neutral organic compound by the cation radical of a triarylamine will in general produce the cation radical of that compound which then undergoes the reactions typical for that particular cation radical under the reaction conditions. In this respect oxidation by triarylamine cation radicals is not expected to differ from other methods of generating the cation radicals. Other examples of cation radical reactions which have been initiated by $Ar_3N^{+\cdot}$ include the intramolecular coupling of 3,3',4,4'-tetra-methoxybibenzyl (Kricka and Ledwith, 1973), the oxidation of hydrazones of aromatic ketones (Handoo and Handoo, 1982) and the dimerization of 1,1-bis($p$-dimethylaminophenyl)ethylene (Bawn et al., 1968). Further applications of triarylamine cation radicals as oxidants include reactions as initiators in cation radical catalyzed reactions and mediators in indirect anodic oxidations.

REDOX CATALYSIS

In conventional anodic oxidation electrons are transferred directly from the substrate to the anode. The current, which is a measure of the rate of the electron-transfer process depends, mainly, on three factors; ($i$) the electrode potential, ($ii$) the so-called exchange current density which is related to the activation energy for the electron-transfer process in the absence of the external electrical field, and ($iii$) the effectiveness of transport of the substrate to the electrode surface. In cases where the third factor is not current limiting the relationship between the electrode potential and the current flowing in one direction is given by (108), where $\beta$ is a symmetry factor often termed

$$i = i_o e^{(1-\beta)(E-E^\circ)F/RT} \tag{108}$$

the transfer coefficient and $i_o$ is the exchange current density. Practical experiments show that the activation energy for electron transfer between an organic molecule and an electrode may often take values of considerable magnitude resulting in correspondingly small values of $i_o$. It is then necessary to operate at electrode potentials displaced far from $E^\circ$, i.e. at high overpotentials, to ensure that appreciable current flows. The origin of high activation energies for electrode reactions may be difficult to find, but it is often associated with the heterogeneous nature of the electron exchange. In such cases it may be possible to circumvent the problem by the addition of a mediator to serve as an electron carrier between the substrate and the electrode. An obvious requirement is that the mediator must have a negligible activation energy for heterogeneous electron transfer. Another requirement is that the cation radical of the mediator be stable to the environment and not participate in other reactions. These requirements are fulfilled by some cation radicals of heteroaromatic molecules properly substituted in positions of high reactivity when necessary. Examples of such compounds are the triarylamines and dibenzodioxin. In addition, 9,10-diphenylanthracene has been used as a mediator. This form of redox catalysis has been treated theoretically in a number of recent papers (Andrieux et al., 1978a,b,c, 1980a,b).

The obvious advantage of applying redox catalysis in electrochemical oxidation is the possibility to use a smaller overpotential to obtain a given current. The technique offers a number of further benefits as well. Since the mediator (Med) is continuously recycled (109–110) only catalytic amounts are necessary in preparative work which facilitates work-up procedures as

$$\text{Med} - e^- \rightleftharpoons \text{Med}^{+\cdot} \quad (109)$$

$$\text{Med}^{+\cdot} + A \rightarrow \text{Med} + A^{+\cdot} \quad (110)$$

$$A^{+\cdot} \rightarrow \text{Products} \quad (111)$$

compared to reactions where stoichiometric amounts of an oxidizing agent have been used. Experience also indicates that problems such as electrode filming, sometimes serious in conventional electrolysis, is in general less severe during redox catalysis.

The anodic oxidation of carboxylate ions is an example of a reaction the outcome of which may depend upon almost any possible experimental parameter (Eberson, 1973). The direct electron transfer from $RCOO^-$ to the anode is a very slow process and considerable overpotentials must be applied. This results in working potentials close to that for the discharge of solvent and supporting electrolyte. When [36] was used as mediator the reactions of aliphatic carboxylates proceeds through steps (112–117) (Schmidt and Steckhan, 1978b). The apparent rate constant, $k = 2 k_{114} K_{113}$, was determined by

$$[36] - e^- \rightleftharpoons [36^{+\cdot}] \quad (112)$$

$$[36^{+\cdot}] + RCOO^- \overset{K_{113}}{\rightleftharpoons} [36] + RCOO\cdot \quad (113)$$

$$RCOO\cdot \overset{k_{114}}{\rightarrow} R\cdot + CO_2 \quad (114)$$

$$R\cdot \rightarrow (R')\cdot \quad (115)$$

$$(R')\cdot + [36^{+\cdot}] \rightarrow (R')^+ + [36] \quad (116)$$

$$RCOO^- + (R')^+ \rightarrow RCOOR' \quad (117)$$

spectroelectrochemical techniques by comparison of experimental data with a theoretical working curve for the mechanism and a value of 60 s$^{-1}$ was observed. Making the assumption that the lifetime of RCOO· is $10^{-10}$s allowed the estimation of $K_{113}$ at $4 \times 10^{-9}$.

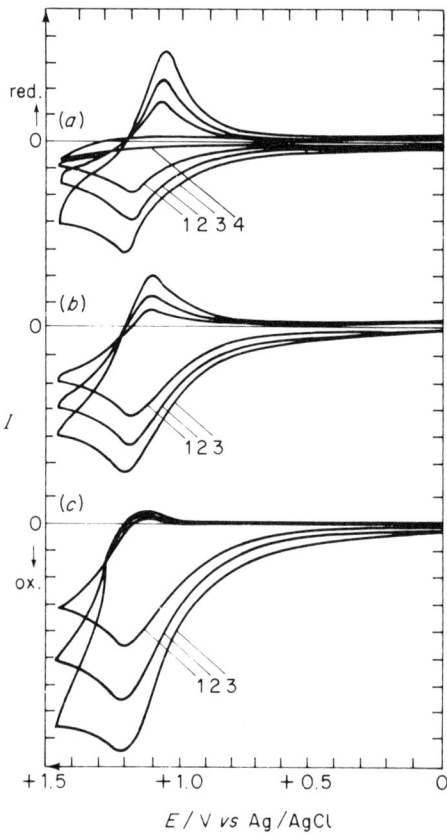

FIG. 11 Cyclic voltammograms of tris-(p-bromophenyl)amine in the presence of tetrabutylammonium (TBA) pelargonate in CH$_3$CN/0.1 M TBAP. 1, 0.294 V s$^{-1}$; 2, 0.588 V s$^{-1}$; 3, 1.176 V s$^{-1}$.

(a) Curves 1, 2 and 3 with 1 mM mediator only; curve 4 with 4 mM TBA-pelargonate only; (b) 1 mM mediator + 2 mM TBA-pelargonate; (c) 1 mM mediator + 4 mM TBA-pelargonate. (Reprinted with permission from Schmidt and Steckhan, 1978b)

The preparative electrolysis of straight-chain aliphatic carboxylates in the presence of [36] resulted in a mixture of products, mainly esters with the general composition RCOOR′ where R′ is an unbranched chain shorter than or equal to R. Fragmentation of R to R' was believed to take place most probably at the radical stage since the reaction of the carbenium ion R$^+$ would be expected to be accompanied by 1,2-hydride shifts and result in branched-chain esters. Such products were not observed and that route was deemed less likely.

Details of redox catalyzed reactions may be revealed by cyclic voltammetry as well. The cyclic voltammograms shown in Fig. 11a were obtained on solutions containing either only the mediator (1)–(3) or only the substrate (4). Only negligible current for the oxidation of substrate can be observed in the potential range where the mediator is oxidized. The effect of adding increasing amounts of the substrate is illustrated by the voltammograms in Figs 11b and 11c. The current due to oxidation of the mediator increases and the corresponding reduction peak disappears. The increase in current during redox catalysis is dependent upon the rate and the concentration of the substrate and may be used quantitatively to evaluate the reaction kinetics (Andrieux et al., 1980b).

Spectroelectrochemical techniques were also used to study the oxidation of 1,1-diphenylethylene [53] catalyzed by the cation radical of dibenzodioxin (Genies et al., 1981). The product of the reaction, 1,2,4,4-tetraphenyl-3-butene-1-one [55], was believed to result from hydrolysis of the dimer

dication [54] accompanied by phenyl migration. Rapid-scan spectrometry was used in the detection of the intermediate cation radical [53$^{\ddot{+}}$]. The disappearance of the dibenzodioxin cation radical, BDO$^{\ddot{+}}$, under pseudo first order conditions was found to be consistent with rate law (118) and $4 k_{120}$ was estimated to be equal to $4.71 \text{ s}^{-1}$ which corresponds to $k_{120}$ equal to

$$d[\text{BDO}^{\ddot{+}}]/dt = 4 k_{120}[\text{BDO}^{\ddot{+}}][53] \tag{118}$$

46 M$^{-1}$s$^{-1}$ under the conditions of the experiments. The experimental technique did not distinguish between coupling of two cation radicals (121) or cation radical–substrate coupling (122–123).

Cyclic voltammetry and voltammetry at rotating electrodes were used in the study of the redox catalyzed oxidative cleavage of dialkyl and diaryl-

$$BDO - e^- \rightleftharpoons BDO^{+\cdot} \tag{119}$$

$$BDO^{+\cdot} + [53] \underset{}{\overset{k_{120}}{\rightleftharpoons}} BDO + [53^{+\cdot}] \tag{120}$$

$$2\,[53^{+\cdot}] \rightarrow Ph_2C{=}CH{-}CH{=}CPh_2 + 2\,H^+ \tag{121}$$

$$\begin{cases} [53^{+\cdot}] + [53] \rightleftharpoons [(53)_2^{+\cdot}] & (122) \\ [(53)_2^{+\cdot}] + BDO^{+\cdot} \rightarrow Ph_2C{=}CH{-}CH{=}CPh_2 + BDO + 2\,H^+ & (123) \end{cases}$$

$$Ph_2C{=}CH{-}CH{=}CPh_2 + 2\,BDO^{+\cdot} \rightarrow [54] + 2\,BDO \tag{124}$$

thioacetals (Martigny and Simonet, 1980). The magnitude of the catalytic current was observed to depend on the nucleophilic properties of the solvent system. In the presence of nucleophiles the catalytic current corresponded to the exchange of two electrons, while one electron behavior was observed when nucleophiles were not added. Reactions (125–128) were proposed to

$$DPA - e^- \rightleftharpoons DPA^{+\cdot} \tag{125}$$

$$DPA^{+\cdot} + R^1R^2C(SR^3)_2 \overset{k_{126}}{\rightarrow} DPA + (R^1R^2CSR^3)^+ + R^3S\cdot \tag{126}$$

$$(R^1R^2CSR^3)^+ + Nu^- \rightarrow R^1R^2C(Nu)SR^3 \tag{127}$$

$$R^1R^2C(Nu)SR^3 \underset{+Nu^-}{\overset{-e^-}{\rightarrow}} R^1R^2C(Nu)_2 + R^3S\cdot \tag{128}$$

account for the observations. The catalysts used were [36] and 9,10-diphenylanthracene (DPA). In the absence of added nucleophiles reaction (127) is slow and the overall number of electrons exchanged corresponded to 1. In the presence of nucleophiles, water or pyridine, (127) is fast and a second electron is exchanged leading ultimately to $R^1R^2C(Nu)_2$ which when $Nu = H_2O$ is a ketone hydrate. Thus, redox catalysis offers a convenient way to eliminate a thioacetal protecting group. Rotating disk electrode voltammetry was used to estimate $k_{126}$ to be of the order of $10^3 M^{-1}s^{-1}$.

The concept of redox catalysis has been extended to the use of electrodes coated with polymer films to which the mediator is attached. The theoretical basis for the application of these electrodes has recently been reported (Andrieux and Savéant, 1978; Anson *et al.*, 1983a,b). The possible application of chemically modified electrodes for the study of cation radical reactions is obvious.

Another extension of redox catalysis involves the photo-stimulated electron transfer between the mediator cation radical and an organic substrate in solution. Irradiation of the anodically generated cation radical of N,N,N',N'-tetraphenyl-*p*-phenylenediamine and benzyl alcohol caused the oxidation of the alcohol to benzaldehyde and regeneration of the parent amine (Moutet and Reverdy, 1982). The photostimulated oxidation of [53] and the two substituted derivatives by six different mediator systems has recently been reported (Moutet and Reverdy, 1983). Of the mediators shown in Scheme 3, only BDO$\overset{+}{\cdot}$ reacts with [53] in the dark. The product distribution in dry

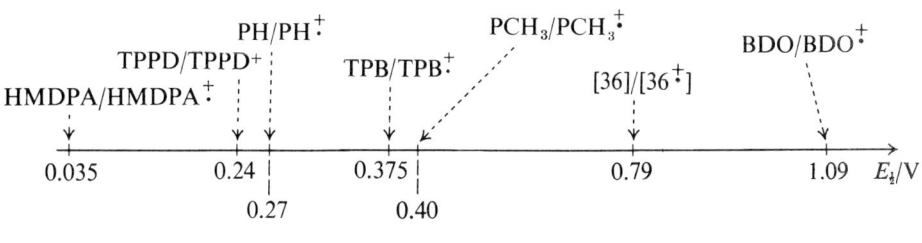

**Scheme 3**

acetonitrile was found to be dependent on the oxidation potential of the mediator. The tetrahydrofuran [56] was only formed to any appreciable extent when the more powerful mediators TPB$\overset{+}{\cdot}$, PCH$_3$$\overset{+}{\cdot}$ and [36$\overset{+}{\cdot}$] were present in solution, while the amount of the butadiene [57] reached a maximum when mediators having intermediate oxidation potentials were used. The dihydronaphthalene derivative [58] was the only product detected when the mediator was HMDPA. The reaction mechanism postulated involves the coupling of $(XC_6H_4)_2C=CH_2$ with the corresponding cation radical as a key step. This reaction is then followed by three competing reactions including

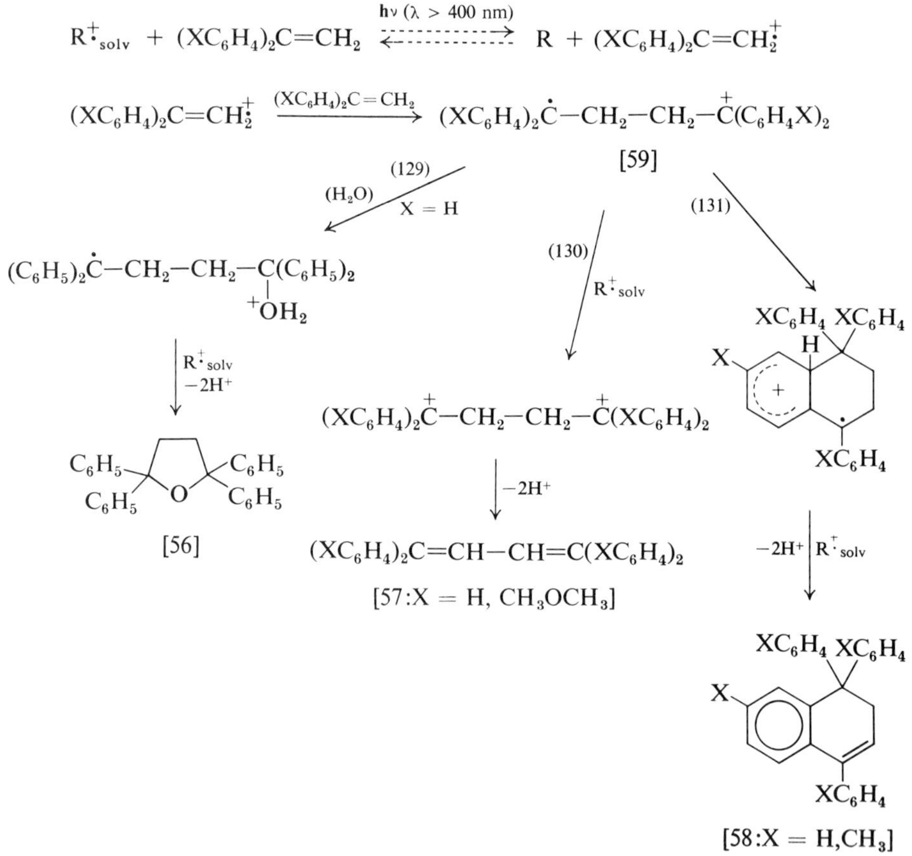

$$R^{+\cdot}_{solv} + (XC_6H_4)_2C=CH_2 \underset{h\nu\ (\lambda > 400\ nm)}{\rightleftharpoons} R + (XC_6H_4)_2C=CH_2^{+\cdot}$$

$$(XC_6H_4)_2C=CH_2^{+\cdot} \xrightarrow{(XC_6H_4)_2C=CH_2} (XC_6H_4)_2\dot{C}-CH_2-CH_2-\overset{+}{C}(C_6H_4X)_2$$
$$[59]$$

(129) (H$_2$O), X = H

(130) R$^{+\cdot}_{solv}$

(131)

$(C_6H_5)_2\dot{C}-CH_2-CH_2-\underset{|}{C}(C_6H_5)_2$
$\phantom{(C_6H_5)_2\dot{C}-CH_2-CH_2-}{}^+OH_2$

R$^{+\cdot}_{solv}$
$-2H^+$

[56] tetrahydrofuran with C$_6$H$_5$, C$_6$H$_5$, C$_6$H$_5$, C$_6$H$_5$ substituents

$(XC_6H_4)_2\overset{+}{C}-CH_2-CH_2-\overset{+}{C}(XC_6H_4)_2$

$\Big| -2H^+$

$(XC_6H_4)_2C=CH-CH=C(XC_6H_4)_2$

[57: X = H, CH$_3$OCH$_3$]

Dihydronaphthalene-type structure with X, XC$_6$H$_4$, XC$_6$H$_4$, XC$_6$H$_4$ substituents (cation radical)

$-2H^+ \Big| R^{+\cdot}_{solv}$

Naphthalene-type structure with X, XC$_6$H$_4$, XC$_6$H$_4$, XC$_6$H$_4$ substituents

[58: X = H, CH$_3$]

(130), a homogeneous electron transfer between cation radical [59] and the mediator cation radical. The dependence of the product distribution on the oxidation potential of the mediator was attributed to the competition between electron transfer (130) and reactions (129) and (131).

CATALYSIS BY CATION RADICALS

A variety of aspects of electron-transfer catalysis in general have recently received considerable attention (Alder, 1980; Savéant, 1980; Chanon, 1982; Chanon and Tobe, 1982). Although most of the examples known so far originate from inorganic chemistry and the chemistry of anion radicals, the concept has already had significant influence on the way of thinking in cation radical chemistry as well. Electron-transfer catalysis is a chain process, the prototype of which can be illustrated by reactions (132–134) for the catalyzed conversion of A to B. Reaction (133) may be a rearrangement or a more

$$A - e^- \rightleftharpoons A^{+\cdot} \tag{132}$$

$$A^{+\cdot} \rightarrow B^{+\cdot} \tag{133}$$

$$B^{+\cdot} + A \rightarrow A^{+\cdot} + B \tag{134}$$

complicated chemical process. In any case, as long as this reaction is sufficiently exothermic, A will effectively be converted to B once the reaction has been initiated independent of whether the homogeneous electron transfer between $B^{+\cdot}$ and A is energetically favorable or not. The initial oxidation of A to $A^{+\cdot}$ may be carried out electrochemically, photochemically, or by conventional homogeneous electron transfer.

*The $S_{ON}2$ reaction*

Aromatic substitution initiated by electron transfer to the substrate resulting in the intermediate formation of an unstable anion radical is known as the $S_{RN}1$ reaction (Bunnett, 1978) and the mechanism is outlined below. Inspection of the four steps (135–138) reveals that this is an anion radical chain

$$ArX + e^- \rightleftharpoons ArX^{-\cdot} \tag{135}$$

$$ArX^{-\cdot} \rightarrow Ar\cdot + X^- \tag{136}$$

$$Ar\cdot + Nu^- \rightarrow ArNu^{-\cdot} \tag{137}$$

$$ArNu^{-\cdot} + ArX \rightarrow ArNu + ArX^{-\cdot} \tag{138}$$

process, which in the ideal case does not require the net consumption of charge. The observation that 4-fluoroanisole in the presence of acetate ion oxidizing conditions could be converted to 4-acetoxyanisole with current yields exceeding 100% indicated that a similar process involving cation radical intermediates is feasible (Nyberg and Wistrand, 1976; Eberson and Jönsson, 1980). Careful mechanistic analysis has provided evidence that this so-called $S_{ON}2$ reaction does indeed involve a cation radical chain (Scheme 4) (Eberson and Jönsson, 1981; Eberson *et al.*, 1982). A similar mechanism has earlier been discussed for aromatic photosubstitution, but the cation radical chain character has not been established in that case. Quantum yields for photosubstitution rarely exceed 0.5 (Havinga and Cornelisse, 1976; Heijer *et al.*, 1977; Parkányi, 1983).

An important intermediate in the $S_{ON}2$ mechanism is the radical [60]. That such radicals in fact are involved was demonstrated by showing that acetoxylation could be initiated by benzoyloxy radicals formed *in situ* by

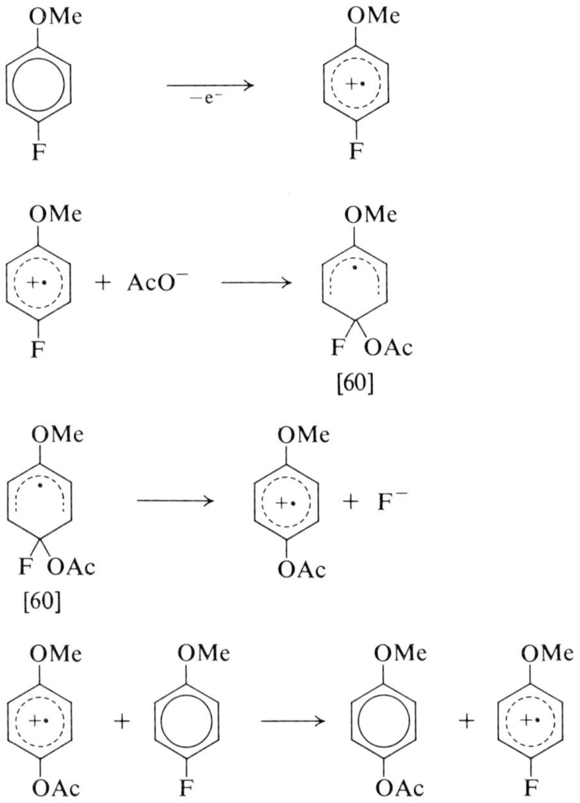

**Scheme 4**

decomposition of dibenzoyl peroxide (Scheme 5) (Eberson and Jönsson, 1981). An additional benefit of the use of this type of initiation was that the ratio between the yields of 4-benzoyloxyanisole formed in the first cycle and 4-acetoxyanisole formed in subsequent cycles could be directly related to the chain length of the reaction, which was typically found to be around 5. The mechanism assignment was supported by thermochemical calculations which indicated that [60] is 4 kcal mol$^{-1}$ less stable than 4-fluoroanisole cation radical plus acetate ion and that the latter are about 12 kcal mol$^{-1}$ less stable than 4-acetoxyanisole cation radical plus fluoride ion. Other examples of substitution reactions which may possibly involve the $S_{ON}2$ pathway were briefly discussed (Eberson et al., 1982). A common feature of these reactions is that only difficultly oxidized nucleophiles are involved. This is a necessary prerequisite for the mechanisms which requires oxidation of substrate to the cation radical *in the presence* of the nucleophile.

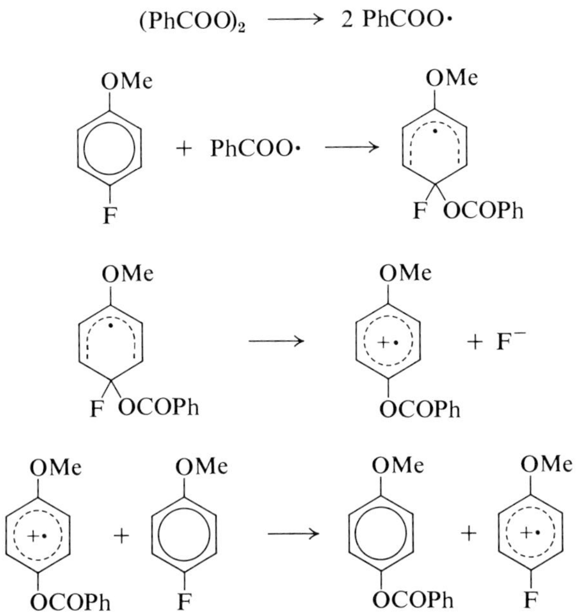

$(PhCOO)_2 \longrightarrow 2\ PhCOO\cdot$

**Scheme 5**

*Formation of 1,2-dioxetans*

The role of cation radical intermediates in the photoinitiated formation of 1,2-dioxetans by irradiation of suitable olefins in the presence of oxygen has aroused considerable interest (Barton *et al.*, 1975; Eriksen and Foote, 1980; Zaklika *et al.*, 1980; Schaap *et al.*, 1980). A key point in these discussions is the importance of the direct reaction of olefin cation radicals with triplet oxygen compared to the fast bond formation between olefin cation radicals and superoxide ion. It has been argued (Schaap *et al.*, 1980) that the longer reaction times and reduced yields of dioxetans in the presence of superoxide ion makes it unlikely that a major part of the dioxetan is formed in a cation radical chain process involving triplet oxygen.

Unequivocal evidence that olefin cation radicals can react rapidly with oxygen emerged from voltammetric and preparative studies of the oxidation of adamantylideneadamantane [61] in the presence and in the absence of molecular oxygen (Nelsen and Akaba, 1981; Clennan *et al.*, 1981). Cyclic voltammetry studies of [61] at low voltage sweep rates in acetonitrile in the absence of molecular oxygen was indicative of the formation of a long-lived cation radical (Fig. 12, top). On the other hand, introduction of oxygen into the cell brought about a great reduction in current (Fig. 12, bottom). This is indicative of an electron-transfer initiated process which consumes

[61]

[62]

[63]

[64]

substrate. Preparative electrolysis (Clennan et al., 1981) or homogeneous oxidation with [36$^{+\cdot}$]SbCl$_6^-$, NO$^+$PF$_6^-$ or NO$_2^+$PF$_6^-$ (Nelsen and Akaba, 1981) demonstrated the formation of oxygenated products among which the 1,2-dioxetan [62] could be detected as the major product when the oxidation was carried out electrochemically (yield = 85%) or chemically, providing

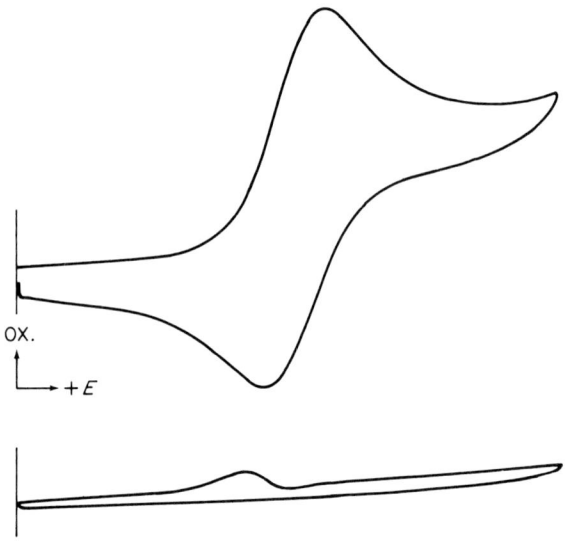

FIG. 12 Cyclic voltammograms of [61] in the absence of oxygen (above) and in oxygen-saturated (below) acetonitrile containing 0.1 M tetra-n-butylammonium perchlorate, scanned from 1.15 to 1.75 V vs SCE at 200 mV s$^{-1}$. (Reprinted with permission from Nelsen and Akaba, 1981)

that the oxidant was only present in catalytic amounts. Higher concentrations of $NO^+PF_6^-$ or $NO_2^+PF_6^-$ resulted in the formation of the epoxide [63] and the spiroketone [64], the latter presumably being formed by pinacol rearrangement of [63]. It was suggested that the formation of [63] might be related to the NO· formed during the chemical oxidation, but no conclusive evidence was presented. The most likely mechanism for the cation radical chain reaction is shown in Scheme 6.

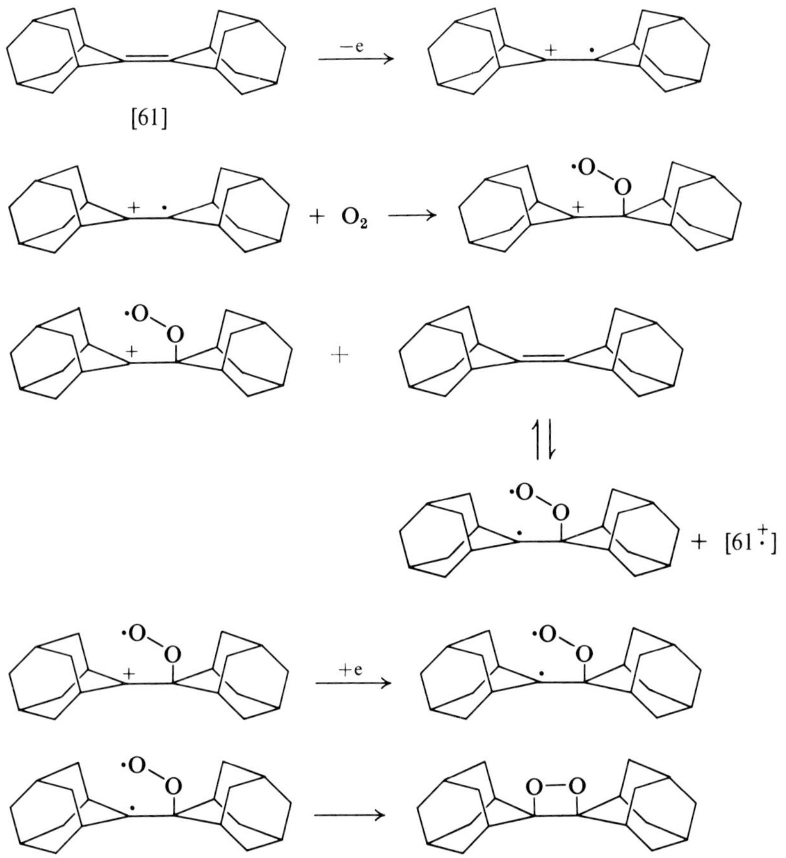

**Scheme 6**

*The tetramerization of N-benzylaziridines*

Anodic oxidation of N-benzylaziridine has been reported to result in oligomerization in high yield with charge consumption varying between 0.05 and 0.25 F mol$^{-1}$ when the oxidation was carried out in dichloromethane (Kossai *et al.*, 1980, 1982). The reaction was also carried out using redox catalysis

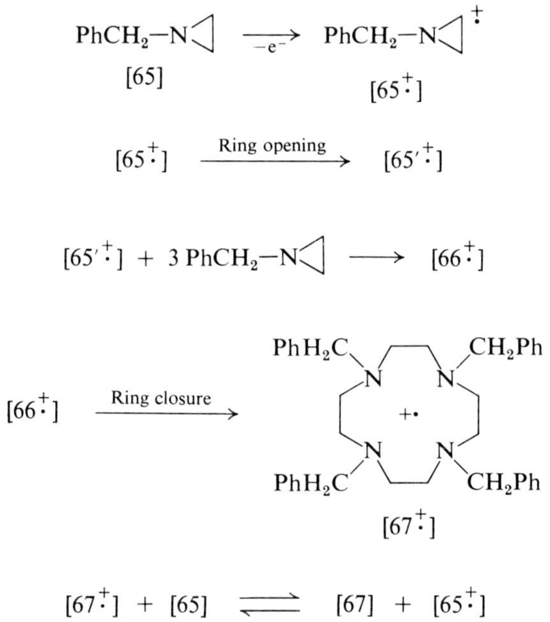

$$\text{PhCH}_2\text{−N}\!\!< \quad \xrightarrow{-e^-} \quad \text{PhCH}_2\text{−N}\!\!<^{+\cdot}$$
$$[65] \qquad\qquad\qquad [65^{+\cdot}]$$

$$[65^{+\cdot}] \xrightarrow{\text{Ring opening}} [65'^{+\cdot}]$$

$$[65'^{+\cdot}] + 3\,\text{PhCH}_2\text{−N}\!\!< \longrightarrow [66^{+\cdot}]$$

$$[66^{+\cdot}] \xrightarrow{\text{Ring closure}}$$

PhH$_2$C–N    N–CH$_2$Ph
   \        /
    (ring, +·)
   /        \
PhH$_2$C–N    N–CH$_2$Ph

$$[67^{+\cdot}]$$

$$[67^{+\cdot}] + [65] \rightleftharpoons [67] + [65^{+\cdot}]$$

with [36] or thianthrene as mediators. In the latter case the current consumption was somewhat higher (0.54 F mol$^{-1}$) than that during direct electrolysis. A solution of the aziridine containing *p*-toluenesulfonic acid showed no signs of oligomerization after six hours and this was interpreted to indicate that protons liberated in a side reaction under oxidative conditions could not be responsible for the conversion to the tetramer. A cation radical chain mechanism was proposed to account for the results. The cation radical of substrate [65$^{+\cdot}$] was assumed to undergo fast ring opening to a species which could attack unoxidized starting material and cause a sequence of reactions resulting in the tetramer in its ring opened form. Ring closure would result in [67$^{+\cdot}$] which was assumed to exchange electrons with the starting material in the propagation step. This example is noteworthy in that the amine cation radical did not undergo either of the two reactions typical of these substances, deprotonation and hydrogen atom abstraction.

*The decomposition of diaryldiazomethanes*

One of the most thoroughly studied ion radical chain reactions in organic chemistry is the decomposition of diaryldiazomethanes catalyzed by oxidants (Ho *et al.*, 1974; Bethell *et al.*, 1977, 1979a; Oshima *et al.*, 1978; Benati *et al.*, 1981) or induced by anodic oxidation (Jugelt and Pragst, 1968a,b; Pragst and Jugelt, 1970a,b). The major products of the decomposition of diphenyldiazomethane in dry aprotic solvents are tetraphenylethylene (TPE) and benzophenone azine (BPA), the proportions of which depend on several experimental factors, most important of which is the concentration of the oxidant.

Kinetic experiments in non-aqueous acetonitrile (Bethell et al., 1977, 1979a) showed that the reaction catalyzed by [36$\overset{+}{\cdot}$] or Cu(II) perchlorate in the strict absence of Cu(I) followed a simple first order expression and the observed rate constant was found to be independent of the nature of the oxidant. These observations were taken to be evidence that the first step of the reaction is fast electron transfer followed by a rate-determining step involving $Ph_2CN_2$ and its cation radical. This view was further supported by the results of product analysis, which demonstrated that the ratio between TPA and BPA formed was likewise independent of the identity of the catalyst. Unimolecular decomposition of $PhCN_2\overset{+}{\cdot}$ to $Ph_2C\overset{+}{\cdot}$ and $N_2$ was deemed unlikely since formation of a high energy species such as $Ph_2C\overset{+}{\cdot}$ would be expected to be rate-determining and thus the overall reaction rate was expected to be independent of the substrate concentration which was inconsistent with the first order dependence observed.

The decomposition of $Ph_2CN_2$ has been observed to be catalyzed by Cu(I) as well, but in a reaction significantly slower than the oxidatively initiated process (Bethell et al., 1979b and reference cited). However, the inclusion of Cu(I) in the Cu(II) catalyzed reaction had a rate retarding effect and increasing the concentrations of Cu(I) caused the reaction order in $Ph_2CN_2$ to change from one to two. This effect of Cu(I) indicates that the initial electron transfer is reversible and application of the steady state principle on $Ph_2CN_2\overset{+}{\cdot}$ led to a rate law, (139), in agreement with all of the kinetic

$$\text{Rate} = kK[Ph_2CN_2]^2[Cu(II)]/([Cu(I)] + K[Ph_2CN_2]) \qquad (139)$$

evidence. The rate constant ($k$) and equilibrium constant ($K$) refer to the mechanism illustrated in Scheme 7.

The rate-determining bimolecular reaction between $Ph_2CN_2$ and $Ph_2CN_2\overset{+}{\cdot}$ may lead to the formation of the cation radicals of either TPE or BPA, both of which are sufficiently strongly oxidizing to abstract an electron from substrate to initiate a new cycle.

Low temperature esr spectrometry evidence supported the mechanistic proposal. At about —40°C a five times seven line spectrum was detected which was attributed to the cation radical of $Ph_2CN_2$. Upon warming, the spectrum changed to that of $BPA\overset{+}{\cdot}$ (Ho et al., 1974).

TPE was also the product formed by the anodic oxidation of $Ph_2CN_2$ (Jugelt and Pragst, 1968a,b; Pragst and Jugelt, 1970a,b) and a cation radical

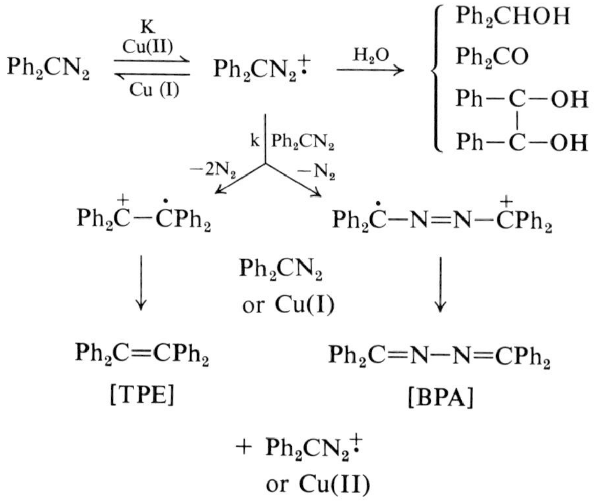

$$Ph_2CN_2 \underset{Cu(I)}{\overset{K \atop Cu(II)}{\rightleftarrows}} Ph_2CN_2^{+\cdot} \xrightarrow{H_2O} \begin{cases} Ph_2CHOH \\ Ph_2CO \\ Ph-C-OH \\ \quad | \\ Ph-C-OH \end{cases}$$

$$\overset{-2N_2}{\swarrow} \overset{k | Ph_2CN_2}{\underset{-N_2}{\searrow}}$$

$$Ph_2\overset{+}{C}-\overset{\cdot}{C}Ph_2 \qquad\qquad Ph_2\overset{\cdot}{C}-N=N-\overset{+}{C}Ph_2$$

$$\downarrow \qquad\qquad Ph_2CN_2 \qquad\qquad \downarrow$$
$$\qquad\qquad \text{or } Cu(I)$$

$$Ph_2C=CPh_2 \qquad\qquad Ph_2C=N-N=CPh_2$$
$$\text{[TPE]} \qquad\qquad\qquad \text{[BPA]}$$

$$+ Ph_2CN_2^{+\cdot}$$
$$\text{or } Cu(II)$$

**Scheme 7**

chain reaction was proposed to account for this. The proportions of TPE and BPA were found to vary with the identity of the aprotic solvent and in some cases was independent of whether or not oxygen was excluded during the reaction. However, no convincing evidence has been offered to explain these aspects of the $Ph_2CN_2$ decomposition. A similar dependence on oxygen was noted during a study of the decomposition of $Ph_2CN_2$ catalyzed by various organic sulfur compounds. In the absence of oxygen, first order decay of starting material was observed, while an apparent autocatalytic process was observed when oxygen was present (Benati et al., 1981). When water was present in the solvent during the Cu(II) catalyzed process a major portion of the reaction took a completely different course and increasing amounts of $Ph_2CHOH$, $Ph_2CO$ and $(PhCOH)_2$ were detected with increasing water concentration (Bethell et al., 1979a). These products could be demonstrated to result from competing acid catalysis of the decomposition by protons liberated during the reaction between $Ph_2CN_2^{+\cdot}$ and water.

Kinetic analysis of the reaction between m- and p-substituted diphenyldiazodiphenylmethanes and tetracyanoethylene (TCNE) resulted in a linear Hammett relationship (140) (Oshima et al., 1978) when $\sigma^+$ values were used and this together with the negative $\rho$ ($-2.67$) indicated that appreciable charge was developing at the methane carbon on the way to the transition

$$\log k = -2.67\sigma^+ - 1.23 \qquad (r = 0.989) \qquad (140)$$

state in agreement with an initial electron-transfer reaction. In this particular case the product of the reaction was found to be a cyclopropane derivative analogous to that expected as a result of a carbene cycloaddition reaction.

*Intramolecular release of strain* via *cation radicals*

Organic chemists have long been fascinated by strained small ring molecules. The study of these substances has resulted in a deeper insight into the fundamental nature of the chemical bond (Greenberg and Liebman, 1978). Due to the violation of the "normal" bond angles molecules belonging to this class are as a rule thermodynamically unstable and considerable energy is expected and observed to be released during the conversion to less strained isomers. As an example relevant to the discussion to follow, the strain energy of quadricyclane [68] is 96 kcal mol$^{-1}$ (Kabakoff *et al.*, 1975) of which 27 kcal mol$^{-1}$ is released during the rearrangement to norbornadiene [69] (Wiberg and Connon, 1976). However, in spite of the intrinsic instability of these compounds surprisingly many of them show only low reactivities and limited tendencies to undergo isomerization. The reasons for this are primarily that the direct conversion from one isomer to the other can be inhibited by a

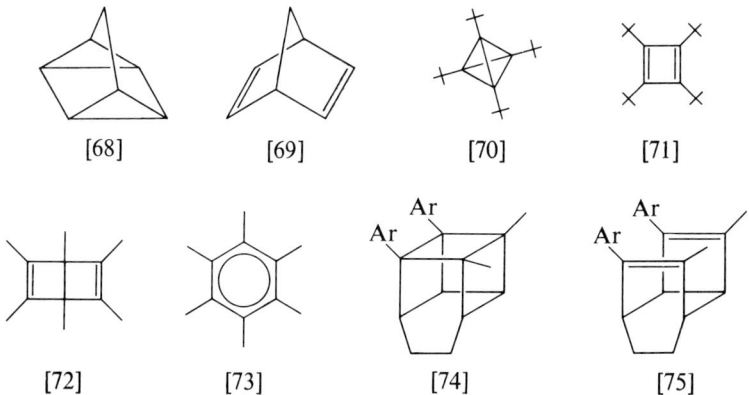

barrier imposed by orbital symmetry and that substituents, especially bulky ones like t-butyl, may sterically hinder the interconversion, i.e. the "corset effect". An example of the latter is found in the isomerization of tetra-t-butyltetrahedrane [70] to tetra-t-butylcyclobutadiene [71], a process which does not take place at temperatures lower than the melting point of [70] at 135°C (Maier *et al.*, 1981a) because of the repulsion between the t-butyl groups in the transition state. The free energy of activation was found to be equal to 28.6 kcal mol$^{-1}$ by nmr measurements (Maier *et al.*, 1981b).

The thermal stability of these polycyclic compounds is in sharp contrast to the instantaneous isomerization that is observed during photolysis. The quantum yields in polar solvents such as acetonitrile are as a rule greater than 1. This along with the recognition that ion radical pairs may be formed during the photolysis provided the background for the now generally accepted cation radical chain mechanism for the photoisomerization in polar solvents (141–143). Typical examples include the isomerization of hexamethyl dewar

$$D + A \xrightarrow{h\nu} D^{+\cdot} + A^{-\cdot} \qquad (141)$$

$$D^{+\cdot} \rightarrow (D')^{+\cdot} \qquad (142)$$

$$(D')^{+\cdot} + D \rightarrow D' + D^{+\cdot} \qquad (143)$$

benzene [72] to hexamethylbenzene [73] (T. R. Evans et al., 1973; Jones and Becker, 1983), quadricyclane [68] to norbornadiene [69] (Roth and Shilling, 1981) and the aryl substituted homocubane [74] to the corresponding diene [75] (Mukai et al., 1981; Okada et al., 1981).

More direct evidence for the participation of cation radicals in steps (142) and (143) has been obtained through electrochemical studies and by homogeneous chemical oxidation by tris(p-bromophenyl) ammoniumyl [$36^{+\cdot}$] and $AlCl_3/CH_2Cl_2$. Anodic oxidation of [72] in acetonitrile was shown to result in complete conversion to [73] after the consumption of 0.1 F mol$^{-1}$ (T. R. Evans et al., 1973). No other products were detected. It was estimated that the chain length was about 10 under the conditions employed. More exhaustive electrolysis caused further oxidation of [73]. The oxidation of derivatives of [68] by [$36^{+\cdot}$] was demonstrated to bring about the same conversion as was previously observed during photochemical experiments (Hoffmann and Barth, 1983). Reaction times varied from 10 minutes to 10 days depending upon the nature of the substituents. It was concluded that the redox catalyzed reaction was most satisfactorily explained by an initial oxidation of the quadricylane derivative to the cation radical by the catalyst followed by reactions (142) and (143). The observation of the relatively long reaction times is presumably connected to the rate of the initial electron-transfer step. The conversion of [$68^{+\cdot}$] to [$69^{+\cdot}$] is known to be an exceedingly fast process. It was not even possible to detect the cation radical during γ-radiation of [68] in an electron scavenging matrix at 77 K (Haselbach et al., 1979). Only [$69^{+\cdot}$] was observed just as in the direct irradiation of [69]. A fast conversion was also observed when [70] was treated with $AlCl_3/CH_2Cl_2$ (Bock et al., 1980). Attempts to obtain esr evidence for the transient existence of [$70^{+\cdot}$] failed. The only cation radical that could be found in solution was [$71^{+\cdot}$] which could also be prepared directly by the action of the oxidizing solution on [71]. Attempts to detect [$70^{+\cdot}$] by cyclic voltammetry at low temperature failed as well (Fox et al., 1982). The voltammogram recorded in acetonitrile at $-70°C$ (!) is shown in Fig. 13. A single irreversible oxidation peak was observed at $+0.50$ V on the anodic going scan and no reduction current due to the cation radical could be found on the reverse scan but a new redox couple appeared at about $+0.15$ V which was demonstrated to be due to [71]/[$71^{+\cdot}$].

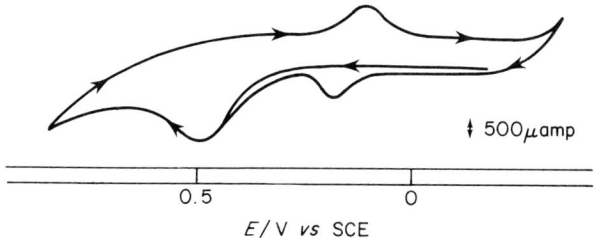

FIG. 13 Cyclic voltammogram of [70] ($5 \times 10^{-3}$M; dry degassed $CH_3CN$ containing 0.1 M $Bu_4NBF_4$; glassy carbon, $-70°C$; scan rate 50 mV s$^{-1}$). (Reprinted with permission from Fox et al., 1982)

A significant reduction of the activation barrier for isomerization in passing from neutral substances to the cation radicals is in agreement with theoretical predictions (Haselbach et al., 1979; Bock et al., 1980; Chu and Lee, 1982). The results of open and closed shell MNDO calculations for the sterically less hindered tetramethyltetrahedrane are illustrated in Fig. 14. The barrier for the isomerization of the neutral compound was estimated to be 41 kcal mol$^{-1}$ while the cation radical does not reside in an energy well and is expected to isomerize spontaneously to the cyclobutadiene derivative.

Although there appears to be little doubt that (142) is a strongly exothermic reaction, it is more difficult to predict whether the electron transfer (143) is a favourable process or not. Oxidation potentials for [68] and [69] and a series of related compounds have been measured (Gassman et al., 1978; Gassman and Yamaguchi, 1982). In spite of the fact that reversible potentials for the oxidation of [68] and its analogues could not be obtained due to the high reactivity of the corresponding cation radicals, the differences in potentials observed for the two series were sufficiently large to allow the conclusion to be drawn that the more strained compounds are much easier to oxidize than their rearranged isomers (Table 4). However, it was emphasized that this result cannot be generalized to the extent that the thermodynamically more stable compounds will also have higher oxidation potentials. There does not appear to be a direct relationship between the energies of the HOMO, the orbitals from which the electrons are presumed to originate, and the thermodynamic stability or strain energy for a given series of compounds. This was demonstrated by a comparison of cubane [76] with the isomer [77]. Although [77] is more easily oxidized than [76] it is thermodynamically more stable and may be obtained from [76] by Ag$^+$ catalyzed isomerization. The conversion of [70] to [71] is another example of the same phenomenon. Thus, it can be concluded that the driving force in these isomerizations is thermodynamically favorable reaction (142) while (143) may be either endergonic or exergonic.

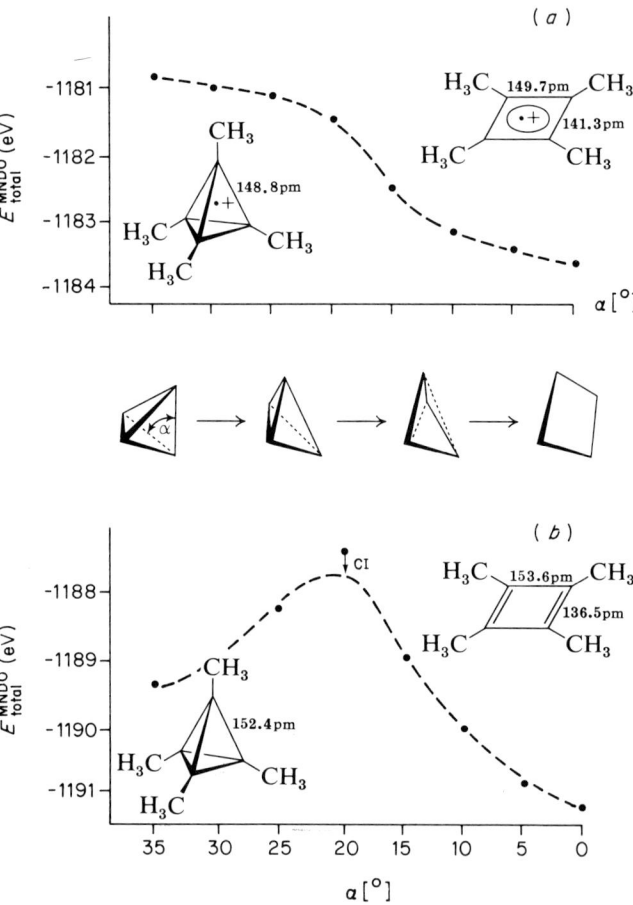

FIG. 14 Energy profiles calculated by open shell (a) and closed shell (b) MNDO for the valence isomerization of tetramethyltetrahedrane. (Reprinted with permission from Bock et al., 1980)

[76]   [77]

# TABLE 4

Half-wave oxidation potentials for a series of quadricyclane type compounds and their diene isomers[a]

| Compound | $E_{1/2}$ V | Compound | $E_{1/2}$ V |
|---|---|---|---|
| quadricyclane | 0.91 | norbornadiene | 1.56 |
| bis-CF$_3$ quadricyclane | 2.19 | bis-CF$_3$ norbornadiene | 2.51 |
| bis-CO$_2$CH$_3$ quadricyclane | 1.64 | bis-CO$_2$CH$_3$ norbornadiene | 2.06 |
| O-bridged bis-CO$_2$CH$_3$ quadricyclane | 1.95 | O-bridged bis-CO$_2$CH$_3$ norbornadiene | 2.26 |
| N-Ts bis-CO$_2$CH$_3$ quadricyclane | 1.95, 2.36 | N-Ts bis-CO$_2$CH$_3$ norbornadiene | 2.20  2.37 |
| nortricyclane | 2.12 | norbornene | 1.95 |

[a] Gassman and Yamaguchi, 1982

*Cycloaddition reactions*

The metal ion and photochemically catalyzed 2 + 2 cycloadditions of N-vinylcarbazole [78], 4,4′-dimethylamino-1,1-diphenylethylene [79] (Bell *et al.*, 1969a; Crellin *et al.*, 1970; Ledwith, 1972) and indene [80] (Farid and Shealer, 1973) demonstrated that cation radicals may play an important role in this area of chemistry. Knowledge that the effect of metal ions is a genuine catalytic one emerged from the observation that [79] was converted

[78]   [79]   [80]

to the corresponding cyclobutadiene derivative (head-to-tail) in 90% yield in the presence of only 1/1000th of the stoichiometric amount of Cu(II) ion (Bell *et al.*, 1969a). The cation radical chain mechanism suggested for the photochemical conversion of [78] and [80] to the corresponding cyclodimers was based on the observation of quantum yields significantly greater than unity (Crellin *et al.*, 1970) and the fact that triplet quenchers had no effect on the reaction (Farid and Shealer, 1973). Related photochemical studies have recently been reviewed (Mattes and Farid, 1982). The report that [82] was formed in low yield during anodic oxidation of [81] gave further evidence for the cation radical chain mechanism, illustrated in detail for the anodically stimulated reaction in Scheme 8 (Cedheim and Eberson, 1976).

The recent discovery that the Diels–Alder reaction is subject to cation radical catalysis which results in enormous rate enhancements is of fundamental importance (Bellville *et al.*, 1981). It is a well-known fact that the Diels–Alder reaction proceeds most efficiently when the dienophile is electron deficient and that electron-rich olefins require the addition of Lewis acids in order to give synthetically useful yields of the adduct. Conversion of an electron-rich dienophile to the corresponding electron-deficient cation radical might be expected to compensate for this inherent lack of reactivity of the neutral compound. This was indeed found to be the case. Four points of interest originate from the initial studies. (*i*) Comparison of the reaction conditions and times for the dimerization of 1,3-cyclohexadiene [83] demonstrated the superiority of the cation radical catalyzed reaction as compared to the conventional Diels–Alder reaction. The dimer [84] was found in 70%

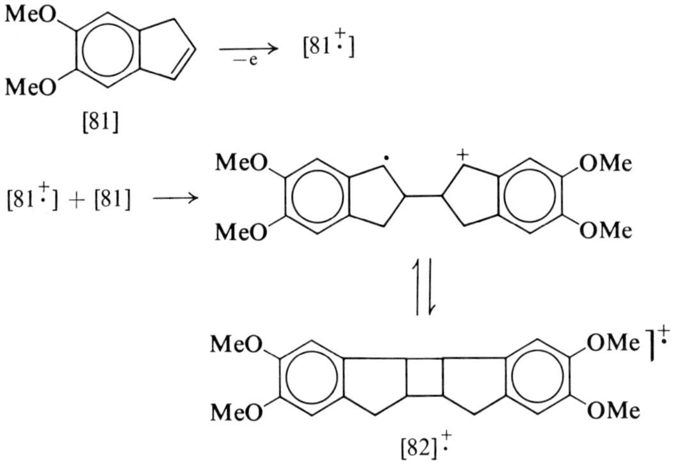

ORGANIC CATION RADICALS

[Scheme showing 5,6-dimethoxyindene [81] → [81⁺·] via −e]

[81⁺·] + [81] → [dimer radical cation] ⇌ [82]⁺·

[82⁺·] + [81] → [82] + [81⁺·]

**Scheme 8**

$2$ [83] → [84] + *exo*

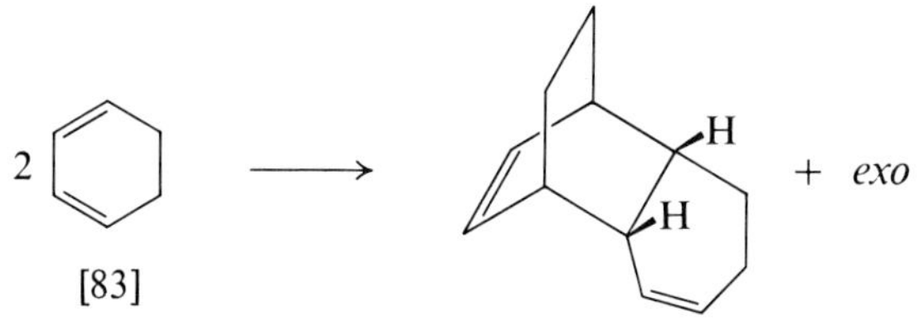

A: 200°C; 20 h; 30%; *endo/exo* = 4/1
B: [36⁺·]; 0°C; 15 min; 70%; *endo/exo* = 5/1

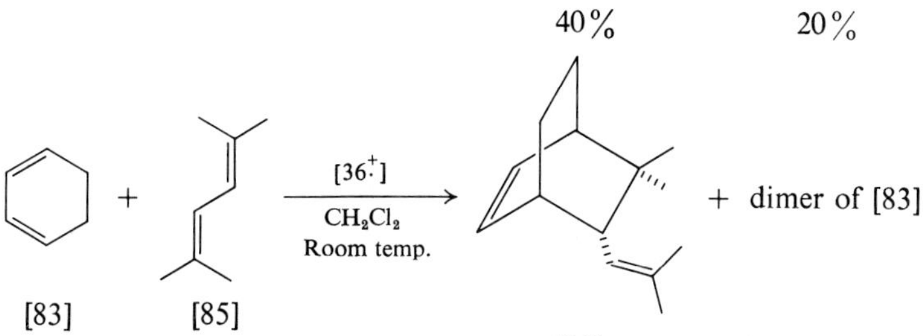

[83] + [85] $\xrightarrow[\text{CH}_2\text{Cl}_2]{[36^{+\cdot}]}$ [86] (40%) + (20%) + dimer of [83]
Room temp.

$\dfrac{endo}{exo} = \dfrac{4}{3}$

yield after 15 min in the catalyzed process while only a modest 30% yield was obtained after 20 h under the more drastic conditions of the uncatalyzed process. (*ii*) Steric hindrance may be a serious obstacle in the normal reaction but is much easier to overcome by application of redox catalysis. The product [86] containing a quaternary carbon (C–5) was obtained in satisfactory (40%) yield together with the dimer of [83]. (*iii*) The suprafacial stereospecificity of the conventional Diels–Alder reaction is retained in the catalytic process as shown by the results of the reaction between [83] and the three isomers of [87]. A single stereoisomer was obtained from *trans,trans*-[87], the *trans,endo*

adduct [88], while *cis,cis*-[87] gave rise to the *cis,endo* and *cis,exo* compounds [89] and [90]. The *cis,trans* isomer of [87] gave three products with the *cis,endo* [91], the *trans,endo* [92] and the *trans,exo* [93] configurations. *Exo* isomers were only formed when the propenyl moiety had the *cis*-configuration, apparently because of steric destabilization of the otherwise favored transition state for the formation of the *endo* isomer. (*iv*) That even dienes which do not have rigidly co-planar double bonds react smoothly was borne out by the dimerization of [94] to [95].

It is noteworthy that [83] competed more favourably as the dienophile in the reaction with *trans,trans*-[87] than in that with [85], with observed yields of self-adducts being 40 and 20%, respectively. This result was found to be consistent with the view that the catalyzed process most probably

[94] $\xrightarrow{[36^{\dotplus}]}{CH_2Cl_2, 0°C \\ 10 \text{ min}}$ [95] 50% endo + exo

involves reaction between the cation radical of the dienophile and the diene, a 4 + 1 rather than a 3 + 2 addition. The more highly substituted diene [85] is expected to form the cation radical more easily than is *trans,trans*-[87] and therefore compete efficiently with the formation of [83$\dotplus$]. This conception of the electronic distribution between the partners of the catalyzed process was found to be in accord with orbital correlation diagrams for the two possibilities, which indicated that the reaction between s-*cis*-1,3-butadiene and ethylene cation radical, the 4 + 1 addition, is an allowed process while the "role-inverted" 3 + 2 addition is forbidden (Bellville and Bauld, 1982). A detailed analysis of cation radical pericyclic reactions in general (Bauld et al., 1983) led to the conclusion that the activation energy for the cation radical catalyzed processes are lowered considerably compared to the electrically neutral reactions and that barrier reductions as great as 60 kcal mol$^{-1}$ might be expected. It was proposed that the catalyzed processes involve three steps; (*i*) formation of the cation radical, (*ii*) pericyclic reaction, and (*iii*) electron uptake to form the neutral adduct.

The synthetic utility of the cation radical initiated Diels–Alder reaction has been further developed by the design of a cation radical carrying polymer, which in practical work may be used in a way operationally similar to that of the Merrifield amino acid synthesis (Bauld et al., 1982).

## 5 Fragmentation reactions of cation radicals

It has been demonstrated by mass spectrometry and similar techniques that considerable bond weakening may be associated with the removal of an electron from an organic molecule. Fragmentation is a common reaction of cation radicals in the gas phase and related reactions should obviously be expected to take place in solution even though the presence of the solvent would be expected to influence both the energetics and the nature of the products. One major difference between gas phase and solution chemistry of cation radicals is to be expected. Deprotonation of cation radicals in the gas phase is a high energy process due to the formation of a "naked" proton. In solution the large solvation energy of the proton may provide the driving force for cation radical deprotonation and this reaction is expected to play an important role in the chemistry of a large range of cation radicals.

ACIDITY OF CATION RADICALS

The acidities of the cation radicals of phenols, amines and hydrazines are of magnitudes which allow the direct experimental determination of $pK_a$-values or related parameters. The general approach that has been used in these determinations is to generate the cation radical or the corresponding base, a neutral radical, in solvent systems of differing acidities so that the extent of equilibria (144–146) can be monitored by uv/visible and esr spectroscopy or in some cases by conductivity measurements. The cation radicals and

$$Ar-OH^{+\cdot} + B \rightleftharpoons Ar-O\cdot + BH^+ \tag{144}$$

$$R_2NH^{+\cdot} + B \rightleftharpoons R_2N\cdot + BH^+ \tag{145}$$

$$R_2N-NHR^{+\cdot} + B \rightleftharpoons R_2N-\dot{N}R + BH^+ \tag{146}$$

neutral radicals involved are most conveniently generated for this purpose by such methods as flash photolysis (Land and Porter, 1960, 1963; Land et al., 1961), radiolysis or pulse radiolysis (Simic and Hayon, 1971; Fessenden and Neta, 1972; Nelsen et al., 1980b) or by flowing solutions of substrate against solutions of a suitable oxidant such as Ce(IV) (Dixon and Murphy, 1976; Holton and Murphy, 1979). Typical $pK_a$-values are listed in Table 5.

TABLE 5

Values of $pK_a$ for $RH^{+\cdot}$

| Substrate | $pK_a$ | Method | Ref. |
|---|---|---|---|
| 96 | 10.4 | Pulse radiolysis + conductivity | Nelsen et al., 1980b |
| 97 | 7.0 | ,, ,, ,, | ,, ,, |
| 98 | 6.5–7.5 | Radiolysis + esr spectr. | Fessenden and Neta, 1972 |
| 99 | 7.0 | Flash photolysis + uv/vis spectr. | Land and Porter, 1963 |
| 100 | 2.9 | Pulse radiolysis + uv spectr. | Simic and Hayon, 1971 |
| 101 | −0.8 | Ce(IV) oxidation + flow esr spectr. | Dixon and Murphy, 1976 |
| 102 | −1.6 | ,, ,, | ,, ,, |
| 103 | −2.0 | ,, ,, | ,, ,, |
| 104 | ~−3[a] | Thermochem. calc. | Nicholas and Arnold, 1982 |
| 105 | ~−11[a] | ,, ,, | ,, ,, |

[a] Cf. Table 6

Direct measurements of $pK_a$-values of cation radicals of benzene and substituted benzenes have not been reported and it appears that these cation radicals are either too acidic (those from alkylbenzenes) or too susceptible to nucleophilic attack (those from benzene) for this approach to succeed.

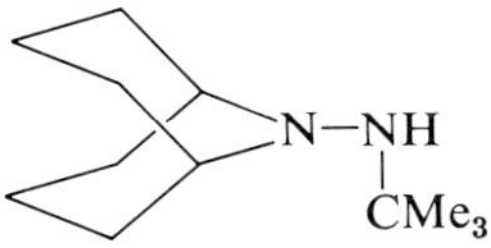

[96] 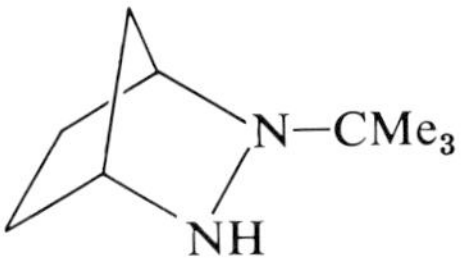 [97] [98] [99]

[100] MeONH$_2$ [101] HO–C$_6$H$_4$–OH [102] Me–C$_6$H$_4$–OH [103] C$_6$H$_5$–OH

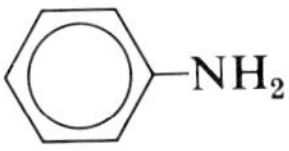

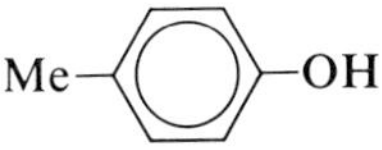

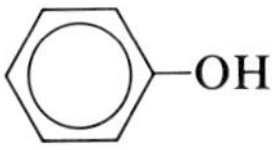

[104] 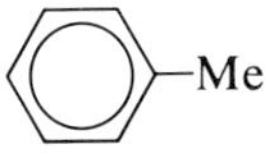 [105]

The alternative is to resort to thermochemical calculations. A detailed analysis of the problem has recently been made (Nicholas and Arnold, 1982). Three different methods were used to estimate the $pK_a$-values for the cation radicals of benzene and toluene. The first (method I) is based on (147) which can be derived using a thermochemical cycle making the assumption that the difference in solvation energy of RH and R· is negligible. In (147)

$$pK_{a(RH^{\ddagger})sol} = (-FE°_{RH} + \Delta G°_{tr(H^+)sol} + \Delta G°_{BDE(RH)g} - \Delta G°_{f(H)g})/2.303RT \quad (147)$$

$E°_{RH}$ is the reversible potential for the formation of the cation radical, $\Delta G°_{tr(H^+)sol}$ is the free energy associated with the transfer of a proton from water to the solvent of interest, $\Delta G°_{BDE(RH)g}$ is the bond dissociation energy of RH in the gas phase and $\Delta G°_{f(H)g}$ is the standard free energy of formation of the hydrogen atom. The free energies have either been calculated or are known and the $pK_a$-value can be estimated providing that the reversible potential can be determined. In the second method (II) the proton transfer of the cation radical to a base B⁻ is considered (148). In this case the calculation requires the knowledge of $\delta E°$, which is the difference between the reversible

$$pK_{a(RH^{\ddagger})sol} = (F\delta E° + \Delta G°_{BDE})/2.303RT + pK_{a(BH)sol} \quad (148)$$

potentials for the oxidation of RH and BH, respectively, and $\Delta G°_{BDE}$ the difference between the gas phase dissociation energies of R—H, and B—H,

TABLE 6
Estimates of the p$K_a$-values of the cation radicals of benzene and toluene at 300K[a]

| Parent compound | Method I solvent = | | Method II solvent = MeCN | | | | Method III solvent = | |
|---|---|---|---|---|---|---|---|---|
| | MeCN | $H_2O$ | (BH = HCl) | (BH = HBr) | | $H_2O$ | MeCN | $H_2O$ |
| Benzene | −3.8 ± 2 | −3 ± 4 | −2.7 ± 3 | −1.9 ± 3 | | −3 ± 4 | −2.3 ± 2 | <2.0 ± 3 |
| Toluene | −13 ± 2 | — | −11 ± 3 | −10 ± 3 | | — | −10 ± 2 | <−6.0 ± 3 |

[a] Nicholas and Arnold, 1982

and p$K_{a(BH)sol}$. Information regarding the gas phase acidity of $RH^+$ is required in the third method (III) which involves (149). It was pointed out that the method can have limited applications since the appropriate solvation

$$pK_{a(RH^{+\cdot})sol} = pK_{a(RH^{+\cdot})g} + \Sigma \Delta G°_{solv}/2.303RT \qquad (149)$$

energies are usually not known. The results obtained for benzene and toluene using all three methods are summarized in Table 6.

Inspection of the data in Table 5 reveals that the cation radicals of aliphatic amines and hydrazines are as a rule only moderately acidic. Deprotonation can be effectively suppressed in solvents like TFA or $H_2SO_4$ in acetic acid as shown by the applications of ammoniumyl radical in the Hofmann–Löffler reaction and homolytic aromatic amination (c.f. Section 6). In these reactions advantage is taken of the improved electrophilicity of the protonated species compared to the neutral radical. The almost identical $pK_a$-values found for the cation radicals of [98] and [99] may seem surprising, especially when comparison is made with the values for protonated substrates, which for the two compounds are 10.7 and 4.6, respectively. In other words, anilinium ion is a stronger acid than the cation radical of aniline, while the opposite order is observed for [98—$H^+$] and [98$^{+\cdot}$] and also for related inorganic systems. This relative difference of approximately six p$K$-units most probably reflects the resonance stabilization of the aniline cation radical, an effect which is absent in the anilinium ion.

Phenolic cation radicals show considerable acidities, the $pK_a$-values being typically around $-2$. This is reflected in the electrochemical oxidation of phenol derivatives, which proceeds in a chemically irreversible reaction unless very strongly acidic conditions are employed. Suppression of the deprotonation of these species requires acids like $FSO_3H$ in the presence of which reversible potentials may be measured at low temperature (Hammerich et al., 1976). Stable solutions of hydroquinone cation radical may be prepared

$$\text{HO-C}_6\text{H}_4\text{-OH} + \text{HO-C}_6\text{H}_4\text{-OH}^{2+} \rightleftharpoons 2\ \text{HO-C}_6\text{H}_4\text{-OH}^{+\cdot} \qquad (150)$$

at $-50°C$ in $CH_2Cl_2/FSO_3H$ through electron transfer between hydroquinone and diprotonated benzoquinone (150) (Hammerich and Parker, 1982c). Potential-pH diagrams for the region between pH $= -1$ and 15 have recently been published for hydroquinone, phenol and 2,4,6-tri-t-butylphenol (Bailey et al., 1983).

The calculated $pK_a$-values of benzene and toluene cation radicals are very small and accordingly both species ought to show pronounced acidity.

Deprotonation of benzene cation radicals has in fact been proposed in an early electrochemical study of benzene oxidation to account for the formation of biphenyl (Osa et al., 1969). However, the formation of biphenyl does not require that benzene cation radical deprotonate before coupling, and the mechanism of this reaction is still not known. So far, there has been no unambiguous demonstration of deprotonation of benzene cation radical and it is unlikely that it can be observed, due to the competition of attack by nucleophilic impurities in the solvent system or by the solvent system itself. It is most probable that a solvent system of sufficiently low nucleophilicity would be too acidic to allow the deprotonation to take place. In this context it should be mentioned that the rate constant for reaction between benzene cation radical and water has been estimated to be higher than $10^7 M^{-1}s^{-1}$ (Neta et al., 1977; Sehested and Holcman, 1978). The fate of the hydroxycyclohexadienyl radical produced may be further oxidation to the cation followed by deprotonation to phenol or dissociation to benzene and hydroxyl radicals (Eberhardt, 1981; Eberhardt et al., 1982). The extremely high acidity of toluene cation radical compared to that of benzene cation radical reflects the difference in stability of the corresponding bases, phenyl and benzyl radical, respectively. While the former is a σ-radical, the latter is a highly resonance stabilized π-system. This difference is reflected in the value of $\delta\Delta G°_{BDE}$ which amounts to approximately 21 kcal mol$^{-1}$. Nicholas and Arnold (1982) pointed out that a difference of approximately 2.4 kcal mol$^{-1}$ in bond dissociation energies corresponds to 2 p$K$-units for closely related systems. For the comparison of benzene with toluene, the 21 kcal mol$^{-1}$ should roughly be responsible for a difference of 16–17 p$K$-units. However, some of this difference is counterbalanced by the lower oxidation potential of toluene as compared to benzene.

Rate constants for deprotonation of cation radicals have only occasionally been reported. The decay of the methylbenzene cation radicals has been studied in weakly acidic aqueous solution (Sehested and Holcman, 1978). The rate of deprotonation was found to be highly dependent on the number of methyl groups and second order rate constants varying from $10^7 M^{-1}s^{-1}$ (toluene) to $1.6 \times 10^4 M^{-1}s^{-1}$ (pentamethylbenzene) were found. It is of interest to note that in neutral or slightly alkaline solutions the reaction with $H_2O$ or $OH^-$ may compete favorably with proton loss. Rate constants for the reaction with hydroxide ion approach the diffusion controlled limit. Attempts have been made to estimate the activation energies for deprotonation of the methylbenzene cation radicals from thermochemical calculations similar to those already discussed in Section 3 and based on the same assumption (Eberson et al., 1978a). The data in Table 7 may be of limited importance for work carried out under nonaqueous conditions and in the presence of often poorly defined bases different from water. The thermochemical calculations showed that the activation energy for deprotonation is strongly dependent on the nature of the solvent and obviously the base.

TABLE 7

Rate constants for deprotonation of methylbenzene cation radicals in water[a]

|  | Proton loss reaction $k/s^{-1}$ | Reaction with $H_2O$ $k/M^{-1}s^{-1}$ | Reaction with $OH^-$ $k/M^{-1}s^{-1}$ |
|---|---|---|---|
| Toluene | $1.0 \pm 0.5 \times 10^7$ | $> 2 \times 10^7$ |  |
| o-Xylene | $2.0 \pm 0.5 \times 10^6$ | $8 \pm 4 \times 10^5$ |  |
| m-Xylene | $2.0 \pm 0.5 \times 10^6$ | $1.5 \pm 0.5 \times 10^6$ |  |
| p-Xylene | $1.4 \pm 0.2 \times 10^6$ | $5.0 \pm 2.0 \times 10^5$ |  |
| Mesitylene | $1.5 \pm 0.2 \times 10^6$ | $1.0 \pm 0.5 \times 10^6$ |  |
| Hemimillitene | $1.5 \pm 0.3 \times 10^6$ | $5.0 \pm 2.0 \times 10^4$ | $\sim 1 \times 10^{10}$ |
| Pseudocumene | $2.0 \pm 0.2 \times 10^5$ | $3.0 \pm 1.0 \times 10^3$ | $3.5 \pm 0.5 \times 10^9$ |
| Isodurene | $1.0 \pm 0.2 \times 10^5$ | $1.0 \pm 0.3 \times 10^3$ | $1.2 \pm 0.2 \times 10^9$ |
| Prehnitene | $2.5 \pm 0.2 \times 10^5$ | $4 \pm 2 \times 10^2$ | $6.0 \pm 1.0 \times 10^8$ |
| Durene | $2.7 \pm 0.5 \times 10^4$ | $6 \pm 2$ | $1.5 \pm 0.3 \times 10^8$ |
| Pentamethylbenzene | $1.6 \pm 0.3 \times 10^4$ | $<4$ | $1.0 \pm 0.2 \times 10^8$ |

[a] Sehested and Holcman, 1978

The acidity of tertiary ammoniumyl radicals of the general structure $RCH_2\overset{+\cdot}{N}R'_2$ has not been measured, but α-deprotonation to the corresponding resonance stabilized radical [106] is a commonly observed process in photoredox reactions (Chow et al., 1978; Lewis et al., 1982) and has been proposed as an important step in the electrochemical oxidation of tertiary amines. The neutral radical may be oxidized further to the iminium ion under the latter conditions (151) and the synthetic utility of this reaction has been reported (Shono et al., 1982). Convincing evidence for the ability of tertiary

$$RCH_2\overset{+\cdot}{-}NR_2 \xrightarrow{-H^+} R\overset{-}{C}H\overset{+\cdot}{-}NR'_2 \leftrightarrow R\overset{\cdot}{C}H\overset{\cdot\cdot}{-}NR'_2 \xrightarrow{-e^-} RCH\overset{+}{=}NR'_2 \quad (151)$$
$$[106]$$

ammoniumyl radicals to participate in fast, reversible acid-base reactions resulted from the nuclear polarization pattern observed when triethylamine was irradiated in the presence of naphthalene directly in the nmr probe (Gardini and Bargon, 1980). The evidence presented consisted of the following points. (i) The formation of an α-carbon-centered radical as an intermediate could be excluded. Back electron transfer in the ion radical pair resulting from irradiation caused only enhanced absorption of the quartet due to the α-$CH_2$ group. An α-centered carbon radical would be expected to affect the β-protons as well. (ii) In the presence of $H_2O$ or an alcohol, emission was observed for the OH-protons, which is consistent with fast proton exchange between the alcohol or water and ammoniumyl radicals that have escaped from the solvent cage and carry the complementary negative polarization at their α-protons. (*iii*) Polarization could not be observed when the α-perdeuterated amine was used. (*iv*) Irradiation of Me$_3$N in the presence of D$_2$O or CD$_3$OD caused the CH$_3$-singlet to split into a doublet-like signal and at the same time negative polarization of OH was observed. This observation was found to be in agreement with D—H exchange between the amine and the hydroxylic compounds. (*v*) In the presence of a strong base like MeO$^-$ in CD$_3$CN as the solvent, emission was observed in a quintet assigned to the solvent, indicating that the small amounts of CD$_2$CN$^-$ could also compete for the α-protons of the amine.

The acidity of tertiary ammoniumyl radicals should be seen in the light of the resonance stabilization of the corresponding base [106]. For this energy gain to be effective it is required that a coplanar arrangement of the σ-bonds can be achieved to gain maximum overlap of the p-orbitals forming the "three-electron π-bond". Steric interactions or a rigid carbon framework may hinder a favorable geometry of the C—N bond with destabilization of the radical as the result, which again will cause the acidity of the ammoniumyl radical to be reduced. This Bredt's rule stabilization has a pronounced influence on the stability of ammoniumyl radicals as evidenced by several studies by Nelsen and coworkers (Nelsen, 1979; Nelsen *et al.*, 1980a; Nelsen and Gannett, 1982). As a result of this stabilization the reversible oxidation potential of the "simple" amine [107] could be measured, resulting in the value +0.63 V *vs* SCE.

$$\text{N-C-Me structure}$$

[107]

The effective bases during deprotonation of cation radicals are in general not known. Only few studies have been carried out in the presence of a deliberately added base or buffer system and the most likely candidate under less controlled conditions is residual water, always present in the solvent unless special precautions are taken. An exception to this general statement is found

$$\text{RCH}_2\text{—}\overset{+\cdot}{\text{NR}'_2} + \text{RCH}_2\text{—NR}'_2 \longrightarrow \text{RC}\overset{\cdot\cdot}{\text{H}}\text{—NR}'_2 + \text{RCH}_2\text{—}\overset{+}{\text{N}}\text{HR}'_2 \qquad (152)$$
$$\underset{-e^-}{\longrightarrow} \text{RCH}=\overset{+}{\text{NR}'_2}$$

in studies of ammoniumyl radicals in which case the unoxidized amine, if present and in the absence of a stronger base, is capable of taking up the liberated protons (152). Another example is illustrated by results obtained

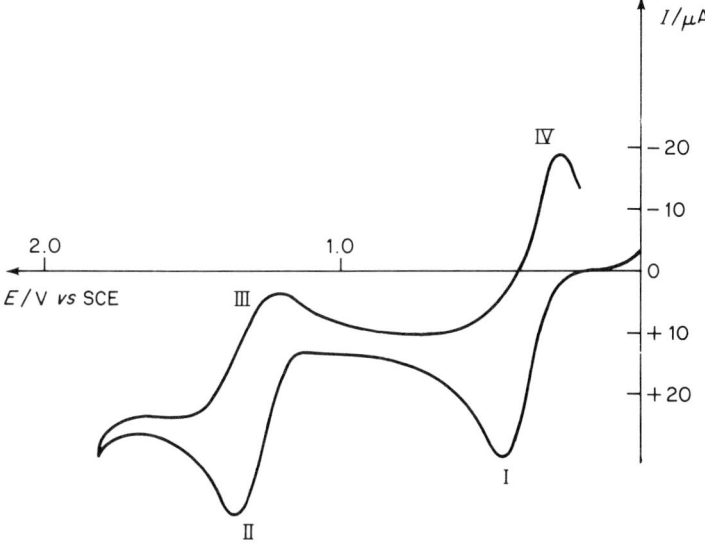

FIG. 15 Cyclic voltammogram of [108]. (Reprinted with permission from Speiser and Rieker, 1979)

during a thorough study of the voltammetric oxidation of 2,6-di-t-butyl-4-(4-dimethylaminophenyl)phenol [108] (Speiser and Rieker, 1977, 1978, 1979, 1980). The cyclic voltammogram of [108] is shown in Fig. 15. The first oxidation peak (I) corresponds to the conversion of [108] to [110] the reduction of which takes place at peak (IV). The peak height of (I) is roughly that of a one-electron process although the formation of [110] obviously requires that two electrons per molecule are transferred. This apparent contradiction can be explained by considering the processes taking place at peak (II), which was shown to be due to the oxidation of protonated [108] formed in the acid-base reaction (153). Thus, approximately half of the substrate is consumed by protons during step (153), giving rise to the overall process summarized in (154), and which is consistent with the observation of an apparent one-electron process at peak (I). Addition of a non-electroactive base like 2,6-lutidine caused the peak height of (I) to grow at the expense of that of (II) in agreement with the above interpretation. The analogous situation, protonation of an electrochemically formed anion radical by a weakly acidic substrate, is frequently encountered as well and gives rise to lower peak currents.

$$2\ [108] \xrightarrow{-2\,e^-} [110] + [108-H^+] \tag{154}$$

PRODUCTS OF PHENOLIC CATION RADICAL REACTIONS

The anodic oxidation of phenol and especially of derivatives substituted in one or more of the positions 2, 4 and 6 has been the subject of extensive studies (cf. e.g. Hammerich, 1983). Oxidation may take place within a broad

ORGANIC CATION RADICALS 133

range of potentials limited approximately by the $E^\circ$-values of the two redox couples [111]/[112] and [113]/[114] which are typically +1.6 V and −0.3 V vs SCE, respectively. Since [111] and [113] are connected through an acid-base reaction, the actual value of the working potential will depend on the acidity of the electrolysis solution. Even under very acidic conditions the predominant reaction of [112] will be loss of a proton to form the phenoxyl redical [114], which is easily explained by the $pK_a$-values of −2 for phenolic cation radicals. Although it is conceivable that [112] may participate in other reactions, for example couplings and attack by nucleophiles, other products than [114] cannot unambiguously be traced back to the cation radical. The oxidation of [114] to [115] takes place at approximately +1 V vs SCE and is not intrinsically dependent on the proton concentration in solution. Whether this further oxidation will take place does, however, indirectly depend on the acidity of the solvent through the influence on the relative proportions of [111] and [113], and thus on the actual working potential.

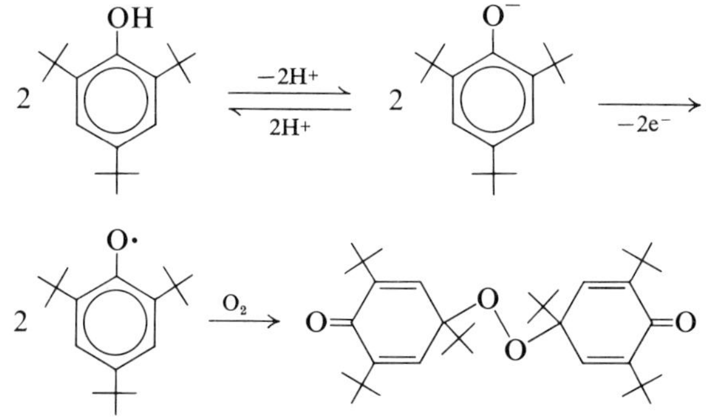

(155)

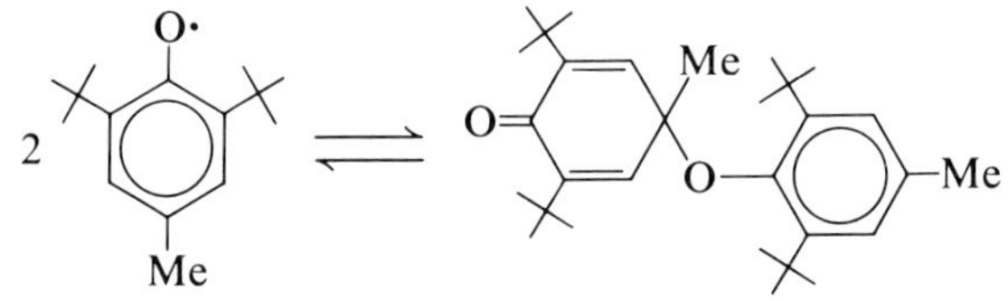

(156)

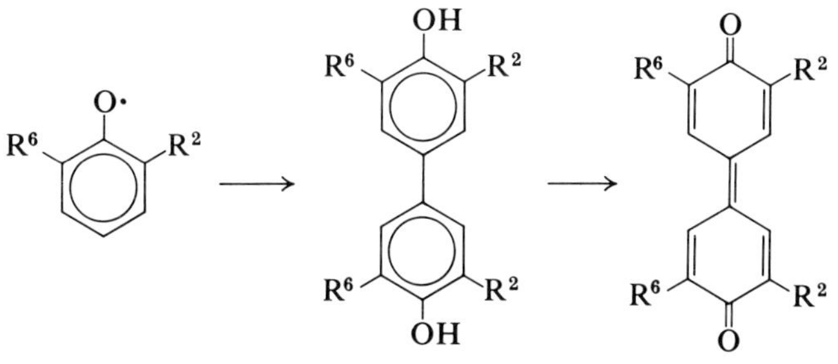

(157)

When appropriately substituted, e.g. by t-butyl groups in the 2,4 and 6-positions, solutions of [114] prepared by oxidation of [113] are stable if protected from oxygen in the presence of which peroxides are formed (155) (Richards *et al.*, 1975; Suttie, 1969). When the 4-substituent is less bulky, 4-methyl or 4-ethyl coupling (156) may take place (Richards and Evans, 1977). Finally, when the 4-position is unsubstituted coupling may lead to the ultimate formation of diphenoquinones (157) (Nilsson *et al.*, 1973, 1978). However, the formation of dimeric species like [116] does not necessarily require radical intermediates, since the coupled products may also arise by nucleophilic attack of substrate on the phenoxonium ion [115].

The major reaction of phenoxonium ions [115] is nucleophilic attack, especially by residual or deliberately added water or by alcohols, resulting in the formation of cyclohexa-2,5-dienone derivatives (158) (Nilsson *et al.*, 1973; Ronlán and Parker, 1971). The preparation of cyclohexa-2,5-dienones by anodic oxidation has later been developed to the practical synthetic level (Rieker *et al.*, 1978).

$$\underset{R^4}{\underset{|}{\overset{O}{\underset{\|}{C_6H_2R^6R^2}}}}^+ \xrightarrow{Nu^-} \underset{R^4\ Nu}{\underset{|}{\overset{O}{\underset{\|}{C_6H_2R^6R^2}}}} \qquad (158)$$

## SYNTHETIC AND MECHANISTIC ASPECTS OF ALKYLBENZENE OXIDATIONS

The deprotonation of alkylbenzene cation radicals has been studied in detail in nonaqueous solvents both under aprotic and acidic conditions. As a result of investigations in acetonitrile by electrochemical techniques like cyclic voltammetry, chronoamperometry and chronopotentiometry as well as spectroelectrochemistry it was concluded that the mechanism for deprotonation included steps (159)–(162) with (160) being rate-determining and irreversible (Bewick *et al.*, 1975, 1976, 1977, 1978, 1980c). However, more recent studies of durene and hexamethylbenzene [117] have provided results

$$Ar\text{—}CH_3 \underset{e^-}{\overset{-e^-}{\rightleftharpoons}} Ar\text{—}CH_3^{+\cdot} \qquad (159)$$

$$Ar\text{—}CH_3^{+\cdot} \rightarrow Ar\text{—}CH_2^{\cdot} + H^+ \qquad (160)$$

$$Ar\text{—}CH_2^{\cdot} + Ar\text{—}CH_3^{+\cdot} \rightleftharpoons Ar\text{—}CH_2^+ + Ar\text{—}CH_3 \qquad (161)$$

$$Ar\text{—}CH_2^+ + Nu^{(-)} \rightarrow Products \qquad (162)$$

inconsistent with this view on the mechanism. Analysis of the reaction mechanism for [117] by double potential step chronoamperometry in $CH_2Cl_2$ containing TFA indicated that the reaction is second order in cation radical, and not first order as required by steps (159)–(162), and it was demonstrated that the half-life of ($117^{\stackrel{+}{\cdot}}$) was proportional to the TFA concentration (Barek et al., 1980). Together, these two results demand that (160) is a true equilibrium and that (161) has to be considered as a possible rate-determining

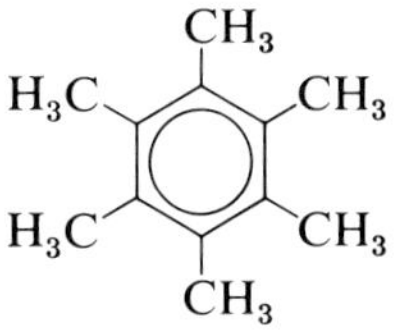

[117] 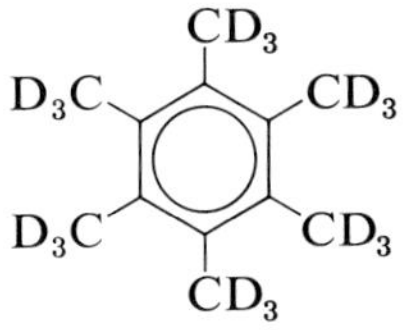    [118] 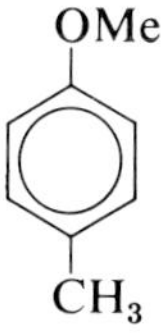    [119]

[120] 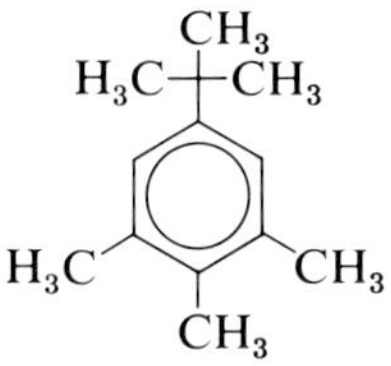    [121] 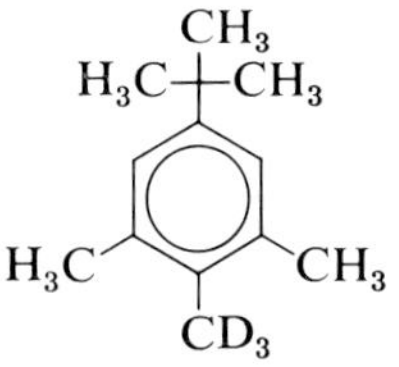    [122]

step under the experimental conditions. Further complexities in both neutral and acidic acetonitrile were revealed by a linear sweep voltammetry study, the results of which could only be interpreted if it was assumed that (162) participated in determination of the overall rate of cation radical decay as well (Schmid Baumberger and Parker, 1980). The experimental results were found to be consistent with rate law (163), where B and BH+ represent the base and conjugate acid in (164). The complexity of the reaction was further

$$\text{Rate} = k_{\text{app}}[\text{Ar—CH}_3^{\stackrel{+}{\cdot}}]^2[\text{Nu}^{(-)}][\text{B}]/[\text{Ar—CH}_3][\text{BH}^+] \quad (163)$$

$$\text{Ar—CH}_3^{\stackrel{+}{\cdot}} + \text{B} \rightleftharpoons \text{Ar—CH}_2^{\cdot} + \text{BH}^+ \quad (164)$$

substantiated by results obtained from derivative cyclic voltammetric analysis which demonstrated that the apparent rate constants for the disappearance of [117] and the $d_{18}$-derivative of [118] were concentration dependent and that the overall process had a negative activation energy (Parker, 1981a). The kinetic deuterium isotope effect could be estimated to be between 4.0 and 4.9 depending on temperature. The temperature effect clearly indicated that the rate-determining step follows a reversible reaction which is displaced to the left by increasing temperature. It was considered most likely that (160) responded to temperature changes in this way due to the large differences in solvation associated with this step while the temperature dependence of (161) was expected to be moderate.

The intrinsic complexity of reaction mechanism (159)–(164)–(161)–(162) may obviously be encountered as well in studies of metal-ion oxidation of alkylbenzenes in the same or related solvents. Experience has shown that the primary event in these reactions apparently may be described as anything between one-electron transfer followed by deprotonation, analogous to the first two steps in anodic oxidation, and hydrogen atom abstraction leading directly to Ar—CH$_2^\cdot$ dependent on the nature of the metal ion and its immediate surroundings. Thus, in metal ion oxidation the first step may add further complications to the ones already discussed. On the other hand, the reactions sometimes appear to be kinetically more simple with apparent rate constants reflecting only the electron transfer and the subsequent proton transfer as in the cases discussed below. Although other aspects certainly are interesting in themselves, we shall restrict our treatment to the cases where it has been demonstrated beyond any reasonable doubt that cation radicals are involved as reactive intermediates. The reader interested in details concerning the factors influencing the nature of the first step is referred to the chapter by Eberson (1982) as well as several reviews and books cited in a very recent paper by the same author (Eberson, 1983).

The primary evidence for an initial step involving electron transfer has been provided by kinetic studies. A classic example is the oxidation of $p$-methoxytoluene [119] by Mn(III) acetate in acetic acid (Andrulis et al., 1966). The reaction, which predominantly yields anisyl acetate, was found to be satisfactorily described by rate law (165). Furthermore, experiments carried

$$-\frac{d[\text{Mn(III)}]}{dt} = \frac{k[\text{Mn(III)}][119]}{[\text{Mn(II)}]} \qquad (165)$$

out at 70°C showed that the apparent rate constant was insensitive to changes in the concentration of acetate ion. The inverse dependence of the rate on the Mn(II) concentration suggested a reversible electron-transfer reaction as the first step followed by rate determining conversion of the resulting cation radical of [119] to products. Oxidation of the $\alpha,\alpha,\alpha$-$d_3$-derivative [120] resulted in a deuterium kinetic isotope effect of approximately 5 for the reaction, which indicated that the rate determining step involved deprotonation of [119$^{\cdot+}$] to the neutral $p$-methoxybenzyl radical. This, together with the identity of the final product and rate law (165) provided the necessary background to suggest mechanism (166)–(169) for the overall reaction, where the product $k'K$ equals $k$ in the rate law. However, the apparent

$$Ar-CH_3 + Mn(III) \stackrel{K}{\rightleftharpoons} Ar-CH_3^{+\cdot} + Mn(II) \quad (166)$$

[119]
$$Ar-CH_3^{+\cdot} \stackrel{k'}{\rightarrow} Ar-CH_2^{\cdot} + H^+ \quad (167)$$

$$Ar-CH_2^{\cdot} + Mn(III) \stackrel{fast}{\rightarrow} Ar-CH_2^{+} + Mn(II) \quad (168)$$

$$Ar-CH_2^{+} + AcOH \stackrel{fast}{\rightarrow} Ar-CH_2OAc + H^+ \quad (169)$$

inconsistency of a rate-determining deprotonation step insensitive to the concentration of acetate ion remained to be explained. The low dissociation constants for metal-ion acetates in acetic acid suggested a modification of the first two steps to (170)–(171). Thus, the formation of [119$^{+\cdot}$] is accompanied

$$Ar-CH_3 + Mn(OAc)_3 \stackrel{K}{\rightleftharpoons} [Ar-CH_3^{+\cdot}AcO^-] + Mn(OAc)_2 \quad (170)$$

$$[Ar-CH_3^{+\cdot}AcO^-] \stackrel{k'}{\rightarrow} Ar-CH_2^{\cdot} + AcOH \quad (171)$$

by liberation of an acetate ion and provided that these two species exist for some time as a caged ion pair, proton transfer may take place within the solvent cage rendering the irreversible step essentially independent of the presence of added base. The nature of the counter ion may obviously depend on the exact composition of the oxidizing solution and accordingly the transition state for proton transfer may be placed at different positions along the reaction co-ordinate. This was considered the most likely explanation of the differences in the *intra*molecular selectivity observed for the oxidation of *p*-ethyltoluene and isodurene by cobaltic acetate, manganic acetate/$H_2SO_4$ and ceric ammonium nitrate in acetic acid (Baciocchi *et al.*, 1980a).

Determination of deuterium kinetic isotope effects for the $(NH_4)_2Ce(NO_3)_6$ oxidation of [117], [119] and [121] and the specifically labelled derivatives [118], [120] and [122] resulted in values varying between 1.0 and 6.2 dependent on the identity of the substrate and the concentration of Ce(III) (Baciocchi *et al.*, 1980b). For [121]–[122] both $k_H$ and $k_D$ increased slightly with increasing concentration of Ce(III), which was attributed to a salt effect, but the ratio $k_H/k_D$ was only insignificantly affected (Table 8). However, a quantitative analysis of the side chain substituted products (nitrates and acetates) demonstrated that the introduction of deuterium into the 2-methyl group reduced substitution in this position. The absence of a significant deuterium kinetic isotope effect indicated rate-determining electron transfer followed by fast deprotonation, the rate of which obviously depends on whether deuterium is present or not as indicated by the relative amounts of products formed in the two cases. The more easily oxidized compounds

TABLE 8

Second order rate constants ($k$) and deuterium kinetic isotope effects ($k_H/k_D$) for Ce(IV) oxidations of some alkylaromatic compounds in acetic acid at 25°C[a]

| Substrate | [Ce(III)],[b] $M \times 10^3$ | $k_H$,[c] $s^{-1}M^{-1}$ | $k_D$,[c] $s^{-1}M^{-1}$ | $k_H/k_D$[d] |
|---|---|---|---|---|
| [121]–[122] | | $4.3 \times 10^{-4}$ | $4.3 \times 10^{-4}$ | 1.0 |
| | 1.07 | $6.0 \times 10^{-4}$ | $4.9 \times 10^{-4}$ | 1.2 |
| | 2.62 | $9.3 \times 10^{-4}$ | $7.6 \times 10^{-4}$ | 1.2 |
| [117]–[118] | | 1.83 | 1.16 | 1.6 |
| | 0.315 | 1.77 | 0.63 | 2.8 |
| | 1.07 | 1.65 | 0.33 | 5.0 |
| | 2.62 | 1.31 | 0.24 | 5.4 |
| [119]–[120] | | $4.7 \times 10^{-1}$ | $1.34 \times 10^{-1}$ | 3.5 |
| | 0.317 | $1.28 \times 10^{-1}$ | $2.4 \times 10^{-2}$ | 5.3 |
| | 1.04 | $7.1 \times 10^{-2}$ | $1.22 \times 10^{-2}$ | 5.8 |
| | 2.64 | $5.8 \times 10^{-2}$ | $9.4 \times 10^{-3}$ | 6.2 |

[a] Baciocchi et al., 1980b; $[Ce(IV)]_0 = 1.05 \times 10^{-3}M$
[b] Concentration of added Ce(III), as $Ce(NO_3)_3 \cdot 6H_2O$
[c] The average error is $\pm 5$–$7\%$
[d] The average error is $\pm 10\%$

[119]–[120] resulted in a $k_H/k_D$-value larger than one. Introduction of increasing concentrations of Ce(III) in the solution caused the second order rate constants to decrease, $k_D$ more than $k_H$, and accordingly the ratio $k_H/k_D$ to increase. The same trend was observed for [117]–[118]. The variation of the $k_H/k_D$-value with the concentration of Ce(III) indicated that $k_{173}$ for these two pairs of compounds is of a magnitude comparable to $k_{-172}$ [Ce(III)] suggesting that rate control had partly been taken over by step (173).

$$Ar\text{—}CH_3 + Ce(IV) \underset{k_{-172}}{\overset{k_{172}}{\rightleftharpoons}} Ar\text{—}CH_3^{+\cdot} + Ce(III) \quad (172)$$

$$Ar\text{—}CH_3^{+\cdot} \overset{k_{173}}{\to} Ar\text{—}CH_2^{\cdot} + H^+ \quad (173)$$

A study of the *inter*molecular selectivity in Ce(IV) oxidations of methylbenzenes had earlier indicated that cation radicals were likely products in the first step (Baciocchi et al., 1976). A good linear correlation between $\ln(k_{relative})$ (to durene) and the charge-transfer transition energy, $h\nu_{CT'}$ of the corresponding TCNE complexes was observed. The value of the slope, 0.7, indicated that significant charge transfer had taken place in the transition state suggesting that a cation radical is the primary intermediate.[1] The reaction order in Ce(IV) was found to be somewhat larger than one, which might point to the participation of Ce(IV) associates in the electron-transfer process. The observation of reaction orders higher than one for the metal-ion oxidant had previously also been observed for Co(III) acetate (Sakota *et al.*, 1969) and the importance of dimeric associates discussed (Hanotier *et al.*, 1973; Hanotier and Hanotier-Bridoux, 1973). However, the nature of the first step in Co(III) oxidations in acetic acid is suspected not to be a simple electron-transfer step in general as indicated by the positional selectivity and the deuterium kinetic isotope effects observed in oxidations of [121], [122] and 1,2,3-trimethylbenzene (Baciocchi *et al.*, 1982). Reaction with cobaltic acetate was compared to anodic and ceric ammonium nitrate oxidations as well as hydrogen atom abstraction by N-bromosuccinimide (NBS). The results obtained for anodic and ceric ammonium nitrate oxidations were very similar, once again emphasizing the electron-transfer properties of Ce(IV), while Co(III) exhibited a behavior much more similar to that of NBS suggesting a mechanism possibly involving hydrogen atom abstraction.

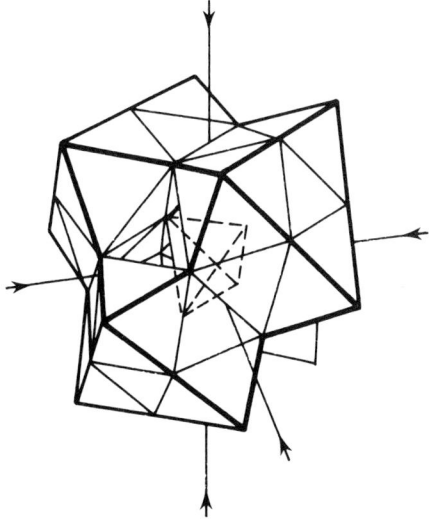

FIG. 16 The structure of the 12-wolframocobalt(III)ate ion (the Keggin structure). (Reprinted with permission from Eberson and Wistrand, 1980)

The tendency of for example Co(III) to participate in association equilibria and ligand exchange reactions may render interpretation of the observed kinetics unnecessarily complicated. A solution to this problem has been found in the application of the 12-tungstocobalt(III)ate (Co(III)W) (Fig. 16)

[1] Cf. also the discussion of aromatic electrophilic substitution in Section 6.

in which the cobaltic ion is protected from direct interactions with solvent and substrate by the screening shell of the $WO_6$ octahedra (Eberson and Wistrand, 1980; Eberson, 1983). Alkylaromatics oxidized by this species in acetic acid undergo α-acetoxylation in the absence of added acetate ion, in the presence of which nuclear acetoxylation is also observed. In this respect the results are very similar to those obtained by anodic oxidation. A thorough kinetic study of the oxidation of [119] in acetic acid containing 45% (w/w) water, in order to minimize possible ion-pairing effects, under pseudo first order conditions ([119] in excess) provided rate data in excellent agreement with mechanism (174)–(175). The deuterium kinetic isotope effect was determined to be between 5 and 7 dependent on the actual conditions of the measurements. The rate constants $k_{174}$ and $k_{-174}$ were estimated from Marcus theory to be equal to 2–10 $M^{-1}s^{-1}$ and $>10^{10} M^{-1}s^{-1}$, respectively. The rate

$$Co(III)W + [119] \underset{k_{-174}}{\overset{k_{174}}{\rightleftharpoons}} Co(II)W + [119^{+\cdot}] \qquad (174)$$

$$[119^{+\cdot}] + base \overset{k_{175}}{\rightarrow} Ar\text{-}CH_2^{\cdot} + base\ H^+ \qquad (175)$$

constant for deprotonation of the cation radical of [119] could be estimated to be 80 $M^{-1}s^{-1}$ (base = water) and $3.6 \times 10^5 M^{-1}s^{-1}$ (base = acetate ion).

A Hammett ρ-value of −3.4 resulted from the competitive oxidation of a series of 4-substituted anisoles and a $k_H/k_D$-value of 5.9 for [119] were found in agreement with formation of a cation radical in the first step (Nyberg and Wistrand, 1978). Recent results from oxidations by peroxydisulphate, via $SO_4^{-\cdot}$, have indicated electron transfer as the primary event also for this species (Jönsson and Wistrand, 1979; Minisci et al., 1983).

Oxidation of alkylbenzenes either anodically or by metal ions is an important synthetic route to a wealth of both side chain and nuclear substituted products. In the light of the results already discussed it is not surprising that the competition between side chain and nuclear substitution is heavily dependent on the deliberate addition of nucleophiles to the oxidation solution, for example addition of acetate ion to acetic acid. Increasing concentrations of the nucleophile tend in general to favor the products derived from nuclear substitution. Details concerning anodic substitution of alkylbenzenes may be found in a review by Eberson and Nyberg (1976) and a summary of oxidative side chain substitution of toluene, durene, hexamethylbenzene and p-methoxytoluene has been reported (Eberson et al., 1978a).

The benzyl radicals formed by deprotonation of alkylbenzene cation radicals may be trapped by quinones which provides a useful synthetic route to 2-substituted benzoquinones (Scheme 9) (Citterio, 1980). Yields were reported to be typically around 85%.

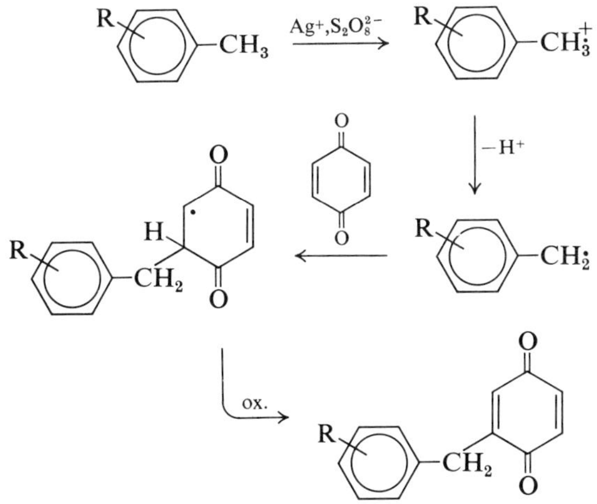

**Scheme 9**

CLEAVAGE OF CARBON—CARBON BONDS

The fragmentation of cation radicals through cleavage of a carbon—carbon bond is beyond doubt a very common and important process in cation radical chemistry in solution as it is in the gas phase (mass spectroscopy). However, despite the numerous cases where product analysis shows that cleavage of a carbon—carbon bond has taken place during an oxidation reaction, it has only been possible to establish unambiguously that the bond breaking actually does take place at the cation-radical state in very few studies. This lack of mechanistic evidence is connected to the fact that the initially formed cation radical in such reactions reacts so rapidly that a direct characterization of the intermediate, as well as kinetic and mechanism analysis of its decay is precluded. Thus, except for very rare cases, indirect evidence for the details of the reaction must be relied upon.

Examples of both older and more recent work, which possibly involve carbon—carbon bond cleavage of a cation radical, include the oxidation of 1,2-diarylethanes by Ce(IV) (Trahanovsky and Brixius, 1973), the oxidative cleavage of alkylphenylcarbinols (Trahanovsky and Cramer, 1971; Mayeda, 1975) and tetraphenyloxirane (Mayeda et al., 1972). Cleavage reactions related to the latter process are the photoinduced electron-transfer ring opening of 1,2-diphenyloxirane (Albini and Arnold, 1978) and the electrochemical oxidative cleavage of the 1,2 carbon—carbon bond in 1,1,2,2-tetraphenylcyclopropane (Wayner and Arnold, 1982).

Reduction of tropylium ion generates the neutral radical, which dimerizes in a fast irreversible step to bitropyl. The reverse reaction, cleavage of the bitropyl to tropylium ion, may be accomplished for example electrochemically. Similar reductive dimerizations and oxidative cleavages have been reported for numerous related systems including substituted pyrylium ions (cf. Siskin, 1972). Anodic oxidation of the dimer [123] in the presence of an aromatic hydrocarbon like rubrene or 9,10-dimethylanthracene was found to be accompanied by emission of light, which could be demonstrated to arise from the excited singlet of the aromatic hydrocarbon (Pragst and Ziebig, 1978). Magnetic field effects and energy calculations indicated that the excited singlet was formed by electron transfer between the radical [124] and the cation radical of the aromatic hydrocarbon [125$\overset{+}{\cdot}$] followed by triplet–triplet annihilation. The emission intensity was observed to depend on the electrode potential in a characteristic way as shown in Fig. 17 together with voltammograms of [123], the hydrocarbon rubrene [125] and the sum of the two. The high intensity peak is observed at a potential where the major electrochemical reaction is the direct electron transfer from [123] to the anode corresponding to the reaction steps (176)–(179). In the potential range where rubrene also undergoes diffusion controlled electron transfer, a broad

$$[123] \xrightarrow{-e^-} [123\overset{+}{\cdot}] \quad (176)$$

$$[123\overset{+}{\cdot}] \xrightarrow{\text{Fast}} [124] + [126] \quad (177)$$

$$[125] \underset{e^-}{\overset{-e^-}{\rightleftarrows}} [125\overset{+}{\cdot}] \quad (178)$$

$$[125\overset{+}{\cdot}] + [124] \longrightarrow {}^3[125]^* + [126] \quad (179)$$

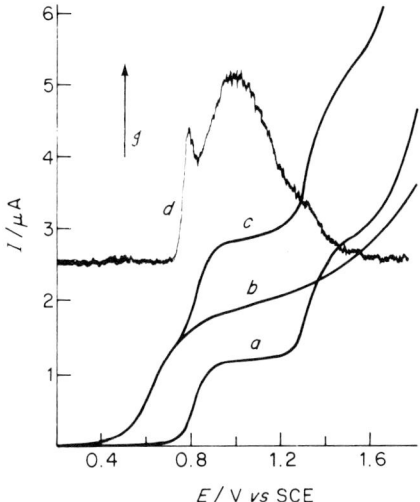

FIG. 17 Voltammograms at the rotating disk electrode ($\omega = 157\ s^{-1}$) of rubrene [125] (curve $a$, $c = 5 \times 10^4 M^{-1}$), the bipyrane [123] (curve $b$, $c = 5 \times 10^{-4} M^{-1}$) and of the mixture of [125] and [123] (curve $c$) and ECL intensity ($\mathscr{I}$) curve ($d$) of the system [123]–[125] in 0.1 M $N(C_2H_5)_4ClO_4/MeCN/C_6H_6$. (Reprinted with permission from Pragst and Ziebig, 1978)

intensity curve was observed, which was due to the indirect oxidation of [123] by [125$^{+}_{\cdot}$] as illustrated by the steps (180)–(181)–(179). The dip in the

$$2\ [125] \underset{2\ e^-}{\overset{-2\ e^-}{\rightleftharpoons}} 2\ [125^{+}_{\cdot}] \qquad (180)$$

$$[125^{+}_{\cdot}] + [123] \rightarrow [125] + [124] + [126] \qquad (181)$$

intensity curve was explained as being due to a diminished concentration of [125$^{+}_{\cdot}$] due to the homogeneous electron-transfer step (181). This study supported the view that the oxidative cleavage of dimers like [123] does indeed proceed *via* the intermediate formation of the cation radical [123$^{+}_{\cdot}$].

## CLEAVAGE OF BONDS BETWEEN CARBON AND HETEROATOMS

Anodic cleavage of benzyl ethers has received considerable attention, mainly because of the common application of benzyl ethers as alcohol protecting groups. Conventional deblocking is carried out under reducing conditions, and thus the mild oxidative cleavage offers an attractive alternative in cases where reducing conditions have to be avoided. The reaction proceeds as illustrated in (182) and may be carried out either by direct anodic oxidation

$$\text{ArCH}_2\text{OR} \xrightarrow[\text{H}_2\text{O},\ -2\text{H}^+]{-2\ e^-} \text{ArCHO} + \text{ROH} \tag{182}$$

(Miller et al., 1971; Mayeda et al., 1972; Weinreb et al., 1975) or indirectly by redox catalysis (cf. Section 4). The mechanism of the anodic cleavage has recently been under vigorous debate. The conclusion arrived at by Mayeda et al. (1972) was that the initially formed cation radical was converted to a phenyl-alkoxy carbenium ion by proton loss and subsequent electron

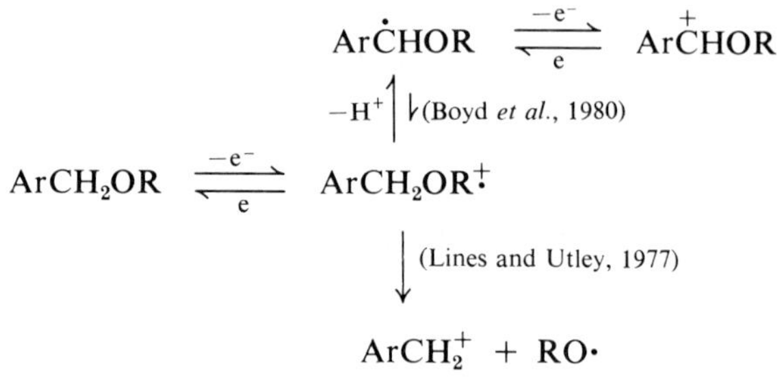

$$\text{Ar}\overset{\cdot}{\text{C}}\text{HOR} \underset{e}{\overset{-e^-}{\rightleftarrows}} \text{Ar}\overset{+}{\text{C}}\text{HOR}$$

$$-\text{H}^+ \uparrow \downarrow \text{(Boyd et al., 1980)}$$

$$\text{ArCH}_2\text{OR} \underset{e}{\overset{-e^-}{\rightleftarrows}} \text{ArCH}_2\text{OR}^{+\cdot}$$

$$\downarrow \text{(Lines and Utley, 1977)}$$

$$\text{ArCH}_2^+ + \text{RO}\cdot$$

**Scheme 10**

transfer (Scheme 10). Hydrolysis of this ion would then give the observed products directly. This interpretation was criticized (Lines and Utley, 1977) based on results obtained from a study of the anodic cleavage of dibenzyl ether, and a mechanism was proposed involving unimolecular cleavage of the C—O bond of the cation radical during which a benzyl cation and a benzyloxy radical is formed. Some of the experimental evidence supporting this conclusion was the observation that exhaustive oxidation of substrate required only one F mol$^{-1}$ and not two as demanded by (182). Furthermore, it was observed that in acetonitrile containing TFA, the major products were N-benzylacetamide and benzaldehyde. The latter was believed to arise from disproportionation of the benzyloxy radicals accompanied by the formation of benzyl alcohol. N-Benzylacetamide was considered to most likely arise via reaction between the benzyl cation and the solvent acetonitrile, which would result in formation of a nitrilium ion as an intermediate. Attack of benzyl alcohol on this ion could then lead to the N-benzylamide. Anodic oxidation of $^{18}$O-dibenzyl ether supported this explanation in that it was shown that the oxygen in both products, benzaldehyde and N-benzylacetamide, came from the ether. In a rebuttal of this mechanism (Boyd et al., 1980) it was pointed out that the discrepancy concerning the coulometric n-values had its origin in differences in the experimental conditions, and that a competing non-electrochemical destruction of substrate during the very dry conditions employed by Lines and Utley was responsible for the low charge consumption. The strongest evidence against the C—O bond cleavage mechanism was the observation of a primary deuterium kinetic isotope effect close to 1.9 for the oxidation of $PhCH_2OCD_2Ph$ indicating a rate-determining step involving proton transfer. The cation radical of dibenzyl ether could not be observed directly in either study, thus precluding more direct kinetic evidence for the mechanism of the cation radical decay.

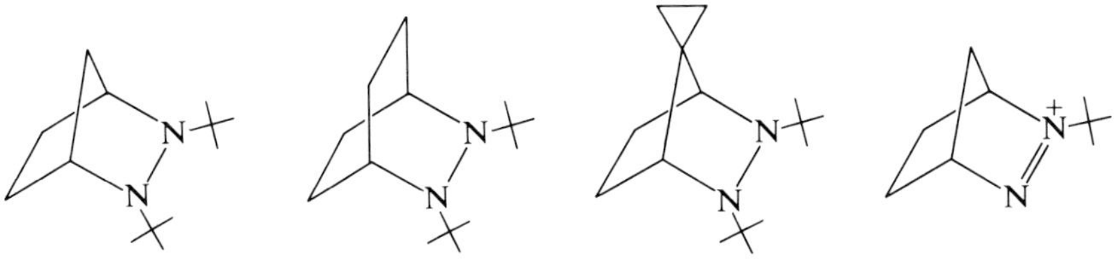

[127]    [128]    [129]    [130]

Carbon—nitrogen bond cleavage has been observed for the cation radicals of the t-butyl substituted hydrazine derivatives [127], [128] and [129] (Nelsen and Parmelee, 1981). Cyclic voltammetry at −25°C of [127] demonstrated that the lifetime of the corresponding cation radical was several seconds, while the rate of decomposition at room temperature matched first order rate constant of $1\ s^{-1}$. The cation radicals of [128] and [129] were found to decompose approximately 10 times faster and 10 times slower, respectively. The fate of all three cation radicals was loss of one t-butyl group resulting in formation of the corresponding diazenium ion, for example [130]. It was not possible to distinguish between loss of t-Bu· and t-Bu⁺.

Anodic oxidation of p-iodo-N,N-dimethylaniline [131] has been reported to result in cleavage of the carbon—iodine bond and the ultimate formation

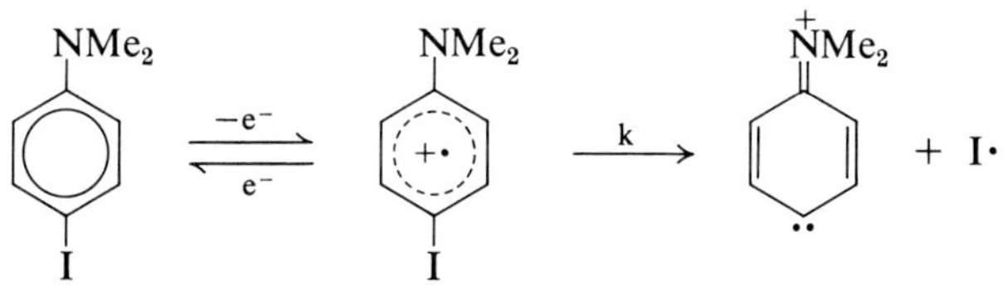

[131]    [134]    [133]

$2\ I· \longrightarrow I_2$

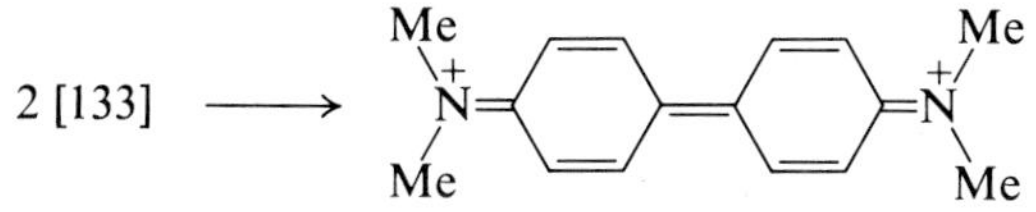

$2\ [133] \longrightarrow$ [132]

of $I_2$ and the N,N,N',N'-tetramethylbenzidine dication [132] (Hand et al., 1971). Coulometry showed that the process required the passage of one F mol$^{-1}$ and a mechanism was proposed, which involved rate-determining elimination of an iodine atom from the initially formed cation radical. Dimerization of the resulting carbene [133] was proposed to account for the formation of [132]. The peak current ratio, $i_p^c/i_p^a$, showed little dependence on the concentration of substrate indicating that the cleavage followed first order kinetics. The latter result could not be reproduced during a reinvestigation (Ahlberg et al., 1980). Data from cyclic voltammetry indicated that a second order rate law was followed and approximate second order rate constants were estimated from a working curve for the EC(dim) mechanism. The values 3.6 ($\pm 1.5$) × $10^3$ and 1.9 ($\pm 0.7$) × $10^4 M^{-1}s^{-1}$ for the p-bromo and p-iodo compound, respectively, resulted from this treatment. However, attempts to fit experimental data obtained by double potential step chronoamperometry (DPSC) to a theoretical working curve for a simple dimerization reaction failed, and this together with the finding that the overall reaction rate was almost insensitive to temperature variations suggested a more complicated reaction scheme. Simple comparison of $E°$-values demonstrated that

$$2 \text{ Me}_2\text{N-}\langle+\cdot\rangle\text{-X} \underset{k_{-183}}{\overset{k_{183}}{\rightleftarrows}} \text{[135]} \tag{183}$$

[134]

$$\text{[135]} \xrightarrow{k_{134}} \text{[132]} + X_2 \tag{184}$$

$$\text{[132]} + \text{[131]} \underset{k_{-185}}{\overset{k_{185}}{\rightleftarrows}} \text{[134]} + \text{[Me}_2\text{N-}\langle\rangle\text{-}\langle\rangle\text{-NMe}_2]^{+\cdot} \tag{185}$$

the homogeneous electron transfer equilibrium (185) is an exothermic reaction, and taking this into account in the calculations produced new theoretical data which gave a satisfactory fit to those obtained by experiment. The second order rate constants estimated in this way were equal to 4.47($\pm 1.02$) × $10^3$ and 2.23($\pm 0.20$) × $10^4 M^{-1}s^{-1}$ in good agreement with the values obtained by cyclic voltammetry. However, the observed effect of temperature variation was not explained by incorporation of (185) in the reaction mechanism. The lack of response to temperature changes indicated an Arrhenius activation energy close to zero for the overall process, which could only be rationalized by the assumption that the dimerization is reversible (183) and is followed by irreversible loss of halogen in the rate-determining step (184).

ORGANIC CATION RADICALS

Increasing temperature could cause (183) to be displaced in favor of the cation radical [134] and almost counterbalance the increasing rate of step (184). Thus, a mechanism including all three steps (183)–(185) was found to be consistent with the experimental observations.

CLEAVAGE OF SULFUR—SULFUR AND SELENIUM—SELENIUM BONDS

After anodic oxidation of disubstituted disulfides or diselenides in acetonitrile in the presence of an alkene, products were isolated which suggested the intermediate formation of $RS^+$ and $RSe^+$ ions, respectively (186) (Bewick et al., 1980a, 1980b). Investigations by conventional electrochemical techniques supplemented by modulated specular reflectance spectroscopy later

$$RSSR + \;\;\;C=C\;\;\; \xrightarrow[CH_3CN,H_2O]{-2e^-} RS-\underset{|}{\overset{|}{C}}-\underset{|}{\overset{|}{C}}-NHCOCH_3 \qquad (186)$$
(RSeSeR) \hspace{5em} (RSe)

demonstrated that $RS^+$ and $RSe^+$ are in fact the most likely intermediates, and a mechanism was suggested which involves the one-electron oxidation of RSSR (RSeSeR) to the corresponding cation radical (187) followed by rapid cleavage of the S—S (Se—Se) bond (188) (Bewick et al., 1983). During step (188) equimolar amounts of $RS^+$ ($RSe^+$) and RS· (RSe·) are generated and the latter is further oxidized to $RS^+$ ($RSe^+$) under the reaction conditions (189). The overall reaction can be summarized as in Scheme 11. It was proposed that in the absence of added alkene $RS^+$ is most probably trapped by acetonitrile.

$$R-S-S-R \underset{e^-}{\overset{-e^-}{\rightleftarrows}} R-S-S-R^{+\cdot} \qquad (187)$$

$$R-S-S-R^{+\cdot} \xrightarrow{Fast} R-S^+ + R-S\cdot \qquad (188)$$

$$R-S\cdot \xrightarrow{ox.} R-S^+ \qquad (189)$$

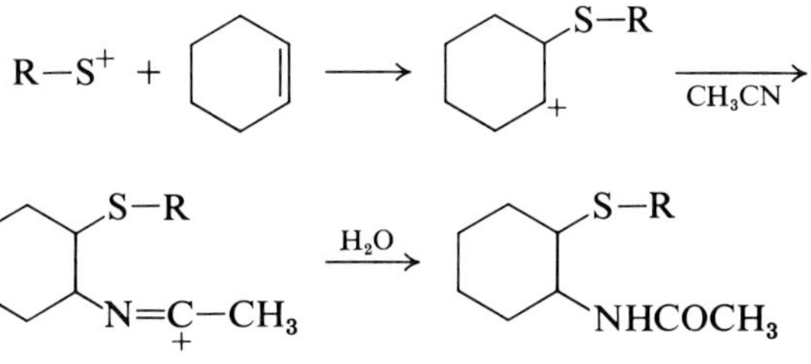

**Scheme 11**

A related reaction involving the indirect anodic oxidation of diphenyl diselenide together with a suitable alkene in methanol or acetic acid results in the formation of α-methoxy or acetoxyselenides (Torii et al., 1980a,b). Mechanistic details were not reported.

The treatment of diaryldisulfides with $AlCl_3/CH_2Cl_2$ resulted in the formation of the corresponding thianthrene cation radicals (Bock et al., 1982; Giordan and Bock, 1982). The mechanism of the conversion is unknown but may involve the cleavage of an S—S bond of the disulfide cation radical.

CLEAVAGE OF NITROGEN—NITROGEN BONDS

Cation radicals of arylsubstituted hydrazines have been reported to undergo a variety of rearrangements, the outcome of which are highly dependent on the structure of the hydrazine and the reaction conditions. In weakly acidic $CH_2Cl_2$ (5% TFA) tetraphenylhydrazine cation radical [136$^{+\cdot}$] was found to rearrange to the N,N'-diphenylbenzidine dication [137$^{2+}$] (Svanholm et al., 1972). An oxidation agent is obviously required, but the identity of this species could not be clearly established. The neutral N,N'-diphenylbenzidine [138] did not oxidize to the dication under the reaction conditions indicating that the rearrangement indeed took place in an oxidized state. The same

[136$^{+\cdot}$] →(ox.) [137$^{2+}$] →(red.) [138]

[136] + 2 H$^+$ → 2 [139$^{+\cdot}$]

[136] + [139$^{+\cdot}$] + H$^+$ ⇌ [140] + [136$^{+\cdot}$]

final product [138] could be isolated after treatment of unoxidized tetraphenylhydrazine [136] with TFA in $CH_2Cl_2$, in this case, however, accompanied by equimolar amounts of protonated diphenylamine [140]. These observations were rationalized by a reaction scheme which included the formation of diphenylamine cation radical [139$\overset{+}{\cdot}$] from protonated [136]. The experimental results did not allow to distinguish between the mono- and diprotonated form of [136] as being the species undergoing cleavage to [139$\overset{+}{\cdot}$]. However, it is conceivable that the reactive species is the N,N'-diprotonated form of [136] which is formed when [136] is treated with TFA-antimony pentafluoride at low temperature and which undergoes rearrangement to [137$^{2+}$] at room temperature (Svanholm and Parker, 1972b).

Related studies have demonstrated that substituted benzidines are only one out of several types of products which may be formed under similar conditions from arylhydrazine cation radicals. When tetra-arylhydrazines carrying electron donating substituents like methyl and methoxy [141] are

dissolved in liquid $SO_2$ spontaneous oxidation to the cation-radicals takes place followed by a slow transformation to the cation-radicals of the corresponding 5,10-dihydrophenazine derivative [142$\overset{+}{\cdot}$] (Nojima et al., 1976). Treatment of [141$\overset{+}{\cdot}$; R = OMe] with TFA results in the formation of the same product together with dianisylamine cation-radical analogous to the reaction described for [136]. In this case [141$\overset{+}{\cdot}$] was detected as an intermediate. Several other hydrazines were investigated and the results are summarized in Scheme 12. The corresponding benzidine was the product of oxidation only in one case. Surprisingly, it was found that irradiation of [136], which did not oxidize spontaneously in liquid $SO_2$, after the formation of [136$\overset{+}{\cdot}$], resulted in conversion to [144$\overset{+}{\cdot}$] and [145] rather than [136] and [137$^{2+}$] which are formed when photolysis of [136$\overset{+}{\cdot}$] is carried out in dichloromethane solution (Svanholm and Parker, 1972a). The observation that [136$\overset{+}{\cdot}$] may be

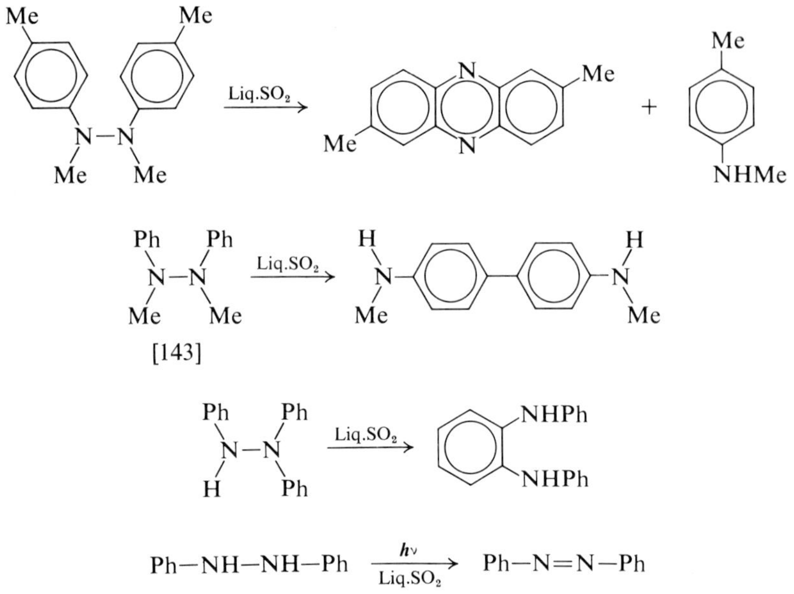

**Scheme 12**

transformed into [137²⁺] again raised the question of the importance of cation-radical intermediates in the benzidine rearrangement. However, benzidines are apparently the final products of arylhydrazine cation-radical reactions only in few cases, if the results in for example liquid SO₂ may be

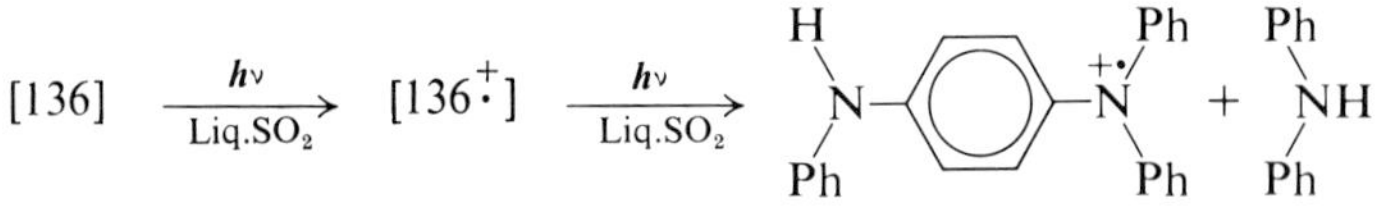

$$[136] \xrightarrow[\text{Liq.SO}_2]{h\nu} [136^{+\cdot}] \xrightarrow[\text{Liq.SO}_2]{h\nu}$$

[144⁺˙]    [145]

generalized, and the different factors governing which of many possible routes the reaction will follow under a given set of conditions are all far from being known. An important point in this connection is the lack of firm knowledge as to whether the cation radical reactions are of *inter-* or *intra-*molecular nature. The diversity of products formed from very similar substrates and the fact that even the same substrate cation-radical may give different products under similar conditions supports the view that at least some of the reactions are *inter*molecular. Another point of interest is the role of the protons liberated during some of the reactions. Under conditions initially aprotic, the acidity increases during the reaction, the mechanistic influence of which is unknown. This is, however, a point which cannot be neglected in the light of the apparent very delicate balance between different possible pathways and the knowledge that the acidity of the solvent is an important factor in these rearrangements.

Thus, independent of whether or not the results emerging from studies of arylhydrazine cation-radical reactions may have a direct bearing on the view on the benzidine rearrangement mechanism, the details of the cleavage of the N—N bond in these species are certainly interesting enough to merit further attention.

In this connection attention should be directed towards a recent study of the prototype benzidine rearrangement of hydrazobenzene based on $^{15}N$, $^{13}C$, $^{14}C$ and deuterium kinetic isotope effects (Shine et al., 1982). In acidic aqueous ethanol the reaction is second order in $H_3O^+$ and proceeds essentially to a 70/30 mixture of benzidine [146] and diphenyline [147]. Four substrates specifically $^{15}N$, $^{15}N'$, $4\text{-}^{14}C$, $4,4'\text{-}^{13}C_2$ and $4,4'\text{-}d_2$ labelled were studied. The results of the heavy atom kinetic isotope effects were especially illuminating. Concerning the $^{15}N$ effect, it was found that formation of both products was accompanied by substantial values of $k(^{14}N)/k(^{15}N)$ indicating that N—N bond weakening had taken place in the transition states. However, a significant difference in the magnitudes of the kinetic isotope effects was observed, that for [146] being 1.0222, while that found for [147] amounted to not less than 1.063, which suggests that formation of [146] and [147] passes through different transition states. This suggestion was supported by the carbon kinetic isotope effects which were of a considerable magnitude for formation of [146], but vanishingly small for formation of [147]. It was concluded that the formation of [146] is a concerted process, while that for [147] is not, and that none of the previous theories for the mechanism of the benzidine rearrangement provide a satisfactory overall description of the reaction. The new data suggest that the formation of [146] is best described as being a consequence of the concerted polar transition state originally proposed by Banthorpe et al. (1964) and that formation of [147] may proceed either via a π-complex (Dewar and Marchand, 1965) or a caged radical pair.

## 6 Cation radicals as intermediates in conventional organic reactions

The mechanisms of a number of well-known reactions in organic chemistry are currently undergoing significant revision. Evidence provided by modern experimental techniques including esr spectroscopy, CIDNP and various electroanalytical methods is incompatible with the prevailing view in conventional organic chemistry that electrons are transferred pairwise, and the role of ion radical intermediates has accordingly been emphasized. Electron transfer as a microscopic step in organic reactions has been considered occasionally, especially by Russian workers (Bilevich and Okhlobystin, 1968; Pokhodenko et al., 1975; Todres, 1978; Zefirov and Makhon'kov, 1982), but it is only in recent years that the concept has been generally recognized, now to the extent that it is difficult to open a new issue of one of the major journals without finding papers devoted to this subject.

Ion radicals, and in the present context especially cation radicals, may arise during a chemical reaction not only by elementary electron transfer, (190)–(192), but also in a number of other ways, some of which are summarized below, (193)–(198). Reactions (195) and (196) have already been illustrated

$$DH + AH \rightarrow DH^{\cdot +} + AH^{\cdot -} \tag{190}$$

$$DH + E^+ \rightarrow DH^{\cdot +} + E\cdot \tag{191}$$

$$DH + Me^{n+} \rightarrow DH^{\cdot +} + Me^{(n-1)+} \tag{192}$$

$$DH-E^+ \rightarrow DH^{\cdot +} + E\cdot \tag{193}$$

$$D-R + H^+ \rightarrow DH^{\cdot +} + R\cdot \tag{194}$$

by the $S_{ON}2$ mechanism (Scheme 4) and the rearrangement of [136] in acidic solution, respectively. Examples of the other reactions will be given in this section.

$$\overset{\cdot}{D}\!\!\underset{X}{\overset{Nu}{\diagdown}} \longrightarrow D-Nu^{\cdot +} + X^- \tag{195}$$

$$R-R + 2H^+ \longrightarrow 2RH^{\cdot +} \tag{196}$$

$$R\cdot + E^+ \longrightarrow R-E^{\cdot +} \tag{197}$$

$$RHX^+ \longrightarrow RH^{\cdot +} + X\cdot \tag{198}$$

In (190)–(198), DH represents an electron-rich and AH an electron-deficient molecule, $E^+$ is an electrophilic reagent and $Nu^-$ a nucleophile, $Me^{n+}$ is a metal ion in oxidation state $+n$, $X^-$ a halide ion and finally, $R\cdot$ represents a neutral free radical.

THE NITRAMINE REARRANGEMENT

Treatment of N-nitroanilines (nitramines) with acid results in isomerization to the corresponding o- and p-nitroanilines. Work published in the 1950s pointed towards a reaction pathway involving *intra*molecular rearrangement of protonated substrate (Hughes and Jones, 1950; Brownstein et al., 1956, 1958). Extensive work by White and coworkers confirmed that the reaction is

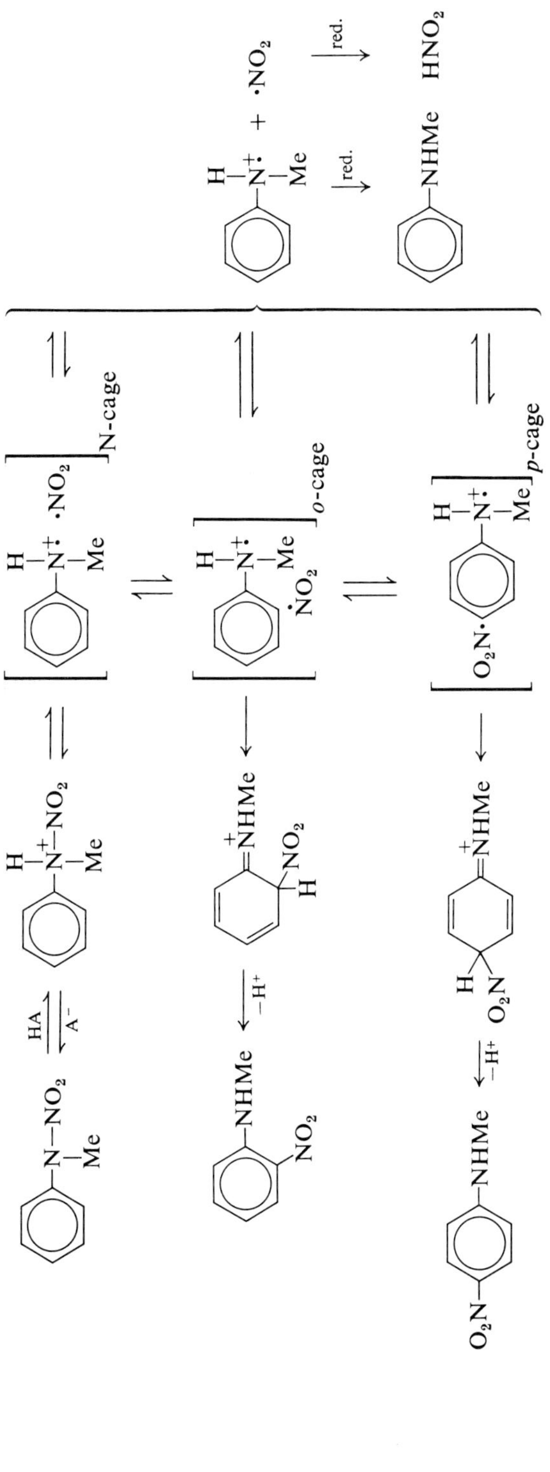

Scheme 13

indeed *intra*molecular in most cases and their experimental results were explained by the mechanism depicted in Scheme 13, which includes rate-determining formation of a caged cation radical-radical pair. The primary evidence for this scheme was (*i*) rearrangement of N-nitro-N-methylaniline in 0.5 N HCl gave *o*- and *p*-nitro-N-methylaniline (52% and 31%, respectively) but *m*-nitro-N-methylaniline could not be detected (<0.1%). Minor amounts of N-methylaniline (10%) and nitrous acid (13%) were formed as well (White et al., 1961, 1964). (*ii*) The reaction was found to be subject to specific acid catalysis and kinetic studies revealed a rate law first order in both substrate and acid. It was concluded that the rearrangement involved reversible protonation of substrate followed by a rate-determining unimolecular step leading to the isomerized product (White et al., 1970a). (*iii*) Hammett treatment ($\sigma^+$) of the rates for 16 substituted N-nitro-N-methylanilines gave a $\rho$-value of $-3.7$ indicating that the substrate becomes increasingly electron deficient on the way to the transition state in agreement with rate-determining formation of [ArNHMe$^{+\cdot}\cdot$NO$_2$]. Formation of [ArNHMe$^{2+}$ NO$_2^-$] was deemed less likely (White and Klink, 1970). (*iv*) The overall rates and product distributions for reactions of N-nitro-N-methylaniline and its 2,6-$d_2$-derivative were identical which eliminated the possibility of the final proton loss being rate-determining (White et al., 1970b). (*v*) Addition of reducing agents like I$^-$, SCN$^-$ and hydroquinone to the reaction mixture lowered the total yield of nitroanilines and augmented the yields of N-methylaniline and HNO$_2$. However, the reaction rates were unaffected which indicated that the diversion of the reaction took place after the rate-determining step. Increasing the concentrations of the reducing agents caused the total yield of nitroanilines to approach a constant value (56%) which is consistent with the idea that the reducing agents did not react directly with substrate or the primary intermediate, the cation radical-radical pair, but with species in equilibrium with it, in this case ArNHMe$^{+\cdot}$ and $\cdot$NO$_2$ (White and White, 1970). (*vi*) The dissociation of the cation radical-radical pair into ArNHMe$^{+\cdot}$ and $\cdot$NO$_2$ suggested that some *inter*molecularity of the reaction might be observed. This was confirmed by the products obtained from reaction of N,2,4-trinitro-N-methylaniline in dilute acid in the presence of 1,4-xylene. 2-Nitro-1,4-xylene was formed together with 2,4-dinitro-N-methylaniline (White and Golden, 1970). (*vii*) The *o/p* product ratio increased with increasing solvent viscosity reflecting the enhanced difficulties of the cation radical-radical pair to undergo the necessary reorientation for the formation of the distant C(4)—N bond (White et al., 1976).

Although the evidence for the proposed reaction mechanism is convincing, no direct indication of the involvement of a cation radical-radical pair was obtained. However, this has recently been achieved through $^{15}$N nmr spectroscopy (Ridd and Sandall, 1982). When $^{15}$N nmr spectra were observed during the rearrangement of $^{15}NO_2$-labelled nitramines, strongly enhanced signals were observed for both starting material and product. The nuclear polarization generated during the reaction was found to be in agreement with the proposal of an intermediate cation radical-radical pair, the fate of which is determined by the competition between dissociation and recombination as illustrated in Scheme 13. The nitramine reaction is an example of cation radical formation by route (194).

## ELECTROPHILIC AROMATIC SUBSTITUTION

The generally accepted mechanism for electrophilic aromatic substitution, (199)–(200), involves a positively charged σ-complex intermediate for which massive and compelling experimental evidence has been presented (Olah, 1971, 1974; Olah and Mo, 1976; Olah et al., 1976). However, the details of its formation are still a matter of intense discussion despite numerous attacks on the problem.

$$\bigcirc + E^+ \longrightarrow \bigcirc\!\!\!<^H_E \qquad (199)$$

$$\bigcirc\!\!\!<^H_E \longrightarrow \bigcirc\!\!-E + H^+ \qquad (200)$$

$$\bigcirc + E^+ \rightleftharpoons [\bigcirc E^+] \qquad (201)$$

$$\bigcirc + E^+ \longrightarrow \bigcirc^{+\cdot} + E\cdot \qquad (202)$$

$$\bigcirc^{+\cdot} + E\cdot \longrightarrow \bigcirc\!\!\!<^H_E \qquad (203)$$

One theory, which assumes that the reaction between substrate and the electrophile, $E^+$, initially leads to the formation of a so-called π-complex, (201), was proposed as a solution to the puzzling observation that the reactivity/selectivity principle seems to be violated during electrophilic aromatic substitution when strongly electrophilic reagents are used (Olah, 1971; see also Ridd, 1971). For example, in the $TiCl_4$-catalyzed competitive benzylation of benzene and toluene with a series of substituted benzyl chlorides it was found that even in cases where the substrate selectivity was low, the reaction exhibited high positional selectivity as measured by the almost total lack of the *m*-isomer among the products (Olah et al., 1970). For this and a number of similar cases it was suggested that the substrate selectivity is determined in the early π-complex, while formation of the σ-complex dictates the positional selectivity. This interpretation has been criticized by showing that rates of attack of the very reactive electrophiles correlate equally well with π- and σ-complex stabilities (Rys et al., 1972) and a recent analysis of no less than 108 electrophilic substitutions demonstrated that there is in fact satisfactory agreement with the reactivity/selectivity relationship with a few notable exceptions, among which aromatic nitration is the most interesting in the present context. A general mechanism which involves a single transition state, of character varying from π-complex-like (early) for reactive electrophiles to σ-complex-like (late) for selective electrophiles, was suggested (Santiago et al., 1979). *Ab initio* STO-3G calculations supported this view.

Another theory takes its starting point in early theoretical work in which the possibility of a mechanism, (202)–(203), involving charge transfer between substrate and the electrophile was analysed (Nagakura and Tanaka, 1954, 1959; Nagakura, 1963; Brown, 1959a,b). It has been pointed out that adoption of the charge-transfer mechanism made it possible to predict the observed relationship (204) between the overall rate of substitution, $k$, and the ionization potential, $I_p$, of the aromatic substrate, and that substitution

$$\log k = \beta I_p + \gamma \tag{204}$$

patterns correlate with the hyperfine coupling constants of the cation radical (Pedersen et al., 1973). Thus, in this case the positional selectivity originates from step (203).

The question of charge transfer in electrophilic aromatic substitution and related reactions in general has recently been taken up (Fukuzumi and Kochi, 1980a,b; 1981a,b; 1983). For example, for electrophilic aromatic substitution it was observed that the relative reactivities of different arenes correlated linearly with the charge-transfer transition energies, $\Delta h\nu_{CT}$, using benzene as the reference compound (Fukuzumi and Kochi, 1981b). Five examples are depicted in Fig. 18. The charge-transfer transition represents an electronic excitation from the ground state to the polar excited state (205) and on the basis of the linear correlations shown in Fig. 18 it was concluded that the formation of an ion pair was tantamount to the activation process for electrophilic aromatic substitution (206). It was suggested that the ion pair is a reasonable approximation to the transition state for the process.

$$[\text{Ar E}] \xrightarrow{h\nu_{CT}} [\text{Ar}^{+\cdot} \text{E}^{-\cdot}]^* \tag{205}$$

$$\text{Ar} + \text{E} \longrightarrow [\text{Ar}^{+\cdot} \text{E}^{-\cdot}] \tag{206}$$

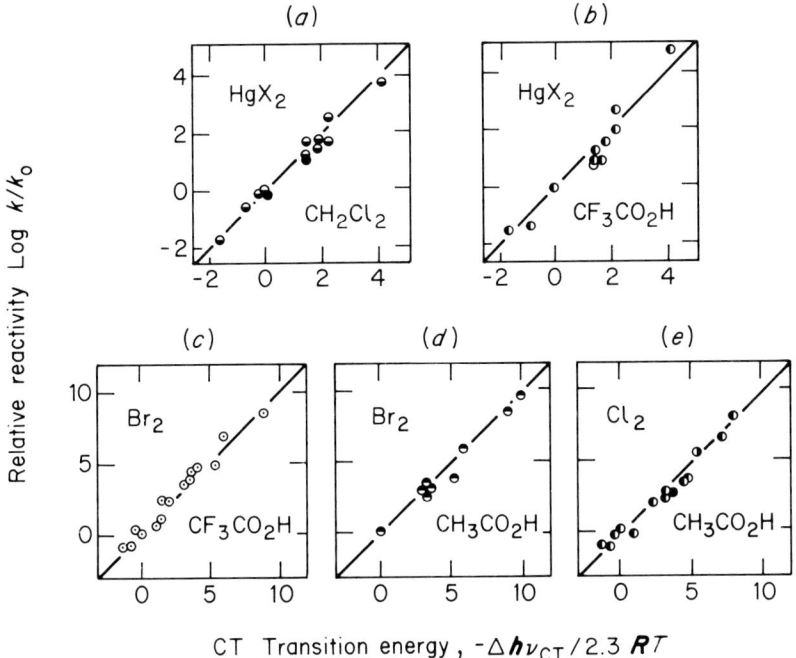

FIG. 18 The correlation of the relative reactivities of arenes with the CT transition energies for (a) $Hg(O_2CCF_3)_2$ in methylene chloride, (b) $Hg(O_2CCF_3)_2$ in trifluoroacetic acid, (c) $Br_2$ in trifluoroacetic acid, (d) $Br_2$ in acetic acid, and (e) $Cl_2$ in acetic acid. (Reprinted with permission from Fukuzumi and Kochi, 1981b)

Formation of the polar ion-pair state in the rate-determining step followed by a fast collapse to the σ-complex corresponds to a separation of the rate-determining step from the product-determining step. The substitution pattern would be expected to reflect the spin density as was suggested earlier (Pedersen et al., 1973). In benzene cation radical the unpaired electron is found in the degenerate $E_{1g}$ HOMO's labelled S for symmetric and A for antisymmetric (Fig. 19). However, the degeneracy is removed upon substitution with the effect that electron withdrawing substituents, Y, cause the energy of S to be lowered considerably while electron donating substituents, X, have the opposite effect. Accordingly, S is the HOMO in Ar—X cation radicals giving rise to the highest spin density in the p-position, while for the cation radicals of Ar—Y, A is the HOMO giving rise to the highest spin density in the m-position. The classification of substituents as electron withdrawing or electron donating could in this way be related directly to the HOMO energy as reflected by the ionization potential instead of as traditionally done by the reactivity, as for example in the Hammett treatment. The

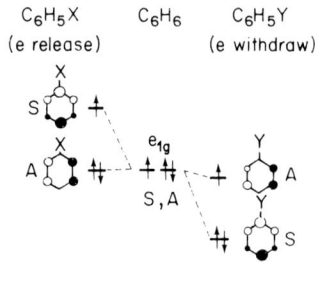

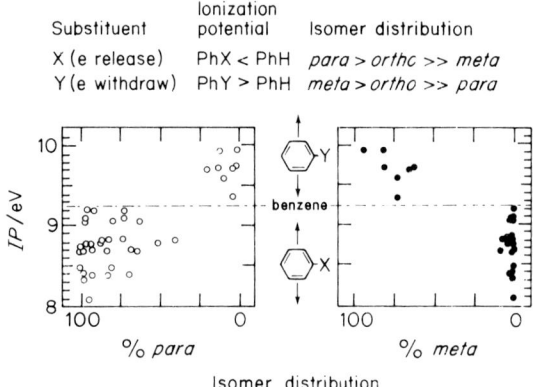

FIG. 19 The product distribution among *para* isomers (left) and *meta* isomers (right) derived from monosubstituted benzenes with electron-releasing (X) and electron-withdrawing (Y) substituents, as indicated by the values of the ionization potential relative to that of benzene (9.23 eV). (Reprinted with permission from Fukuzumi and Kochi, 1981b)

distribution of the $m$-/$p$-isomers predicted in this way is also summarized in Fig. 19. Four features of the plot were emphasized. (*i*) All monosubstituted benzenes fall clearly in one of the two groups with no derivative giving comparable amounts of $m$- and $p$-products. (*ii*) The isomer distribution can be explained solely by the ionization potential and is more or less independent on the nature of the electrophile. (*iii*) The halogens, which traditionally have to be treated in a group of their own, fall clearly in the group containing the electron-releasing substituents. (*iv*) There is a discontinuous change from $p$- to $m$-producing arenes occurring at unsubstituted benzene ($I_p = 9.23$ eV). It was pointed out that the fourth point is not easily accounted for by the types of arguments conventionally used in discussions of aromatic electrophilic substitution, but is nicely borne out by the charge-transfer approach. A similar treatment was reported for the Diels–Alder reaction

(Fukuzumi and Kochi, 1982), a process which recently has been shown to be subject to cation radical catalysis (see Section 4). Essentials of this treatment are also found in a recent paper concerning the valence bond approach to reactivity (Shaik, 1981).

In the following we will restrict the discussion to the question of whether cation radicals are likely intermediates in electrophilic aromatic nitration, a reaction especially well studied with this problem in mind. However, before going into details it is pertinent to emphasize that the conditions of electrophilic aromatic substitution, and of aromatic nitration in particular, are in general oxidizing. Not only because electrophilic reagents as a rule have appreciable electron affinities, but also because of the strongly acidic solvent systems frequently employed. The oxidizing power of $NO_2^+$, $NO^+$ and $X_2$ (X = halogen) is clearly illustrated by the fact that salts like $NO_2^+PF_6^-$ and $NO^+PF_6^-$ (or $NO^+BF_4^-$) are often used to carry out one-electron oxidations, examples of which are found in the conversion of adamantylideneadamantane to the cation radical (see Section 4), the oxidation of O,O,O-trialkylphosphorothiolates and triphenylphosphine sulfide (Blankespoor et al., 1983), and the generation of dimethylsulfide cation radical (Chow and Iwai, 1980). Oxidation by halogens, $X_2$, and nitric acid have likewise been reported to result in the formation of cation radicals of triarylamines, phenols and heterocyclic aromatic compounds (Morkovnik et al., 1980; Koshechko et al., 1981; Pokhodenko et al., 1975). When an easily oxidized aromatic molecule is dissolved in a strong acid such as $H_2SO_4$, small but significant amounts of the corresponding cation radical are often formed. Different mechanisms may be proposed to account for the observation. One involves electron transfer from the unprotonated form of the aromatic molecule to the protonated form (207). Another possibility which cannot be neglected is electron transfer between Ar and protonated oxygen (208). Thus, the mere observations

$$Ar + ArH^+ \rightarrow Ar^{\overset{+}{\cdot}} + ArH\cdot \qquad (207)$$

$$Ar + HO_2^+ \rightarrow Ar^{\overset{+}{\cdot}} + HO_2\cdot \qquad (208)$$

of cation radicals during electrophilic aromatic substitution, as demonstrated by esr-spectroscopy, cannot be taken as unequivocal support for a cation radical mechanism. Further experiments have to render it probable that the cation radicals are in fact intermediates on the way to substituted products. In this context mention should be made of the well-known reactions between cation-radicals derived from aromatic and heteroaromatic compounds and nucleophiles like $NO_2^-$ and $X^-$ which may proceed to nitro- and halogen-substituted aromatics. (See Section 3 for details concerning the reactions of cation radicals with nucleophiles.) Typical examples relevant to electrophilic aromatic substitution include the reactions of perylene and 10-phenylphenoxazine cation radicals with nitrite ion (Ristagno and Shine, 1971;

Shine and Wu, 1979) and of phenothiazine cation radical with chloride and bromide (Shine *et al.*, 1972). Other examples involving cation radicals of sulfur containing compounds may be found in Shine's recent review (1981). Phenothiazinium ion produced by further oxidation and deprotonation of the cation radical has been demonstrated to react with nitrite ion accompanied by the formation of 3-nitrophenothiazine (Morkovnik *et al.*, 1981).

AROMATIC NITRATION

In a much quoted paper, Perrin (1977) pointed out that the proposal of a π-complex prior to formation of the σ-complex does not solve the enigma concerning the reactivity/selectivity paradox, since it is difficult to envisage that $NO_2^+$ in a π-complex has aquired the selectivity which free $NO_2^+$ lacks. Simple thermochemical considerations demonstrated, however, that electron transfer between $NO_2^+$ and toluene, or aromatic compounds more easily oxidized than toluene, is an exothermic process and a mechanism, (209)–(210), which contains the essential elements of the charge-transfer mechanism, was proposed. Perrin's mechanism differs from the previously suggested one

$$Ar + NO_2^+ \rightarrow [Ar^{+\cdot} \cdot NO_2] \quad (209)$$

$$[Ar^{+\cdot} \cdot NO_2] \rightarrow Ar-NO_2^+ \quad (210)$$

in that the electron-transfer step (209) was assumed to give instead of $Ar^{+\cdot}$ and $NO_2^\cdot$, a cation radical-radical pair which subsequently collapses to the σ-complex. The lack of intermolecular selectivity arises because the initial electron transfer is assumed to be diffusion controlled and the positional selectivity because of the non-uniform spin density in the cation radical, an argument similar to that set forth by Pedersen *et al.* (1973). A crucial experiment was the demonstration that controlled potential electrolysis of naphthalene in acetonitrile in the presence of $NO_2$ at a potential, 1.3 V *vs* Ag/0.01 M Ag$^+$, where only naphthalene might be expected to be oxidized (to the cation radical) resulted in a mixture of 1- and 2-nitronaphthalene. The ratio of the 1- and 2-isomer was found to be equal to 9.2, which was considered to be identical within the experimental error to the value (10.9) observed by conventional nitration of naphthalene by $HNO_3/H_2SO_4$ and urea in acetonitrile. This was taken as evidence for the presence of the same intermediate, the $[C_{10}H_8^{+\cdot} \cdot NO_2]$ pair, in both reactions. It was finally pointed out that the assumption of a radical pair offered a better explanation of the long distance migration of the nitro group frequently encountered during aromatic nitration. Migration of electrophiles does normally take place *via* 1,2-shifts, but in the radical pair the nitro group is expected to be allowed to move more freely and therefore is also able to reach more distant carbon atoms.

Perrin's paper stimulated more research, the results of which have provided both support of and evidence against the charge-transfer mechanism. Gas-phase studies by flow discharge mass spectroscopy showed that the reaction between a cation-radical derived from an aromatic hydrocarbon and $NO_2$ is indeed a feasible process (Schmitt et al., 1981). The ion resulting from combination of benzene cation-radical and $NO_2$ was demonstrated to have acidic properties being capable of transferring a proton to THF or pyridine as experiments with $C_6D_6^{+\cdot}$ unequivocally revealed. Since a $\pi$-complex would not be expected to be acidic, it was concluded that the $C_6H_6NO_2^+$ ion contained a C—N $\sigma$-bond and probably had the same structure as the $\sigma$-complex in solution. Although solvation obviously may change the relative stabilities of the species involved, these gas-phase studies did show that the reaction between an arene cation radical and $NO_2$ in the presence of a suitable base does in fact give rise to the corresponding nitroarene.

The strongest experimental evidence for Perrin's proposal were the results of electrochemical nitration of naphthalene. However, a more detailed investigation of this process (Eberson et al., 1978b; Eberson and Radner, 1980) led to severe questioning of this evidence. It was demonstrated that the current yield in the early stages of the reaction exceeded 100%, indicating that a homogeneous process was outrunning the anodic nitration, if it did take place at all. Independent experiments showed that the most likely homogeneous process was nitration by $N_2O_4$, catalyzed by protons liberated during the electrochemical follow-up reactions. Thus, no firm conclusion could be drawn regarding the question at hand; does the reaction between naphthalene cation radical and $NO_2$ provide a feasible route to nitronaphthalenes? The hexafluorophosphate of naphthalene cation radical can be isolated as a solid material with the composition $(C_{10}H_8)_2^{+\cdot}PF_6^-$ (Fritz et al., 1978). When this salt was allowed to react with $NO_2/N_2O_4$ under the same conditions as those employed for the electrochemical experiment, nitration did take place, but the ratio between the 1- and 2-isomer was found to be around 40 as opposed to the values close to 15 observed when naphthalene reacts with $NO_2^+BF_4^-$. Furthermore, it was found that naphthalene cation radical reacted with $NO_2$ in an electron-transfer process during which no reaction between the $NO_2^-$ formed as a result of the electron transfer and $C_{10}H_8^{+\cdot}$ could be detected. That this reaction apparently could not compete with the self-coupling of the cation radicals, a process which is known to have a second order rate constant close to $10^8 M^{-1}s^{-1}$, indicated that the formation of the $\sigma$-complex from naphthalene cation radical and $NO_2$ must be a rather slow process. Perrin's mechanism requires, however, that this reaction approaches the diffusion-controlled limit in order to account for the lack of *inter*molecular selectivity. This, together with the differences observed in the substitution pattern, was taken as evidence that the reaction between naphthalene cation radical and $NO_2$ is not an elementary step during the conventional nitration of this hydrocarbon. The same conclusion was reached in a general study devoted to electrochemical nitration of aromatic hydrocarbons and it was reported that dinitration, which required a more positive potential than that for mononitration, most probably takes place *via* electrogenerated $NO_2^+$ (Achord and Hussey, 1981).

The mechanisms of both 1,2- and 1,3-shifts of the nitro group during aromatization of initially formed *ipso*-products have been investigated carefully and evaluated with reference to the cation-radical-radical pair mechanism. For example, solutions of [148] are slowly, *via* a formal 1,3-shift, converted to [149] and [150] at room temperature without any sign of the

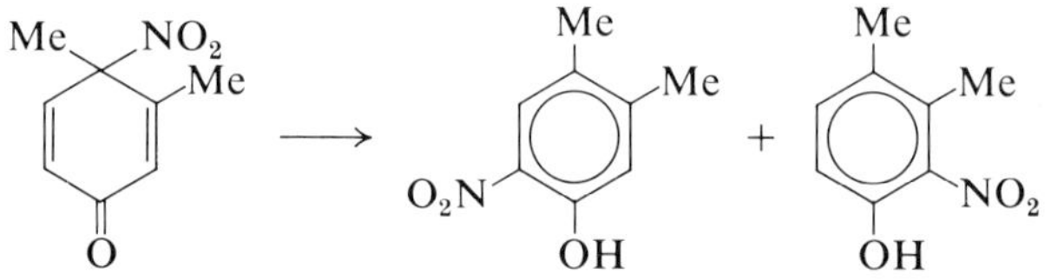

formation of 1,2-shift products (Barnes and Myhre, 1978a). Measurements of kinetics and activation parameters resulted in the proposal of the radical dissociation-recombination mechanism depicted in Scheme 14 for 4-methyl-4-nitrocyclohexa-2,5-dienone, [151], and five reaction characteristics were emphasized as being in support of the mechanism. (*i*) The rate of disappearance of substrate equalled the rate of formation of product. (*ii*) Reaction rates decreased as the polarity or hydrogen bonding ability of the solvent increased, which was attributed to solvent stabilization of [151].

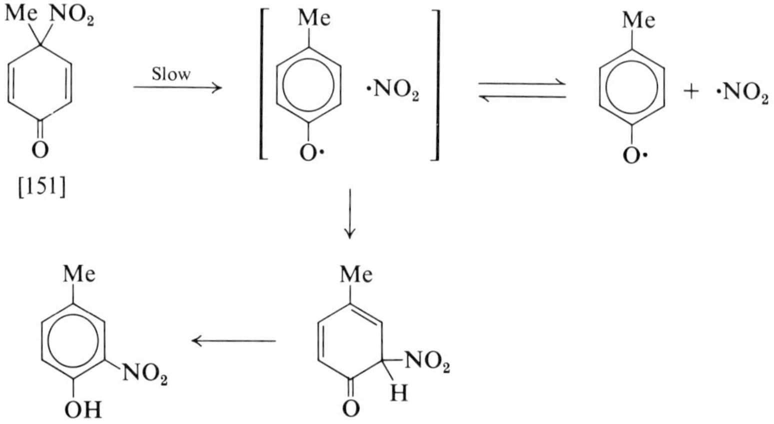

**Scheme 14**

(iii) Activation entropies were small, but positive, consistent with rate-determining dissociation. (iv) Addition of radical scavengers such as hydroquinone did not affect the rate, but did reduce the yields of nitrophenols and increased the yields of alkylphenol products. This result suggests that some diffusion of radicals out of the solvent cage does take place and is similar to that observed in the related nitramine rearrangement. The interpretation was supported by the observation of considerable scrambling during reaction of equimolar quantities of [148] and the 2,6-$d_2$-$^{15}$NO$_2$ derivative. (v) Alkyl substituents reduced the rate of rearomatization, which was attributed to a steric effect. Furthermore it was observed that strong acids catalyzed the reaction and under such conditions it is reasonable to assume that substrate is O-protonated and that the rearomatization corresponds to the cation radical-radical pair pathway proposed by Perrin. This possibility was also considered in a detailed kinetic investigation of the reaction of p-cresol with NO$_2^+$ (Coombes et al., 1979). The results indicated that the product, 4-methyl-2-nitrophenol, was formed by two parallel routes, either directly (211) or via

(211)

the ipso-intermediate [151]. The rate enhancement observed under strongly acidic conditions was found to be consistent with acid catalyzed reversion of [151] to the encounter complex [ep] and it was concluded that the experimental results did not provide compelling evidence for the cation radical-radical pair mechanism.

A 1,2-shift of the nitro group was observed during aromatization of the 1,2-dimethyl-1-nitrocyclohexa-2,5-dienyl cation (Barnes and Myhre, 1978b). The specifically 3,5-$d_2$-labelled isomer [153], which was formed in situ by treatment of [152] with 85% H$_2$SO$_4$, was utilized to investigate the mechanism of the rearrangement. Only [154] and [156] were formed and analysis of the isomer ratio [154]/[156] by low voltage mass spectrometry resulted in the value 1.021 ± 0.014. Application of the steady state principle on [155], assuming that $k_o = k'_o$ and $k_{ipso} = k'_{ipso}$, allowed the evaluation of the rate constant ratio, $k_o/k_{ipso}$, which was estimated to be equal to 0.02. This value indicates that the isotopic isomers [153] and [155] are nearly equilibrated before migration of the nitro group to one of the open positions. If this equilibration is assumed to take place *via* a cation radical-radical pair, the product mixture would also be expected to contain 1,2-dimethyl-4-nitrobenzene, [157], since the spin densities in positions 4 and 5 of the 1,2-dimethylbenzene cation radical are nearly as large as that in positions 1 and 2. However, [157] could not be detected. Furthermore, the spin densities in the positions corresponding to the formation of products actually formed are vanishingly small, and it was accordingly concluded that the nitro group migration took place *via* an *intra*molecular 1,2-shift and not through the radical pair intermediate.

The nitrous acid catalyzed nitration of N,N-dimethylaniline [158] and substituted derivatives is of particular interest in the present context. Attempts to carry out conventional nitration of [158] by $HNO_3$ in $H_2SO_4$ led to complicated product mixtures, the composition of which were highly dependent on the amount of water present in the sulfuric acid (Giffney and Ridd, 1979). The product distribution summarized in Table 9 illustrates this point. The observed kinetics were in agreement with conventional attack by $NO_2^+$ on protonated [158] when the concentration of $H_2SO_4$ was larger than 83%,

TABLE 9

Distribution of major products from nitration of N,N-dimethylaniline [158] by $HNO_3$ in $H_2SO_4$ containing various amounts of water[a]

| % $H_2SO_4$ | % [159] | % [160] | % [161][b] |
|---|---|---|---|
| 87 | 57 | 28 | — |
| 81.8 | 2 | 40 | 26 |
| 74.7 | — | 2 | 63 |

[a]Giffney and Ridd, 1979
[b]The sum of yields of [161] and mono- and dinitrated derivatives

while at lower concentrations the reaction was subject to autocatalysis after a short induction period. It was demonstrated that the catalytic effect was due to NO[+] and that the deliberate addition of nitrous acid to the reaction mixture caused the induction period to vanish. Nitrous acid catalysis of nitration is traditionally explained by initial nitrosation followed by oxidation of the aromatic nitroso compound to the corresponding nitroarene. However, it was shown that the rate of nitrosation under the reaction conditions was much too low for this explanation to be acceptable. The rejection of a nitrosoarene intermediate was supported by an independent study of the nitrous acid catalyzed nitration of phenol (Ross et al., 1980). Here it was observed that nitrosation followed by oxidation led to an isomer distribution different from that resulting from the nitrous acid catalyzed process. The presence of N,N,N',N'-tetramethylbenzidine [161] and nitro derivatives of [161] in the

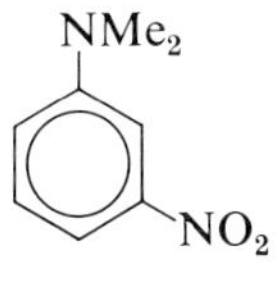

[159] 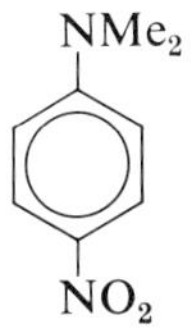 [160] 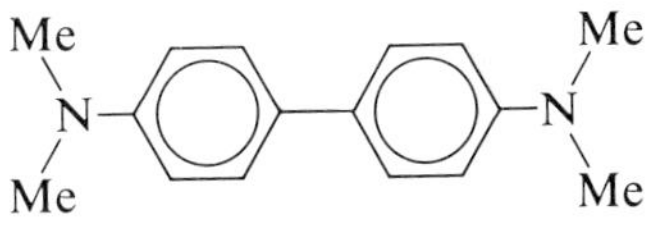 [161]

product mixture from nitration of [158] in dilute sulfuric acid gave an indication of the involvement of N,N-dimethylaniline cation radical and the mechanism in Scheme 15 was suggested. CIDNP studies were inconclusive, but later it was reported that nuclear polarization could be observed for the formation of p-nitro-N,N-dimethylaniline [160] in $H^{15}NO_3$ and 88% $H_2SO_4$, but not for the m-isomer [159] (Ridd and Sandall, 1981). The observed polarization was consistent with a reaction scheme involving the formation of the cation radical-radical pair ([158$\overset{+}{\cdot}$]·$NO_2$) from [158$\overset{+}{\cdot}$] and ·$NO_2$ in solution as indicated in Scheme 15, but not from an electron-transfer reaction between [158] and $NO_2^+$. Thus, it was concluded that at least some of compound [160], but none of [159], was formed through a reaction involving a cation radical-radical pair and that this reaction is subject to nitrous acid

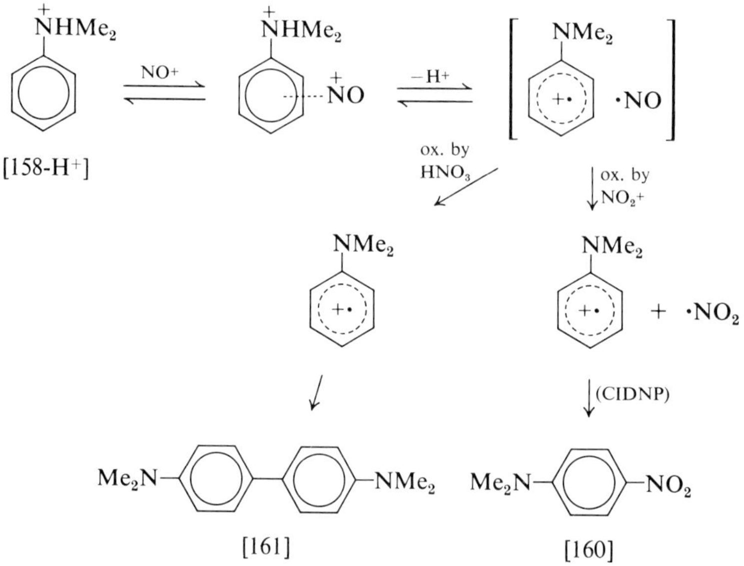

Scheme 15

catalysis. The mechanism change-over in going from concentrated to more dilute sulfuric acid convincingly explains the trend in the product distribution reproduced in Table 9. In concentrated $H_2SO_4$, compound [159] is the major product indicating that conventional attack by $NO_2^+$ on [158—H$^+$] is the predominant reaction pathway, which is supported by both the kinetic and the CIDNP results. Going to more and more dilute $H_2SO_4$ causes the cation radical-radical pair mechanism to become increasingly important, which is accompanied by the disappearance of [159] from the product mixture. The relative proportions of [160] and [161] reflects the competition between oxidation of the cation radical-radical pair by either $NO_2^+$ or $HNO_3$. In the more dilute acid the concentration of $NO_2^+$ is low and thus oxidation by $HNO_3$ competes favorably resulting in the formation of [161] as the major product.

More insight into these reactions was gained through the study of the nitration of *p*-alkyl-N,N-dimethylanilines (Al-Omran et al., 1981). The reaction of [162] with nitric acid in 70% sulfuric acid gave the *o*-nitro derivative, [164], in good yield (78%) *via* formation of the *ipso*-intermediate [163]. The reaction

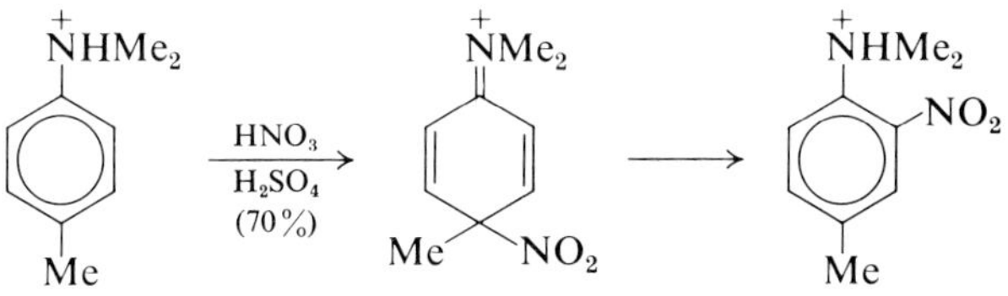

ORGANIC CATION RADICALS

[162-H⁺]    [163]    [164-H⁺]

required nitrous acid catalysis and was accordingly inhibited by the presence of hydrazine. Scheme 16 summarizes the proposed mechanism, which was supported by esr evidence indicating the presence of [162$\overset{+}{\cdot}$] during the

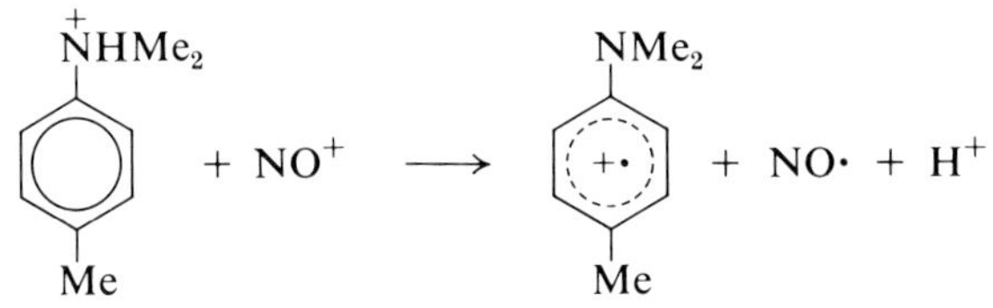

[162-H⁺] + NO⁺ ⟶ [162$\overset{+}{\cdot}$] + NO· + H⁺

NO· + NO$_2^+$ ⟶ NO⁺ + NO$_2^-$

[162$\overset{+}{\cdot}$] + NO$_2^-$ ⟶ [163]

**Scheme 16**

reaction. In this case the presence of the *p*-methyl group hinders the formation of a benzidine derivative by dimerization of the cation radical. The further rearrangement of [163] to [164—H⁺] was found most likely to proceed *via*

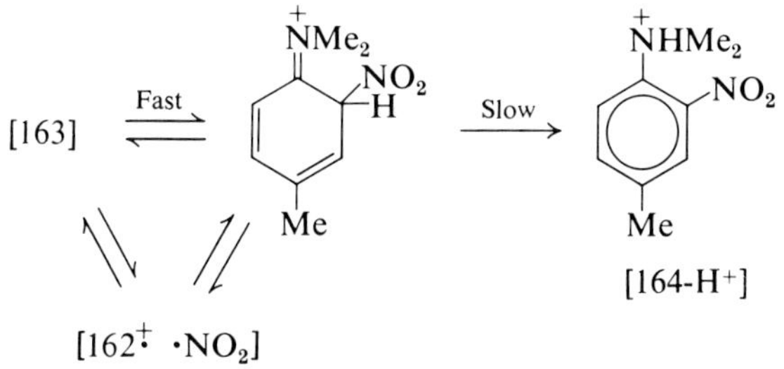

[163] ⇌ (Fast) ... ⟶ (Slow) [164-H⁺]

         [162$\overset{+}{\cdot}$ ·NO₂]

two competitive pathways, one involving a cation radical-radical pair, the other apparently being a direct 1,3-intramolecular shift with the latter as the dominating route. Participation of [162$\overset{+}{\cdot}$ · NO₂] in the aromatization of [163] gained support from CIDNP when the ¹⁵NO₂/¹⁴NO₂ exchange of [165] in nitric acid was studied by ¹⁵N nmr spectroscopy (Helsby *et al.*, 1981).

[165]

Nuclear polarization experiments also indicated the involvement of a cation radical-radical pair in the nitrous acid catalyzed nitration of 4-nitrophenol in TFA (Clemens et al., 1983).

It seems to be too early at present to attempt to give a definitive answer to the question of the importance of cation-radicals in aromatic nuclear nitration. The results from the nitrous acid catalyzed reaction in not too concentrated $H_2SO_4$ strongly indicates that cation radical-radical pairs play a central role in the reaction mechanism. However, in the so-called conventional nitration by $NO_2^+$ in more concentrated acid the picture is a little less clear. Evidently, oxidation of impurities may produce sufficient amounts of $NO^+$ to obscure the mechanistic pattern and render the interpretation of experimental results extremely difficult.

More information pertaining to the role of cation radicals in nitration of aromatic compounds is provided by studies on alkylbenzenes. For these compounds different product distributions may be expected to arise from the reaction between substrate and $NO_2^+$ and the reaction between substrate cation radical and $NO_2$ because of the pronounced acidity of cation radicals of alkylbenzenes (see Section 5). If deprotonation of the cation radical competes favourably with nuclear nitration the predominant products will be derived from side chain nitration while such products are not expected to arise by the conventional reaction pathway. Results from nitration of mesitylene under different reaction conditions (Table 10) provides a nice illustration of this point (Draper and Ridd, 1978). Very clean and efficient nuclear nitration was observed when mesitylene reacted with $HNO_3$ in acetonitrile in the absence of any deliberately added oxidizing agents. On the other hand, nitration by reaction with ceric ammonium nitrate (CAN) gave rise to side chain nitration exclusively, although in an overall modest yield. The latter

TABLE 10

Product compositions from the reactions of mesitylene (0.20–0.21 M) with nitric acid and ceric ammonium nitrate (CAN) in acetonitrile at 65°C[a]

| Time/min | [HNO$_3$]/M | [CAN]/M | [N$_2$O$_4$]/M | % Yield [b,c] | % composition[c] ||
|---|---|---|---|---|---|---|
| | | | | | ArNO$_2$ | ArCH$_2$ONO$_2$ |
| 100 | 3.05 | — | — | 100 | 100 | — |
| 60 | — | 0.26 | — | 35 | — | 100 |
| 30 | — | 0.26 | 0.19 | 5 | 74 | 26 |
| 30 | 0.80 | 0.26 | — | 27 | 67[d] | 33[d] |
| 30 | 0.80 | 0.26 | 0.22 | 38 | 90 | 10 |

[a] Draper and Ridd, 1978
[b] Total yield relative to the initial amount of mesitylene
[c] Determined from the nmr spectrum of the extracted product: errors $c \pm 3\%$
[d] The composition of this product varies with the extent of reaction

reaction could be suppressed to some extent by addition of $HNO_3/N_2O_4$ to the reaction medium and a mixture of both products resulted accordingly. It was emphasized that the absence of side chain products from reaction with $HNO_3$ could not be taken as evidence against the involvement of a cation radical intermediate, which may react further to the σ-complex before deprotonation can take place. However, the observed results do certainly not require that assumption either. An analogous effect of the nitration conditions was recorded during nitration of 2,4,5-trimethylneopentylbenzene by CAN and $HNO_3$ respectively (Suzuki *et al.*, 1980).

THE ABNORMAL WITTIG REACTION

The Wittig reaction is of fundamental importance for the conversion of ketones to unsaturated compounds. The generally accepted mechanism involves the initial formation of a zwitterion [166] which dissociates to products, presumably through a four-membered ring intermediate [167].

Although yields in general are excellent, steric hindrance may cause the reaction pathway to deviate from that depicted. For enolizable ketones steric hindrance can cause self-condensation to take place instead, while sterically hindered ketones which cannot undergo enolization do not react at all under the conditions normally employed, i.e. boiling ether or THF.

During attempts to carry out the reaction between adamantanone [168] and methylenetriphenylphosphorane [169] at higher temperature in boiling toluene, it was unexpectedly observed that adamantanone was reduced to the alcohol in good yield accompanied by the formation of solvent dimers, bibenzyl and o-, m- and p-benzyltoluene (Olah and Krishnamurthy, 1982).

TABLE 11

Products and yields in attempted Wittig reactions[a]

| Ketone | Wittig reagent | Solvent | Products | Yield % |
|---|---|---|---|---|
| [168] | $Ph_3PCH_2$ [169] | Ether | [171] | 80 |
| [168] | $Ph_3PCMe_2$ | Ether | No reaction | |
| [168] | $Ph_3PCH_2$ [169] | Toluene | [170]/[171] $\simeq$ 1/1 | 76 |
| [168] | $Ph_3PCMe_2$ | Toluene | [170] | 82 |
| $Ph_2CO$ | $Ph_3PCMe_2$ | Ether | $Ph_2C=CMe_2$ | 76 |
| $Ph_2CO$ | $Ph_3PCMe_2$ | Toluene | $Ph_2CHOH/Ph_2C=CMe_2 \simeq$ 1/1 | 74 |

[a]Olah and Krishnamurthy, 1982

Other solvents containing easily abstractable hydrogen atoms had the same effect on the course of the reaction and even ketones, which in boiling ether smoothly undergo the Wittig reaction, were found to be reduced to a significant extent in boiling toluene (Table 11). It was proposed that the Wittig reaction at high temperature in toluene and similar solvents, and possibly in general, proceeds *via* the formation of a tight ion pair [172]. When collapse to the zwitterion is hindered sterically, hydrogen atom abstraction from the solvent may compete leading to the observed products, alcohols and solvent dimers. Direct evidence for the ion radical pair by CIDNP experiments could not be obtained, but the involvement of ion radicals in the Wittig reaction has later been verified by esr spectroscopy (Ashby, quoted by Olah and Krishnamurthy, 1982).

ADDITION OF RADICALS TO DIAZONIUM IONS AND PROTONATED HETEROAROMATIC BASES

Azoarenes [173] are frequently detected as byproducts in reactions of arenediazonium ions (Zollinger, 1973, 1978). It was suggested some time ago that the origin of these compounds was addition of aryl radicals to the diazonium ions (212) resulting in the intermediate formation of azoarene cation radicals [174] (Waters, 1942). That this reaction is indeed feasible was later

$$Ar—\overset{+}{N}{\equiv}N + Ar\cdot \rightarrow Ar—N{=}N—Ar\overset{+\cdot}{} \xrightarrow{red.} Ar—N{=}N—Ar \quad (212)$$
$$[174] \qquad [173]$$

confirmed through both kinetic and product studies (Packer et al., 1971, 1974, 1980) and it was suggested that the radical detected by esr spectroscopy during reduction of benzenediazonium ion with dithionite ion (Dixon and Norman, 1964) was in fact also [174] rather than $ArN_2^{\cdot}$ (Bargon and Seifert, 1974; Heighway et al., 1974). The addition of alkyl radicals, $R\cdot$, may likewise give rise to cation radicals, [175] (Heighway et al., 1974).

$$Ar—\overset{+}{N}{\equiv}N + R\cdot \longrightarrow Ar—N{=}N—R\overset{+\cdot}{} \begin{array}{c} \xrightarrow{red.} Ar—N{=}N—R \quad (213) \\ [176] \\ \xrightarrow{?} Ar\cdot + N_2 + R^+ \quad (214) \end{array}$$
$$[175]$$

It has been proposed that the formation of [174] and [175] is reversible and that the oxidation of $R\cdot$ to $R^+$ by diazonium ions proceeds via [175] rather than by a direct electron transfer (214) (Bargon and Seifert, 1974). However, pulse radiolytic studies have revealed that dissociation of [175] into $ArN_2^+$ and $R\cdot$ is a very slow process and that the reduction of $ArN_2^+$ by $R\cdot$ to $Ar\cdot$, $N_2$ and $R^+$ does in fact go via electron transfer (Heighway et al., 1974; Packer et al., 1980). Whether the reaction between $ArN_2^+$ and $R\cdot$ in general gives rise to electron transfer or leads to bond formation seems to be highly dependent on the nature of $R\cdot$. If $R\cdot$ is $Ar\overset{\cdot}{C}H_2$ or highly nucleophilic radicals such as $R\overset{\cdot}{C}HOH$, $R\overset{\cdot}{C}HOR_2'$, $R\overset{\cdot}{C}HNR_2'$ the result is electron transfer (Packer et al., 1980; Werner and Rüchardt, 1969). It is, however, not clear whether this reaction is direct in all cases or may involve [175] as an intermediate (Citterio and Minisci, 1982). If $R\cdot$ is alkyl or aryl, bond formation takes place. Rate constants for the latter case have been measured (Citterio and Minisci, 1982).

$$k(R_3C\cdot) \geqslant 10^8 M^{-1}s^{-1} \text{ and } k(RCH_2^\cdot) \simeq 10^6 M^{-1}s^{-1}$$

The reaction between R· and $ArN_2^+$ has proved to be a synthetically useful procedure for the preparation of aryl alkyl azocompounds provided the reaction is carried out under conditions where [175] is reduced to [176] immediately after its formation in reaction (213) (Citterio and Minisci, 1982). The radical source could be alkyl iodides, ketone peroxides or sulfoxides. In the case of alkyl iodides the overall reaction scheme was believed to include the steps (215)–(218). Yields of R—N=N—Ar were typically between 40 and 80%.

$$Ar—\overset{+}{N}{\equiv}N + Ti(III) \rightarrow Ar\cdot + N_2 + Ti(IV) \tag{215}$$

$$Ar\cdot + RI \rightarrow ArI + R\cdot \tag{216}$$

$$R\cdot + \overset{+}{N}{\equiv}N—Ar \rightarrow R—N{=}N—Ar\overset{+}{\cdot} \tag{217}$$
$$[175]$$

$$R—N{=}N—Ar\overset{+}{\cdot} + Ti(III) \rightarrow R—N{=}N—Ar + Ti(IV) \tag{218}$$
$$[176]$$

Like $ArN_2^+$, protonated heteroaromatic bases are strongly electron deficient and thus excellent reaction partners for nucleophilic carbon centered radicals, for example R· and R—Ċ=O (Minisci, 1976). The key step, which follows the generation of the radical from a suitable source, is addition of the radical to the protonated aromatic base during which a cation radical intermediate is formed. The overall mechanism involves steps (219)–(221) illustrated by the reaction between cyclohexyl radical and protonated isoquinoline (Minisci et al., 1971). The synthetic success of the reaction is highly dependent on the presence of an oxidizing agent ($Ag^{2+}$, $Cu^{2+}$ or $Fe^{3+}$) in step (221), in the absence of which the cation radical [177] may participate in side-reactions. In the general case more than one position is available for substitution, and this gives rise to mixtures of products as was found with protonated pyridine and quinoline, where both 2- and 4-substituted heterocycles were obtained. Furthermore, the acidity of the solvent shows a pronounced influence on the isomer distribution. The ratio between attack in the 2- and 6-position (2:6) of protonated quinoxaline by cyclohexyl radical was found to vary between 15.3 and 0.59 when the concentration of $H_2SO_4$ in acetic acid was raised from 15 to 96% (Caronna et al., 1977). The effect was explained as being due to the increasing concentration of the diprotonated quinoxaline, which is expected to be most susceptible

$$\text{C}_6\text{H}_{11}\text{-COOH} \xrightarrow{\text{Ag}^{2+}} \text{C}_6\text{H}_{11}\cdot + \text{CO}_2 + \text{H}^+ + \text{Ag}^+ \qquad (219)$$

$$\text{(isoquinolinium)} + \text{C}_6\text{H}_{11}\cdot \longrightarrow \text{[177]} \qquad (220)$$

$$\text{[177]} \xrightarrow[\text{ox.}]{-\text{H}^+} \text{(4-cyclohexylisoquinolinium)} \quad (84\%) \qquad (221)$$

to nucleophilic attack in the 6-position in contrast to the monoprotonated form which has the lowest $\pi$-electron density in the 2-position (Citterio et al., 1977). A similar effect was observed when the concentration of $\text{H}_2\text{SO}_4$ in an aqueous solution of the acid was increased (Caronna et al., 1976). The rate constants for the addition of the primary 5-hexenyl radical to several protonated heteroaromatic compounds were measured by competition with the *intra*molecular cyclization of the radical, for which both the rate constant and the activation parameters are known (Citterio et al., 1977). It was found that the pre-exponential factors varied very little with substrate and were close to $10^{9.3}$, while the Arrhenius activation energies were dependent on the nature of the substrate. Values of $E_a$ between 6.9 kcal mol$^{-1}$ (4-methylpyridine) and 2.8 kcal mol$^{-1}$ (quinoxaline) were observed.

REACTIONS OF ALIPHATIC AMMONIUMYL RADICALS

Various aspects of the reactions of aliphatic ammoniumyl radicals have been extensively reviewed during the past twenty years (Wolff, 1963; Kovacic et al., 1970; Neale, 1971; Deno, 1972; Chow, 1973; Minisci, 1973, 1976; Chow et al., 1978). Since very little work has appeared since the exhaustive review by Chow et al. was published, only a short summary of the more important chemical aspects will be given.

Due to the acidic properties of aliphatic ammoniumyl radicals ($pK_a = 3-7$; Simic and Hayon, 1971; Fessenden and Neta, 1972) deprotonation to the corresponding amminyl radicals (222) will be a major, in most cases undesirable, contribution to the reaction scheme unless strongly acidic

$$\begin{matrix} R' \\ R \end{matrix}\!\!>\!\!NH^{+\cdot} \longrightarrow \begin{matrix} R' \\ R \end{matrix}\!\!>\!\!N\cdot + H^+ \qquad (222)$$

conditions are used. For this reason typical solvents for studying the reactions of ammoniumyl radicals include 2N $H_2SO_4$, 4M $H_2SO_4$ in $CH_3COOH$ and neat trifluoroacetic acid, TFA. The ammoniumyl radical sources most frequently employed are N-haloamines and N-nitrosoamines. Protonated N-haloamines are readily converted to ammoniumyl radicals by thermolysis, photolysis or treatment with metal ions like Fe(II). N-Nitrosoamines are cleaved photochemically in dilute acid to ammoniumyl radicals and NO (Chow, 1973).

Due to their strongly electrophilic character ammoniumyl radicals participate readily in hydrogen abstraction reactions and in addition to olefinic and aromatic systems. In cases where both hydrogen abstraction and addition to carbon–carbon double bonds are possible the latter reaction competes favorably as was demonstrated for the cation radical of piperidine in acidic methanol (Chow, 1973). It was calculated that addition to cyclohexene occurred 5000 times faster than hydrogen abstraction from methanol. An *intra*molecular reaction illustrating the same point is found in the cyclization of the ammoniumyl radical [178] (Surzur *et al.*, 1975).

[178]     [179]

Radicals like [179] formed by hydrogen abstraction or addition to a double bond may themselves participate in atom-abstraction reactions. When an N-haloamine is the ammoniumyl radical precursor, a cation radical chain process takes place as illustrated by the intramolecular rearrangement of N-chloro-N-methylcyclo-octylamine (Wawzonek and Thelen, 1950) (Scheme 17). This mechanistic element is a general feature of all the reaction schemes to follow.

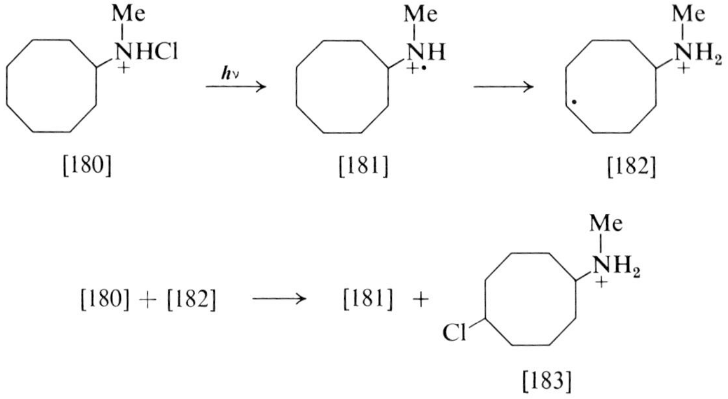

Scheme 17

*Intramolecular hydrogen abstraction: the Hofmann–Löffler reaction*

The conversion of [180] to [183] is an example of the Hofmann–Löffler rearrangement (Wolff, 1963; Deno, 1972), the general features of which are shown in Scheme 18. For acyclic compounds the most favorable transition state for hydrogen abstraction is that giving rise to 1,5-migration which makes the reaction an attractive synthetic procedure for the preparation of pyrrolidines [184]. That the reactive intermediate is in fact the ammoniumyl radical [185] and not the deprotonated form became evident from studies on an

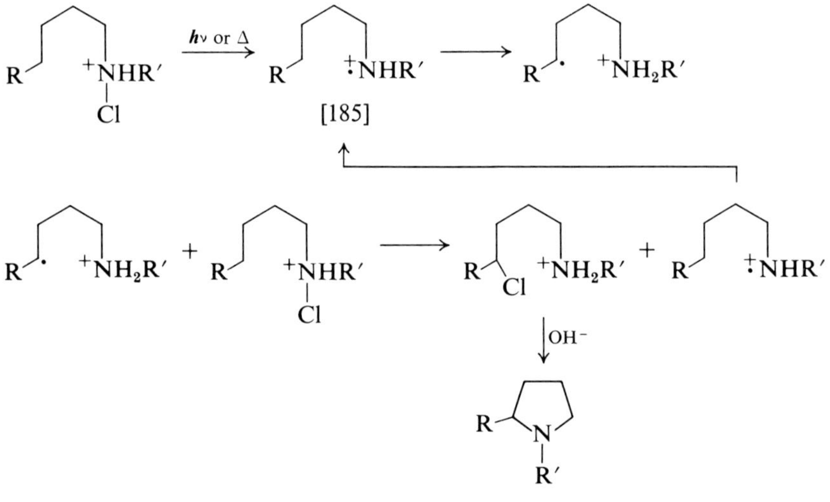

Scheme 18

extensive series of reactions reported by Corey and Hertler (1960). Rearrangement of dibutylchloroamine in acetic acid was investigated as a function of the amount of $H_2SO_4$ added to the solvent system. When no $H_2SO_4$ was added, the formation of the final cyclized product, N-butylpyrrolidine, could not be detected. However, inclusion of increasing amounts of $H_2SO_4$ caused the yield of this compound to increase, reaching a satisfactory 80% yield at 21% $H_2SO_4$.

*Intermolecular hydrogen abstraction*

If the alkyl chain is not sufficiently long for *intra*molecular hydrogen abstraction to take place, the ammoniumyl radical may be trapped by deliberately added reagents (Minisci, 1973). Halogenation carried out in this way (Scheme 19) has been successful for a variety of aliphatic compounds. A striking feature of the *inter*molecular hydrogen abstraction is the predominance of attack in the ω–1 position, probably caused by a minimum of steric hindrance in the transition state for this position. A typical example of the relative yields is given in Scheme 19 for the case where R'H is hexanol and R is i-Pr. The ω–1 isomer accounts for 90% of the chlorinated products in this case (Deno *et al.*, 1971).

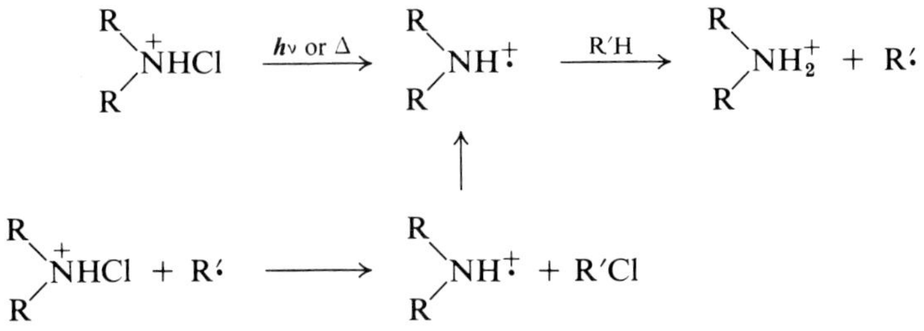

$$R_2\overset{+}{N}HCl \xrightarrow{h\nu \text{ or } \Delta} R_2NH^{+\cdot} \xrightarrow{R'H} R_2\overset{+}{N}H_2 + R'\cdot$$

$$R_2\overset{+}{N}HCl + R'\cdot \longrightarrow R_2NH^{+\cdot} + R'Cl$$

R'H = HO–CH$_2$–CH$_2$–CH$_2$–CH$_2$–$\overset{\omega-1}{CH_2}$–CH$_3$

Relative yields    0    0    2    2    90    6

**Scheme 19**

*Addition to double bonds*

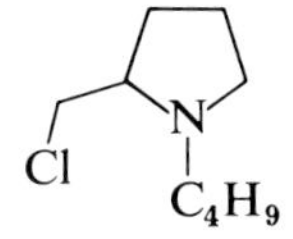

More than one hundred examples of the addition of ammoniumyl radicals to double bonds are offered to the reader in the reviews by Chow *et al.* (1978) and Stella (1983). An example has already been given in the cyclization of [178] to [179] which, after chlorine atom abstraction from the N-haloamine precursor, yielded [186] (55%) (Surzur et al., 1975). The general chain reaction for the intermolecular case is shown in Scheme 20. If the nitrosamine

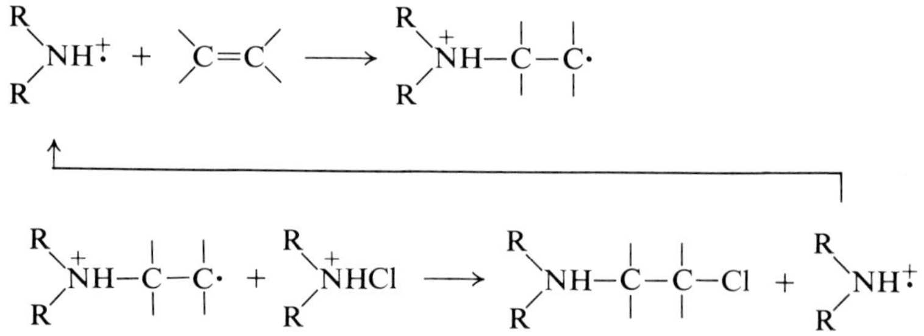

**Scheme 20**

[187] is used as the ammoniumyl radical precursor the reaction may lead to formation of an aliphatic C-nitroso compound [188], which readily undergoes rearrangement to the corresponding oxime [189] (Chow et al., 1967).

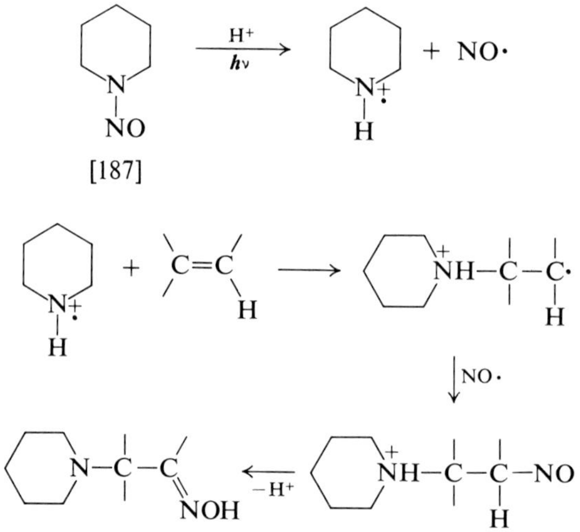

*Homolytic amination of aromatic compounds*

The direct introduction of a dialkylamino group in an aromatic molecule by homolytic aromatic substitution offers a valuable alternative to conventional electrophilic aromatic substitution which would require several steps for the same conversion. The reaction has, seen from a synthetic point of view, a number of virtues among which are simple experimental conditions,

# ORGANIC CATION RADICALS

short reaction times and high substrate and positional selectivity. The mechanism of the reaction, outlined in Scheme 21, follows the same pattern as the previous reactions, in being a cation radical chain process. Due to the strongly

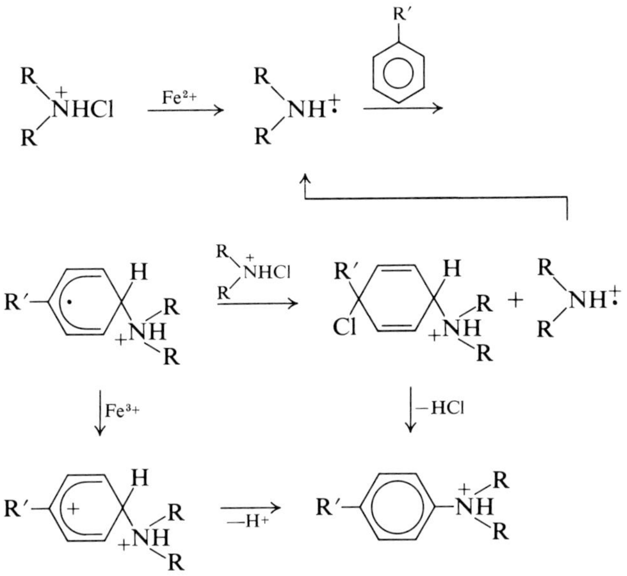

**Scheme 21**

acidic conditions employed, the product amine exists in solution in the protonated form, which protects it from further reaction. The low reactivity of aromatic compounds carrying electron-withdrawing substituents is a common feature of the reaction, while electron-donating substituents facilitate the reaction considerably. The high substrate and positional selectivity has been ascribed to a transition state similar to a charge-transfer complex (223). In

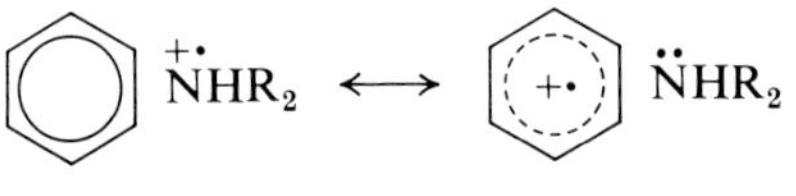

$$\underset{}{\bigcirc}\overset{+\cdot}{N}HR_2 \longleftrightarrow \underset{}{\bigcirc}\overset{..}{N}HR_2 \qquad (223)$$

the extreme, this corresponds to an electron-transfer reaction. Numerous examples of the synthetic utility of homolytic aminations are available through the two reviews by Minisci (1973, 1976) in which other aspects of the reaction are also discussed.

## 7 Concluding remarks

The chemistry of ion radicals can no longer be considered to be a novel area of limited interest. The recent work, some of which has been reviewed here, has provided evidence that ion radicals are reactive intermediates functioning in the mainstream of organic chemistry in much the same way as the more familiar carbenium ions, carbanions, carbenes and free radicals. The recognition that ion radicals play such an important role in organic chemistry in solution is a recent phenomenon. It is probably too early to assess just how great a role is played by these intermediates in reactions which do not involve net changes in oxidation state. As in any other rapidly expanding research area, the initial conclusions may not withstand the assaults of challenging inquiries and it may be found that some of the reactions in which ion radicals are now considered as probable intermediates will be shown not to involve these species at all. On the other hand, it is conceivable that further work will reveal other important roles of ion radicals in organic chemistry. It appears safe to conclude that the reactions of ion radicals will provide an active research arena for physical organic chemists for some time.


## References

Aalstad, B., Ronlán, A. and Parker, V. D. (1981a). *Acta Chem. Scand.* **B35**, 649
Aalstad, B., Ronlán, A. and Parker, V. D. (1981b). *Acta Chem. Scand.* **B35**, 247
Aalstad, B., Ronlán, A. and Parker, V. D. (1982a). *Acta Chem. Scand.* **B36**, 199
Aalstad, B., Ronlán, A. and Parker, V. D. (1982b). *Acta Chem. Scand.* **B36**, 171
Achord, J. M. and Hussey, C. L. (1981). *J. Electrochem. Soc.* **128**, 2556
Adams, R. N. (1966). *Acc. Chem. Res.* **2**, 175
Adams, R. N. (1969). "Electrochemistry at Solid Electrodes". Dekker, New York
Ahlberg, E. and Parker, V. D. (1980). *Acta Chem. Scand.* **B34**, 97
Ahlberg, E., Helgée, B. and Parker, V. D. (1980). *Acta Chem. Scand.* **B34**, 187
Albini, A. and Arnold, D. R. (1978). *Can. J. Chem.* **56**, 2985
Alder, R. W. (1980). *J. Chem. Soc. Chem. Comm.* 1184
Al-Omran, F., Fujiwara, K., Giffney, J. C., Ridd, J. H. and Robinson, S. R. (1981). *J. Chem. Soc. Perkin Trans. 2*, 518
Amatore, C. and Savéant, J. M. (1983). *J. Electroanal. Chem.* **144**, 59
Ambrose, J. F., Carpenter, L. L. and Nelson, R. F. (1975). *J. Electrochem. Soc.* **122**, 876
Amouyal, E., Grand, D., Moradpour, A. and Keller, P. (1982). *Nouveau J. Chim.* **6**, 241
Andrieux, C. P. and Savéant, J. M. (1978). *J. Electroanal. Chem.* **93**, 163
Andrieux, C. P., Dumas-Bouchiat, J. M. and Savéant, J. M. (1978a). *J. Electroanal. Chem.* **87**, 39
Andrieux, C. P., Dumas-Bouchiat, J. M. and Savéant, J. M. (1978b). *J. Electroanal. Chem.* **87**, 55

Andrieux, C. P., Dumas-Bouchiat, J. M. and Savéant, J. M. (1978c). *J. Electroanal. Chem.* **88**, 43
Andrieux, C. P., Dumas-Bouchiat, J. M. and Savéant, J. M. (1980a). *J. Electroanal. Chem.* **113**, 1
Andrieux, C. P., Blocman, C., Dumas-Bouchiat, J. M., M'Halla, F. and Savéant, J. M. (1980b). *J. Electroanal. Chem.* **113**, 19
Andrulis, Jr., P. J., Dewar, M. J. S., Dietz, R. and Hunt, R. L. (1966). *J. Am. Chem. Soc.* **88**, 5473
Anson, F. C., Savéant, J. M. and Shigehara, K. (1983a). *J. Phys. Chem.* **87**, 214
Anson, F. C., Savéant, J. M. and Shigehara, K. (1983b). *J. Am. Chem. Soc.* **105**, 1096
Asmus, K.-D. (1979). *Acc. Chem. Res.* **12**, 436
Baciocchi, E., Mandolini, L. and Rol, C. (1976). *Tetrahedron Lett.* 3343
Baciocchi, E., Mandolini, L. and Rol, C. (1980a). *J. Org. Chem.* **45**, 3906
Baciocchi, E., Rol, C. and Mandolini, L. (1980b). *J. Am. Chem. Soc.* **102**, 7597
Baciocchi, E., Eberson, L. and Rol, C. (1982). *J. Org. Chem.* **47**, 5106
Bailey, S. I., Ritchie, I. M. and Hewgill, F. R. (1983). *J. Chem. Soc. Perkin Trans. 2*, 645
Banthorpe, D. V., Hughes, E. D., Ingold, C. K., Bramley, R. and Thomas, J. A. (1964). *J. Chem. Soc.* 2864
Bard, A. J., Ledwith, A. and Shine, H. J. (1976). *Adv. Phys. Org. Chem.* **12**, 155
Barek, J., Ahlberg, E. and Parker, V. D. (1980). *Acta Chem. Scand.* **B34**, 85
Bargon, J. and Seifert, K.-G. (1974). *Tetrahedron Lett.* 2265
Barnes, G. E. and Myhre, P. C. (1978a). *J. Am. Chem. Soc.* **100**, 973
Barnes, G. E. and Myhre, P. C. (1978b). *J. Am. Chem. Soc.* **100**, 975
Barton, D. H. R., Haynes, R. K., Leclerc, G., Magnus, P. D. and Menzies I. D. (1975). *J. Chem. Soc. Perkin Trans. 1*, 2055
Bauld, N. L., Bellville, D. J., Gardner, S. A., Migron, Y. and Cogswell, G. (1982). *Tetrahedron Lett.* **23**, 825
Bauld, N. L., Bellville, D. J., Pabon, R., Chelsky, R. and Green, G. (1983). *J. Am. Chem. Soc.* **105**, 2378
Bawn, C. E. H., Bell, F. A. and Ledwith, A. (1968). *J. Chem. Soc. Chem. Comm.* 599
Bell, F. A., Crellin, R. A., Fujii, H. and Ledwith, A. (1969a). *J. Chem. Soc. Chem. Comm.* 251
Bell, F. A., Ledwith, A. and Sherrington, D. C. (1969b). *J. Chem. Soc. (C)*, 2719
Bellville, D. J. and Bauld, N. L. (1982). *J. Am. Chem. Soc.* **104**, 2665
Bellville, D. J., Wirth, D. D. and Bauld, N. L. (1981). *J. Am. Chem. Soc.* **103**, 718
Benati, L., Montevecchi, P. C. and Spagnolo, P. (1981). *J. Chem. Soc. Perkin Trans. 2*, 1437
Bethell, D., Handoo, K. L., Fairhurst, S. A. and Sutcliffe, L. H. (1977). *J. Chem. Soc. Chem. Comm.* 326
Bethell, D., Handoo, K. L., Fairhurst, S. A. and Sutcliffe, L. H. (1979a). *J. Chem. Soc. Perkin Trans. 2*, 707
Bethell, D., Eeles, M. F. and Handoo, K. L. (1979b). *J. Chem. Soc. Perkin Trans. 2*, 714
Bewick, A., Edwards, G. J. and Mellor, J. M. (1975). *Tetrahedron Lett.* 4685
Bewick, A., Edwards, G. J. and Mellor, J. M. (1976). *Electrochim. Acta* **21**, 1101
Bewick, A., Edwards, G. J., Mellor, J. M. and Pons, S. (1977). *J. Chem. Soc. Perkin Trans. 2*, 1952
Bewick, A., Mellor, J. M. and Pons, S. (1978). *Electrochim. Acta* **23**, 77

Bewick, A., Coe, D. E., Fuller, G. B. and Mellor, J. M. (1980a). *Tetrahedron Lett.* **21**, 3827
Bewick, A., Coe, D. E., Mellor, J. M. and Walton, D. J. (1980b). *J. Chem. Soc. Chem. Comm.* 51
Bewick, A., Mellor, J. M. and Pons, B. S. (1980c). *Electrochim. Acta* **25**, 931
Bewick, A., Coe, D. E., Libert, M. and Mellor, J. M. (1983). *J. Electroanal. Chem.* **144**, 235
Bilevich, K. A. and Okhlobystin, O. Yu. (1968). *Russ. Chem. Rev.* **37**, 954
Blankespoor, R. L., Doyle, M. P., Smith, D. J., Van Dyke, D. A. and Waldyke, M. J. (1983). *J. Org. Chem.* **48**, 1176
Blount, H. N. (1973). *J. Electroanal. Chem.* **42**, 271
Blum, Z., Cedheim, L. and Eberson, L. (1977). *Acta Chem. Scand.* **B31**, 662
Bock, H. and Kaim, W. (1982). *Acc. Chem. Res.* **15**, 9
Bock, H., Roth, B. and Maier, G. (1980). *Angew. Chem.* **92**, 213
Bock, H., Stein, U. and Rittmeyer, P. (1982). *Angew. Chem.* **94**, 540
Boyd, J. W., Schmalzl, P. W. and Miller, L. L. (1980). *J. Am. Chem. Soc.* **102**, 3856
Brown, R. D. (1959a). *J. Chem. Soc.* 2224
Brown, R. D. (1959b). *J. Chem. Soc.* 2232
Brownstein, S., Bunton, C. A. and Hughes, E. D. (1956). *Chem. Ind.* 981
Brownstein, S., Bunton, C. A. and Hughes, E. D. (1958). *J. Chem. Soc.* 4354
Bunnett, J. F. (1978). *Acc. Chem. Res.* **11**, 413
Burgbacher, C. and Schäfer, H. (1979). *J. Am. Chem. Soc.* **101**, 7590
Caronna, T., Citterio, A., Crolla, T., Ghirardini, M. and Minisci, F. (1976). *J. Heterocycl. Chem.* **13**, 955
Caronna, T., Citterio, A., Crolla, T. and Minisci, F. (1977). *J. Chem. Soc. Perkin Trans. 1*, 865
Cedheim, L. and Eberson, L. (1976). *Acta Chem. Scand.* **B30**, 527
Chanon, M. (1982). *Bull Soc. Chim. France*, II–197
Chanon, M. and Tobe, M. L. (1982). *Angew. Chem. Int. Ed. Engl.* **21**, 1
Cheng, H. Y., Sackett, P. H. and McCreery, R. L. (1978a). *J. Am. Chem. Soc.* **100**, 962
Cheng, H. Y., Sackett, P. H. and McCreery, R. L. (1978b). *J. Med. Chem.* **21**, 948
Chow, Y. L. (1973). *Acc. Chem. Res.* **6**, 354
Chow, Y. L. and Iwai, K. (1980). *J. Chem. Soc. Perkin Trans. 2*, 931
Chow, Y. L., Colon, C. and Chen, S. C. (1967). *J. Org. Chem.* **32**, 2109
Chow, Y. L., Danen, W. C., Nelsen, S. F. and Rosenblatt, D. H. (1978). *Chem. Rev.* **78**, 243
Chu, S.-Y. and Lee, T.-S. (1982). *Nouveau J. Chim.* **6**, 155
Citterio, A. (1980). *Gazz. Chim. Ital.* **110**, 253
Citterio, A. and Minisci, F. (1982). *J. Org. Chem.* **47**, 1759
Citterio, A., Minisci, F., Porta, O. and Sesana, G. (1977). *J. Am. Chem. Soc.* **99**, 7960
Clemens, A. H., Ridd, J. H. and Sandall, J. P. B. (1983). *J. Chem. Soc. Chem. Comm.* 343
Clennan, E. L., Simmons, W. and Almgren, C. W. (1981). *J. Am. Chem. Soc.* **103**, 2098
Corey, E. J. and Hertler, W. R. (1960). *J. Am. Chem. Soc.* **82**, 1657
Coombes, R. G., Golding, J. G. and Hadjigeorgiou, P. (1979). *J. Chem. Soc. Perkin Trans. 2*, 1451
Creason, S. C., Wheeler, J. and Nelson, R. F. (1972). *J. Org. Chem.* **37**, 4440
Crellin, R. A., Lambert, M. C. and Ledwith, A. (1970). *J. Chem. Soc. Chem. Comm.* 682

Dapperheld, S. and Steckhan, E. (1982). *Angew. Chem. Suppl.* 1730
Darwent, J. R., McCubbin, I. and Porter, G. (1982). *J. Chem. Soc. Faraday Trans.* **78**, 903
Davidson, R. S. (1983). *Adv. Phys. Org. Chem.* **19**, 1
Deno, N. C. (1972). *In* "Methods in Free-Radical Chemistry" (E. S. Huyser, ed.) Vol. 3. Dekker, New York, p. 135
Deno, N. C., Billups, W. E., Fishbein, R., Pierson, C., Whalen, R. and Wyckoff, J. C. (1971). *J. Am. Chem. Soc.* **93**, 438
Deronzier, A. and Esposito, F. (1983). *Nouveau J. Chim.* **7**, 15
Dewar, M. J. S. and Marchand, A. P. (1965). *Ann. Rev. Phys. Chem.* **16**, 321
Dixon, W. T. and Norman, R. O. C. (1964). *J. Chem. Soc.* 4857
Dixon, W. T. and Murphy, D. (1976). *J. Chem. Soc. Faraday Trans.* 2 **72**, 1221
do Amaral, L., Bull, H. G. and Cordes, E. H. (1972). *J. Am. Chem. Soc.* **94**, 7579.
Draper, M. R. and Ridd, J. H. (1978). *J. Chem. Soc. Chem. Comm.* 445
Eberhardt, M. K. (1981). *J. Am. Chem. Soc.* **103**, 3876
Eberhardt, M. K., Martinez, G. A., Rivera, J. I. and Fuentes-Aponte, A. (1982). *J. Am. Chem. Soc.* **104**, 7069
Eberson, L. (1973). *In* "Organic Electrochemistry: An Introduction and a Guide" (M. M. Baizer, ed.) Dekker, New York, Ch. XIII
Eberson, L. (1975). *Chem. Commun.* 826
Eberson, L. (1982). *Adv Phys. Org. Chem.* **18**, 79
Eberson, L. (1983). *J. Am. Chem. Soc.* **105**, 3192
Eberson, L. and Nyberg, K. (1976). *Tetrahedron* **32**, 2185
Eberson, L. and Nyberg, K. (1978). *Acta Chem. Scand.* **B32**, 235
Eberson, L. and Jönsson, L. (1980). *J. Chem. Soc. Chem. Comm.* 1187
Eberson, L. and Wistrand, L.-G. (1980). *Acta Chem. Scand.* **B34**, 349
Eberson, L. and Radner, F. (1980). *Acta Chem. Scand.* **B34**, 739
Eberson, L. and Jönsson, L. (1981). *J. Chem. Soc. Chem. Comm.* 133
Eberson, L., Nyberg, K. and Stenerup, H. (1973). *Acta Chem. Scand.* **27**, 1679
Eberson, L., Jönsson, L. and Wistrand, L.-G. (1978a). *Acta Chem. Scand.* **B32**, 520
Eberson, L., Jönsson, L. and Radner, F. (1978b). *Acta Chem. Scand.* **B32**, 749
Eberson, L., Blum, Z., Helgée, B. and Nyberg, K. (1978c). *Tetrahedron* **34**, 731
Eberson, L., Jönsson, L. and Wistrand, L.-G. (1982). *Tetrahedron* **38**, 1087
Elliott, I. W. (1977). *J. Org. Chem.* **42**, 1090
Elliott, I. W. (1979). *J. Org. Chem.* **44**, 1162
Eriksen, J. and Foote, C. S. (1980). *J. Am. Chem. Soc.* **102**, 6083
Evans, D. H. and Nelsen, S. F. (1978). *In* "Characterization of Solutes in Nonaqueous Solvents" (G. Mamantov, ed.). Plenum, New York, p. 131.
Evans, J. F. and Blount, H. N. (1976a). *J. Org. Chem.* **41**, 516
Evans, J. F. and Blount, H. N. (1976b). *J. Phys. Chem.* **80**, 1011
Evans, J. F. and Blount, H. N. (1977a). *J. Org. Chem.* **42**, 976
Evans, J. F. and Blount, H. N. (1977b). *J. Org. Chem.* **42**, 983
Evans, J. F. and Blount, H. N. (1978). *J. Am. Chem. Soc.* **100**, 4191
Evans, J. F. and Blount, H. N. (1979). *J. Phys. Chem.* **83**, 1970
Evans, T. R. and Hurysz, L. F. (1977). *Tetrahedron Lett.* 3103
Evans, T. R., Wake, R. W. and Sifain, M. M. (1973). *Tetrahedron Lett.* 701
Farid, S. and Shealer, S. E. (1973). *J. Chem. Soc. Chem. Comm.* 677
Fessenden, R. W. and Neta, P. (1972). *J. Phys. Chem.* **76**, 2857
Fox, M. A., Campbell, K. A., Hünig, S., Berneth, H., Maier, G., Schneider, K.-A. and Malsch, K.-D. (1982). *J. Org. Chem.* **47**, 3408

Fritz, H. P., Gebauer, H., Friedrich, P., Ecker, P., Artes, R. and Schubert, U. (1978). *Z. Naturforsch.* **B33**, 498
Fukuzumi, S. and Kochi, J. K. (1980a). *J. Am. Chem. Soc.* **102**, 2141
Fukuzumi, S. and Kochi, J. K. (1980b). *J. Phys. Chem.* **84**, 2246
Fukuzumi, S. and Kochi, J. K. (1981a). *J. Am. Chem. Soc.* **103**, 2783
Fukuzumi, S. and Kochi, J. K. (1981b) *J. Am. Chem. Soc.* **103**, 7240
Fukuzumi, S. and Kochi, J. K. (1982). *Tetrahedron* **38**, 1035
Fukuzumi, S. and Kochi, J. K. (1983). *Int. J. Chem. Kinet.* **15**, 249
Gardini, G. P. and Bargon, J. (1980). *J. Chem. Soc. Chem. Comm.* **757**
Gassman, P. G. and Yamaguchi, R. (1982). *Tetrahedron* **38**, 1113
Gassman, P. G., Yamaguchi, R. and Koser, G. F. (1978). *J. Org. Chem.* **43**, 4392
Genies, M., Moutet, J.-C. and Reverdy, G. (1981). *Electrochim, Acta* **26**, 931
Giffney, J. C. and Ridd, J. H. (1979). *J. Chem. Soc. Perkin Trans. 2*, 618
Giordan, J. and Bock, H. (1982). *Chem. Ber.* **115**, 2548
Greenberg, A. and Liebman, J. F. (1978). "Strained Organic Molecules". Academic Press, London and New York
Hammerich, O. (1983). *In* "Organic Electrochemistry, 2nd Revised and Expanded Edition" (M. M. Baizer and H. Lund, eds) Dekker, New York, Ch. 16
Hammerich, O. and Parker, V. D. (1972). *J. Electroanal. Chem.* **38**, App. 9
Hammerich, O. and Parker, V. D. (1973). *Electrochim. Acta* **18**, 537
Hammerich, O. and Parker, V. D. (1974a). *J. Am. Chem. Soc.* **96**, 4289
Hammerich, O. and Parker, V. D. (1974b). *Chem. Commum.*, 245
Hammerich, O. and Parker, V. D. (1981a). *Sulfur Reports* **1**, 317
Hammerich, O. and Parker, V. D. (1981b). *Acta Chem. Scand.* **B35**, 341
Hammerich, O. and Parker, V. D. (1982a). *Acta Chem. Scand.* **B36**, 43
Hammerich, O. and Parker, V. D. (1982b). *Acta Chem. Scand.* **B36**, 59
Hammerich, O. and Parker, V. D. (1982c). *Acta Chem. Scand.* **B36**, 63
Hammerich, O. and Parker, V. D. (1982d). *Acta Chem. Scand.* **B36**, 421
Hammerich, O. and Parker, V. D. (1982e). *Acta Chem. Scand.* **B36**, 519
Hammerich, O. and Parker, V. D. (1983). *Acta Chem. Scand.* **B37**, 303
Hammerich, O., Parker, V. D. and Ronlán, A. (1976). *Acta Chem. Scand.* **B30**, 89
Hand, R., Melicharek, M. Scoggin, D. I., Stotz, R., Carpenter, A. K. and Nelson, R. F. (1971). *Coll. Czech. Chem. Commun.* **36**, 842
Handoo, K. L. and Handoo, S. K. (1982). *Indian J. Chem.* **21B**, 270
Hanotier, J. and Hanotier-Bridoux, M. (1973). *J. Chem. Soc. Perkin Trans. 2*, 1035
Hanotier, J., Hanotier-Bridoux, M. and Radzitzky, P. de (1973). *J. Chem. Soc. Perkin Trans. 2*, 381
Hanson, P. (1980). *Advances in Heterocyclic Chemistry* **27**, 31
Haselbach, E., Bally, T., Lanyiova, Z. and Baertschi, P. (1979). *Helv. Chim. Acta* **62**, 583
Havinga, E. and Cornelisse, J. (1976). *Pure Appl. Chem.* **47**, 1
Heighway, C. J., Packer, J. E. and Richardson, R. K. (1974). *Tetrahedron Lett.* 4441
Heijer, J. den, Shadid, O. B., Cornelisse, J. and Havinga, E. (1977). *Tetrahedron* **33**, 779
Helsby, P., Ridd, J. H. and Sandall, J. P. B (1981). *J. Chem. Soc. Chem. Comm.* 825
Ho, C.-T., Conlin, R. T. and Gaspar, P. P. (1974). *J. Am. Chem. Soc.* **96**, 8109
Hoffmann, R. W. and Barth, W. (1983). *J. Chem. Soc. Chem. Commun.* 345
Holton, D. M. and Murphy, D. (1979). *J. Chem. Soc. Faraday Trans. 2* **75**, 1637
Hughes, E. D. and Jones, G. T. (1950). *J. Chem. Soc.* 2678
Hünig, S. (1980). *Topics Curr. Chem.* **92**, 1
Jensen, B. S. and Parker, V. D. (1973). *Electrochim. Acta* **18**, 665

Jones, C. R. (1981). *J. Org. Chem.* **46**, 3370
Jones, G., II and Becker, W. G. (1983). *J. Am. Chem. Soc.* **105**, 1276
Jugelt, W. and Pragst, F. (1968a). *Angew. Chem.* **80**, 280
Jugelt, W. and Pragst, F. (1968b). *Tetrahedron* **24**, 5123
Jönsson, L. and Wistrand, L.-G. (1979). *J. Chem. Soc. Perkin Trans. 1*, 669
Kabakoff, D. S., Bünzli, J.-C. G., Oth, J. F. M., Hammond, W. B. and Berson, J. A. (1975). *J. Am. Chem. Soc.* **97**, 1510
Kerr, J. B., Jempty, T. C. and Miller, L. L. (1979). *J. Am. Chem. Soc.* **101**, 7338
Kim, K., Hull, V. J. and Shine, H. J. (1974). *J. Org. Chem.* **39**, 2534
Koshechko, V. G., Inozemtsev, A. N. and Pokhodenko, V. D. (1981). *Zh. Org. Khim.* **17**, 2608
Kossai, R., Simonet, J. and Dauphin, G. (1980). *Tetrahedron Lett.* **21**, 3575
Kossai, R., Simonet, J. and Dauphin, G. (1982). *J. Electroanal. Chem.* **139**, 207
Kovacic, P., Lowery, M. K. and Field, K. W. (1970). *Chem. Rev.* **70**, 639
Kricka, L. J. and Ledwith, A. (1973). *J. Chem. Soc. Perkin Trans. 1*, 294
Kupchan, S. M. and Kim, C.-K. (1975). *J. Am. Chem. Soc.* **97**, 5623
Kupchan, S. M., Kameswaran, V., Lynn, J. T., Williams, D. K. and Liepa, A. J. (1975). *J. Am. Chem. Soc.* **97**, 5622
Kupchan, S. M., Dinghra, O. P., Kim, C.-K. and Kameswaran, V. (1978a). *J. Org. Chem.* **43**, 2521
Kupchan, S. M., Dinghra, O. P., Ramachandran, V. and Kim, C.-K. (1978b). *J. Org. Chem.* **43**, 105
Land, E. J. and Porter, G. (1960). *Proc. Chem. Soc.* **84**
Land, E. J. and Porter, G. (1963). *Trans. Faraday Soc.* **59**, 2027
Land, E. J., Porter, G. and Strachan, E. (1961). *Trans. Faraday Soc.* **57**, 1885
Ledwith, A. (1972). *Acc. Chem. Res.* **5**, 133
Lewis, F. D., Ho, T.-I. and Simpson, J. T. (1982). *J. Am. Chem. Soc.* **104**, 1924
Liao, C.-S., Chambers, J. Q., Kapovits, I., and Rabai, J. (1974). *J. Chem. Soc. Chem. Comm.* 149
Lines, R. and Utley, J. H. P. (1977). *J. Chem. Soc. Perkin Trans. 2*, 803
Maier, G., Pfriem, S., Schäfer, U., Malsch, K.-D. and Matusch, R. (1981a). *Chem. Ber.* **114**, 3965
Maier, G., Pfriem, S., Malsch, K.-D., Kalinowski, H. O. and Dehnicke, K. (1981b). *Chem. Ber.* **114**, 3988
Manning, G., Parker, V. D. and Adams, R. N. (1969). *J. Am. Chem. Soc.* **91**, 4584
Marcoux, L. S. (1971). *J. Am. Chem. Soc.* **93**, 537
Martigny, P. and Simonet, J. (1980). *J. Electroanal. Chem.* **111**, 133
Mattes, S. L. and Farid, S. (1982). *Acc. Chem. Res.* **15**, 80
Mayausky, J. S. and McCreery, R. L. (1983). *J. Electroanal. Chem.* **145**, 117
Mayeda, E. A. (1975). *J. Am. Chem. Soc.* **97**, 4012
Mayeda, E. A., Miller, L. L. and Wolf, J. F. (1972). *J. Am. Chem. Soc.* **94**, 6812
McCreery, R. L. and Mayausky, J. S. (1982). *Acta Chem. Scand.* **B36**, 713
McKillop, A., Turrell, A. C. and Taylor, E. C. (1977). *J. Org. Chem.* **42**, 765
Miller, L. L., Wolf, J. F. and Mayeda, E. A. (1971). *J. Am. Chem. Soc.* **93**, 3306
Minisci, F. (1973). *Synthesis* 1
Minisci, F. (1976). *Topics Curr. Chem.* **62**, 1
Minisci, F., Bernadi, R., Bertini, F., Galli, R. and Perchinunno, M. (1971). *Tetrahedron Lett.* 3575
Minisci, F., Citterio, A. and Giordano, C. (1983). *Acc. Chem. Res.* **16**, 27
Morkovnik, A. S., Dobaeva, N. M., Panov, V. B. and Okhlobystin, O. Yu. (1980). *Dokl. Akad. Nauk. SSSR* **251**, 125

Morkovnik, A. S., Dobaeva, N. M. and Okhlobystin, O. Yu. (1981). *Khim. Geterotsikl. Soedin.* 1214
Moutet, J.-C. and Reverdy, G. (1982). *J. Chem. Soc. Chem. Comm.* 654
Moutet, J.-C. and Reverdy, G. (1983). *Nouveau J. Chim.* **7**, 105
Mukai, T., Sato, K. and Yamashita, Y. (1981). *J. Am. Chem. Soc.* **103**, 670
Murata, Y. and Shine, H. J. (1969). *J. Org. Chem.* **34**, 3368
Musker, W. K. (1980). *Acc. Chem. Res.* **13**, 200
Nagakura, S. (1963). *Tetrahedron* **19**, *Suppl. 2*, 361
Nagakura, S. and Tanaka, J. (1954). *J. Chem. Phys.* **22**, 563
Nagakura, S. and Tanaka, J. (1959). *Bull. Chem. Soc. Jpn.* **32**, 734.
Neale, R. S. (1971). *Synthesis* 1
Nelsen, S. F. (1979). *Israel J. Chem.* **18**, 45
Nelsen, S. F. (1981). *Acc. Chem. Res.* **14**, 131
Nelsen, S. F. and Akaba, R. (1981). *J. Am. Chem. Soc.* **103**, 2096
Nelsen, S. F. and Parmelee, W. P. (1981). *J. Org. Chem.* **46**, 3453
Nelsen, S. F. and Gannett, P. M. (1982). *J. Am. Chem. Soc.* **104**, 5292
Nelsen, S. F., Kessel, C. R. and Brien, D. J. (1980a). *J. Am. Chem. Soc.* **102**, 702
Nelsen, S. F., Parmelee, W. P., Göbl, M., Hiller, K.-O., Veltwisch, D. and Asmus, K.-D. (1980b). *J. Am. Chem. Soc.* **102**, 5606
Nelson, R. F. (1974). *In* "Technique of Electroorganic Synthesis" (N. L. Weinberg, ed.) Part I. Wiley, New York, Ch. V
Nelson, R. F. and Feldberg, S. W. (1969). *J. Phys. Chem.* **73**, 2623
Neptune, M. and McCreery, R. L. (1978). *J. Org. Chem.* **43**, 5006
Neta, P., Zemel, H., Madhavan, V. and Fessenden, R. W. (1977). *J. Am. Chem. Soc.* **99**, 163
Nicholas, A. M. de P. and Arnold, D. R. (1982). *Can. J. Chem.* **60**, 2165
Nilsson, A., Ronlán, A. and Parker, V. D. (1973). *J. Chem. Soc. Perkin Trans. 1*, 2337
Nilsson, A., Palmquist, U., Pettersson, T. and Ronlán, A. (1978). *J. Chem. Soc. Perkin Trans. 1*, 696
Nojima, M., Ando, T. and Tokura, N. (1976). *J. Chem. Soc. Perkin Trans. 1*, 1504
Nyberg, K. (1970). *Acta Chem. Scand.* **24**, 1609
Nyberg, K. (1971a). *Acta Chem. Scand.* **25**, 2499
Nyberg, K. (1971b). *Acta Chem. Scand.* **25**, 534
Nyberg, K. (1971c). *Acta Chem. Scand.* **25**, 3770
Nyberg, K. (1973). *Acta Chem. Scand.* **27**, 503
Nyberg, K. (1978). *In* "Encyclopedia of Electrochemistry of the Elements" (A. J. Bard and H. Lund, eds) Vol. XI. Dekker, New York, Ch. XI-1
Nyberg, K. and Trojanek, A. (1975). *Coll. Czech. Chem. Commun.* **40**, 526
Nyberg, K. and Wistrand, L.-G. (1976). *J. Chem. Soc. Chem. Comm.* **898**
Nyberg, K. and Wistrand, L.-G. (1978). *J. Org. Chem.* **43**, 2613
Okada, K., Hisamitsu, K. and Mukai, T. (1981). *Tetrahedron Lett.* **22**, 1251
Okura, I., Kim-Thuan, N. and Takeuchi, M. (1982). *Angew. Chem. Suppl.* 1004
Olah, G. A. (1971). *Acc. Chem. Res.* **4**, 240
Olah, G. A. (1974). "Carbocations and Electrophilic Reactions". Wiley, New York
Olah, G. A. and Mo, Y. K. (1976). *In* "Carbonium Ions" (G. A. Olah and P. van R. Schleyer, eds) *Vol 5*. Wiley, New York, Ch. 36
Olah, G. A. and Krishnamurthy, V. V. (1982). *J. Am. Chem. Soc.* **104**, 3987
Olah, G. A., Tashiro, M. and Kobayashi, S. (1970). *J. Am. Chem. Soc.* **92**, 6369
Olah, G. A., Yu, S. H. and Parker, D. G. (1976). *J. Org. Chem.* **41**, 1983
Osa, T., Yildiz, A. and Kuwana, T. (1969). *J. Am. Chem. Soc.* **91**, 3994

Oshima, T., Yoshioka, A. and Nagai, T. (1978). *J. Chem. Soc. Perkin Trans. 2*, 1283
Packer, J. E., House, D. B. and Rasburn, E. J. (1971). *J. Chem. Soc.* (*B*), 1574
Packer, J. E., Richardson, R. K., Soole, P. J. and Webster, D. R. (1974). *J. Chem. Soc. Perkin Trans. 2*, 1472
Packer, J. E., Heighway, C. J., Miller, H. M. and Dobson, B. C. (1980). *Aust. J. Chem.* **33**, 965
Parkányi, C. (1983). *Pure Appl. Chem.* **55**, 331
Parker, V. D. (1970). *Acta Chem. Scand.* **24**, 2757
Parker, V. D. (1972). *J. Electroanal. Chem.* **38**, App. 8
Parker, V. D. (1980). *Acta Chem. Scand.* **B34**, 359
Parker, V. D. (1981a). *Acta Chem. Scand.* **B35**, 123
Parker, V. D. (1981b). *Acta Chem. Scand.* **B35**, 233
Parker, V. D. (1983a). *Adv. Phys. Org. Chem.* **19**, 131
Parker, V. D. (1983b). *Acta Chem. Scand.* **B37**, 393
Parker, V. D. (1984). *Acta Chem. Scand.* in press
Parker, V. D. and Eberson, L. (1970). *J. Am. Chem. Soc.* **92**, 7488
Parker, V. D. and Ronlán, A. (1975). *J. Am. Chem. Soc.* **97**, 4714
Parker, V. D. and Bethell, D. (1981). *Acta Chem. Scand.* **B35**, 691
Parker, V. D. and Hammerich, O. (1982). *Acta Chem. Scand.* **B36**, 133
Parker, V. D., Palmquist, U. and Ronlán, A. (1974). *Acta Chem. Scand.* **B28**, 1241
Parker, V. D., Sundholm, C., Svanholm, U. and Hammerich, O. (1978). In "Encyclopedia of Electrochemistry of the Elements." (A. J. Bard and H. Lund, eds) Vol. XI. Dekker, New York, Ch. XI–2
Parker, V. D., Aalstad, B. and Ronlán, A. (1983). *Acta Chem. Scand.* **B37**, 467
Pedersen, E. B., Petersen, T. E., Torsell, K. and Lawesson, S.-O. (1973). *Tetrahedron* **29**, 579
Perrin, C. L. (1977). *J. Am. Chem. Soc.* **99**, 5516
Pokhodenko, V. D., Khiznyi, V. A., Koshechko, V. G. and Shkrebtii, O. I. (1975). *Zh. Org. Khim.* **11**, 1873
Pragst, F. and Jugelt, W. (1970a). *Electrochim. Acta* **15**, 1543
Pragst, F. and Jugelt, W. (1970b). *Electrochim. Acta* **15**, 1769
Pragst, F. and Ziebig, R. (1978). *Electrochim. Acta* **23**, 735
Reynolds, R., Line, L. L. and Nelson, R. F. (1974). *J. Am. Chem. Soc.* **96**, 1087
Richards, J. A. and Evans, D. H. (1977). *J. Electroanal. Chem.* **81**, 171
Richards, J. A., Whitson, P. E. and Evans D. H. (1975). *J. Electroanal. Chem.* **63**, 311
Ridd, J. H. (1971). *Acc. Chem. Res.* **4**, 248
Ridd, J. H. and Sandall, J. P. B. (1981). *J. Chem. Soc. Chem. Comm.* 402
Ridd, J. H. and Sandall, J. P. B. (1982). *J. Chem. Soc. Chem. Comm.* 261
Rieker, A., Dreher, E.-L., Geisel, H. and Khalifa, M. H. (1978). *Synthesis* 851
Ristagno, C. V. and Shine, H. J. (1971). *J. Am. Chem. Soc.* **93**, 1811
Ronlán, A. and Parker, V. D. (1971). *J. Chem. Soc.* (*C*), 3214
Ronlán, A. and Parker, V. D. (1974). *J. Org. Chem.* **39**, 1014
Ronlán, A., Bechgaard, K. and Parker, V. D. (1973a). *Acta Chem. Scand.* **27**, 2375
Ronlán, A., Hammerich, O. and Parker, V. D. (1973b). *J. Am. Chem. Soc.* **95**, 7132
Ross, D. S., Hum, G. P. and Blucher, W. G. (1980). *J. Chem. Soc. Chem. Comm.* 532
Roth, H. D. and Schilling, M. L. M. (1981). *J. Am. Chem. Soc.* **103**, 7210
Rys, P., Skrabal, P. and Zollinger, H. (1972). *Angew. Chem. Int. Ed. Engl.* **11**, 874
Sackett, P. H. and McCreery, R. L. (1979). *J. Med. Chem.* **22**, 1447

Sakota, K., Kamiya, Y. and Ohta, N. (1969). *Can. J. Chem.* **47**, 387
Santiago, C., Houk, K. N. and Perrin, C. L. (1979). *J. Am. Chem. Soc.* **101**, 1337
Savéant, J. M. (1980). *Acc. Chem. Res.* **13**, 223
Schaap, A. P., Zaklika, K. A., Kaskar, B. and Fung, L. W.-M. (1980). *J. Am. Chem. Soc.* **102**, 389
Schäfer, H. (1981). *Angew. Chem.* **93**, 978
Schmid Baumberger, R. and Parker, V. D. (1980). *Acta Chem. Scand.* **B34**, 537
Schmidt, W. and Steckhan, E. (1978a). *Angew. Chem.* **90**, 717
Schmidt, W. and Steckhan, E. (1978b). *J. Electroanal. Chem.* **89**, 215
Schmidt, W. and Steckhan, E. (1979a). *Angew. Chem.* **91**, 850
Schmidt, W. and Steckhan, E. (1979b). *Angew. Chem.* **91**, 851
Schmidt, W. and Steckhan, E. (1980). *Chem. Ber.* **113**, 577
Schmitt, R. J., Ross, D. S. and Buttrill, Jr., S. E. (1981). *J. Am. Chem. Soc.* **103**, 5265
Sehested, K. and Holcman, J. (1978). *J. Phys. Chem.* **82**, 651
Seo, E. T., Nelson, R. F., Fritsch, J. M., Marcoux, L. S., Leedy, D. W. and Adams, R. N. (1966). *J. Am. Chem. Soc.* **88**, 3498
Shaik, S. S. (1981). *J. Am. Chem. Soc.* **103**, 3692
Shang, D. T. and Blount, H. N. (1974). *J. Electroanal. Chem.* **54**, 305
Shine, H. J. (1981). *In* "The Chemistry of the Sulfonium Group" (C. J. M. Sterling and S. Patai, eds). Wiley, New York, Ch. 14
Shine, H. J. and Murata, Y. (1969). *J. Am. Chem. Soc.* **91**, 1872
Shine, H. J. and Wu, S.-M. (1979). *J. Org. Chem.* **44**, 3310
Shine, H. J., Silber, J. J., Bussey, R. J. and Okuyama, T. (1972). *J. Org. Chem.* **37**, 2691
Shine, H. J., Zmuda, H., Park, K. H., Kwart, H., Horgan, A. G. and Brechbiel, M. (1982). *J. Am. Chem. Soc.* **104**, 2501
Shono, T., Matsumura, Y., Inoue, K., Ohmizu, H. and Kashimura, S. (1982). *J. Am. Chem. Soc.* **104**, 5753
Silber, J. J. and Shine, H. J. (1971). *J. Org. Chem.* **36**, 2923
Simic, M. and Hayon, E. (1971). *J. Am. Chem. Soc.* **93**, 5982
Siskin, M. (1972). *In* "Methods in Free-Radical Chemistry" (E. S. Huyser, ed.) Vol. 3. Dekker, New York, p. 83
Speiser, B. and Rieker, A. (1977). *J. Chem. Res.(S)*, 314
Speiser, B. and Rieker, A. (1978). *Electrochim. Acta* **23**, 983
Speiser, B. and Rieker, A. (1979). *J. Electroanal. Chem.* **102**, 373
Speiser, B. and Rieker, A. (1980). *J. Electroanal. Chem.* **110**, 231
Steckhan, E. (1978). *J. Am. Chem. Soc.* **100**, 3526
Stella, L. (1983). *Angew. Chem. Int. Ed. Engl.* **22**, 337
Stuart, J. D. and Ohnesorge, W. E. (1971). *J. Am. Chem. Soc.* **93**, 4531
Sullivan, B. P., Dressick, W. J. and Meyer, T. J. (1982). *J. Phys. Chem.* **86**, 1473
Surzur, J. M., Stella, L. and Tordo, P. (1975). *Bull. Soc. Chim. France* 1429
Suttie, A. B. (1969). *Tetrahedron Lett.* 953
Suzuki, H., Nagae, K., Maeda, H. and Osuka, A. (1980). *J. Chem. Soc. Chem. Comm.* 1245
Svanholm, U. and Parker, V. D. (1972a). *J. Am. Chem. Soc.* **94**, 5597
Svanholm, U. and Parker, V. D. (1972b). *J. Chem. Soc. Chem. Comm.* 440
Svanholm, U. and Parker, V. D. (1973). *Acta Chem. Scand.* **27**, 1454
Svanholm, U. and Parker, V. D. (1976a). *J. Am. Chem. Soc.* **98**, 997
Svanholm, U. and Parker, V. D. (1976b). *J. Am. Chem. Soc.* **98**, 2942

Svanholm, U. and Parker, V. D. (1976c). *J. Chem. Soc. Perkin Trans. 2*, 1567
Svanholm, U. and Parker, V. D. (1980). *Acta Chem. Scand.* **B34**, 5
Svanholm, U., Bechgaard, K., Hammerich, O. and Parker, V. D. (1972). *Tetrahedron Lett.* 3675
Svanholm, U., Ronlán, A. and Parker, V. D. (1974). *J. Am. Chem. Soc.* **96**, 5108
Svanholm, U., Hammerich, O. and Parker, V. D. (1975). *J. Am. Chem. Soc.* **97**, 101
Taylor, E. C., Andrade, J. G. and McKillop, A. (1977). *J. Chem. Soc. Chem. Comm.* 538
Todres, Z. V. (1978). *Russ. Chem. Rev.* **47**, 148
Torii, S., Uneyama, K. and Ono, M. (1980a). *Tetrahedron Lett.* **21**, 2653
Torii, S., Uneyama, K. and Ono, M. (1980b). *Tetrahedron Lett.* **21**, 2741
Trahanovsky, W. S. and Cramer, J. (1971). *J. Org. Chem.* **36**, 1890
Trahanovsky, W. S. and Brixius, D. W. (1973). *J. Am. Chem. Soc.* **95**, 6778
Walter, R. I. (1955). *J. Am. Chem. Soc.* **77**, 5999
Waters, W. A. (1942). *J. Chem. Soc.* 266
Wawzonek, S. and Thelen, P. J. (1950). *J. Am. Chem. Soc.* **72**, 2118
Wayner, D. D. M. and Arnold, D. R. (1982). *J. Chem. Soc. Chem. Comm.* 1087
Weinreb, S. M., Epling, G. A., Comi, R. and Reitano, M. (1975). *J. Org. Chem.* **40**, 1356
Werner, R. and Rüchardt, C. (1969). *Tetrahedron Lett.* 2407
White, W. N. and Golden, J. T. (1970). *J. Org. Chem.* **35**, 2759
White, W. N. and Klink, J. R. (1970). *J. Org. Chem.* 35, 965
White, W. N. and White, H. S. (1970). *J. Org. Chem.* **35**, 1803
White, W. N., Klink, J. R., Lazdins, D., Hathaway, C., Golden, J. T. and White, H. S. (1961). *J. Am. Chem. Soc.* **83**, 2024
White, W. N., Lazdins, D. and White, H. S. (1964). *J. Am. Chem. Soc.* **86**, 1517
White, W. N., Hathaway, C. and Huston, D. (1970a). *J. Org. Chem.* **35**, 737
White, W. N., Golden, J. T. and Lazdins, D. (1970b). *J. Org. Chem.* **35**, 2048
White, W. N., White, H. S. and Fentiman, A. (1976). *J. Org. Chem.* **41**, 3166
Wiberg, K. B. and Connon, H. A. (1976). *J. Am. Chem. Soc.* **98**, 5411
Wolff, M. E. (1963). *Chem. Rev.* **63**, 55
Zaklika, K. A., Kaskar, B. and Schaap, A. P. (1980). *J. Am. Chem. Soc.* **102**, 386
Zefirov, N. S. and Makhon'kov, D. I. (1982). *Chem. Rev.* **82**, 615
Zollinger, H. (1973). *Acc. Chem. Res.* **6**, 335
Zollinger, H. (1978). *Angew. Chem. Int. Ed. Engl.* **17**, 141

# The Photochemistry of Aryl Halides and Related Compounds

R. STEPHEN DAVIDSON,[1] JONATHAN W. GOODIN[2] and GRAHAM KEMP[2]

[1] Department of Chemistry, The City University, London, U.K.
[2] Plastics Division, Ciba-Geigy Plastics and Additives Co., Duxford, Cambridge, U.K.

1 Introduction   192
2 Chloroaromatics   196
    Chlorobenzenes   196
    Chlorinated biphenyls   203
    Chlorinated terphenyls   206
    1-Chloronaphthalene and related compounds   206
    9,10-Dichloroanthracene   208
    Reactions of other chlorinated and related compounds of mechanistic interest   209
3 Bromo- and iodoaromatics   212
    Bromo- and iodobenzenes   212
    Bromo- and iodopolycyclic aromatic hydrocarbons   217
4 Assisted dehalogenation of halogenoaromatics   219
    Amine- and sulphide-assisted dehalogenation of halogenoaromatics   219
    Dehalogenation of haloaromatic compounds assisted by sodium borohydride   222
5 Photoinduced nucleophilic substitution   222
    Haloaromatics   222
    Reactions involving vinyl cations   224
    $S_{RN}1$ reactions   226
    Reactions of vinylic halides with aromatic compounds   227
References   229

## 1 Introduction

There have been several earlier reviews on the photochemistry of alkyl and aryl halides (Sammes, 1973; Sharma and Kharasch, 1968; Kharasch, 1969; Grimshaw and de Silva, 1981). Due to the widespread use of substituted aryl halides as pesticides, electrical insulators, moth proofing agents, etc. and the environmental problems which the use of these compounds pose, it seems worthwhile reconsidering the photochemistry of these compounds. Furthermore, the photoinduced decomposition of aryl halides is used extensively in synthesis, and much of this latter work has been recently reviewed (Grimshaw and de Silva, 1981).

It is well established that on photolysis, many aryl halides undergo homolysis (1) to generate aryl radicals and halogen atoms. The question as to

$$Ar - Hal \xrightarrow{h\nu} Ar\cdot + Hal\cdot \qquad (1)$$

whether the excited singlet or triplet state or indeed both excited states are responsible for reaction, is of considerable importance. Thus, if reaction occurs from the excited singlet state, the homolytic process may well be in competition with an electrocyclic reaction. Compound [2] illustrates the point. In principle this compound may photoisomerize to give [1] which may then undergo further photoreactions. The photoisomerization is in competition with the electrocyclic reaction to give [3] and the homolytic reactions which produces [4]. This example also illustrates the way in which conformational factors affect the outcome of these reactions. That haloaromatic compounds can undergo electrocyclic reactions to the exclusion of homolysis is shown by the elegant work of Begley and Grimshaw (1977).

They showed that [5] photocyclizes to give [6] as the only product. Electrochemical reduction of aryl halides results in the generation of aryl radicals; the formation of [7] must involve such species. The absence of [7] from the products of the photochemical reaction indicates that in this case the electrocyclic reaction takes precedence over the homolytic reaction. In a related piece of work (Grimshaw and Haslett, 1980) it was found that irradiation of [8] in cyclohexane solution produces [9] and [10]. Electrochemical reduction of [8] does not produce [9] but gives a range of other products which are related to [10].

[Structures [8], [9], [10] with hv arrow]

The ability of an electrocyclic reaction to compete with a homolytic reaction will depend upon the relative energies of activation for the two processes and the energy available for reaction. Steric factors should be important in electrocyclic reactions since there is likely to be more steric overcrowding in the transition state. The homolytic reaction may well result in the relief of steric overcrowding in the rate-determining step, e.g. the homolysis of 2-halobiphenyls. The competition between electrocyclic and homolytic processes should also depend upon the nature of the halogen since homolysis becomes progressively easier as the substituent is changed F → Cl → Br → I. Thus in the case of [11] irradiation leads to products which were shown, unequivocally, to be derived *via* radical intermediates (Hey *et al.*, 1971, 1972). The formation of cyclized products in this reaction is attributed to the fact that the favoured conformation of the amide is that depicted in [11a] rather than [11b].

The precise mechanism for photoinduced homolysis of aryl—halogen compounds will be dependent upon both the nature of the halogen substituent and the aryl group. However, two facts should be borne in mind. The first of these is the strength of the carbon—halogen bond which is to be homolyzed; some typical values (Egger and Cocks, 1973) are shown in Table 1. The second point is that halogen substituents increase intersystem crossing from both the excited singlet to the triplet state and from the triplet state to the singlet ground state. One may well anticipate that the triplet state of haloaromatics will be efficiently populated and therefore be a likely candidate for

TABLE 1

Bond dissociation energies for some carbon—halogen bonds[a]

| | $D(C-X)^b$ | | $D(C-X)^b$ |
|---|---|---|---|
| Ph—Cl | 94.5 | Ph—Br | 79.2 |
| Ph—I | 64.4 | Ph—F | 123.9 |
| $CH_3CH_2$—Cl | 80.8 | $CH_3CH_2$—Br | 67.6 |
| $CH_3CH_2$—I | 53.2 | | |

[a] Egger and Cocks, 1973
[b] In kcal mol$^{-1}$

the state responsible for homolysis. However, the triplet states of many systems, e.g. biphenyl, naphthyl and anthracenyl have energies below the energy required to homolyze the aryl—chlorine bond. It is therefore convenient to discuss the mechanisms of this reaction by compound type.

## 2 Chloroaromatics

CHLOROBENZENES

Measurement of the fluorescence lifetimes of the dichlorobenzenes in the vapour phase (Shimoda et al., 1979) showed that the chlorine substituents induce fast non-radiative decay of the excited singlet state and that since the aryl—chlorine bond is of lower energy than the first excited singlet state this decay must involve bond homolysis as well as intersystem crossing. The fluorescence lifetimes depend upon the pattern of substitution: 1,2-dichlorobenzene 290 ns, 1,3-dichlorobenzene 280 ns, and 1,4-dichlorobenzene 170 ns. Chlorobenzene has a low fluorescence quantum yield and a high quantum yield for the production of radicals.

The mechanism of bond homolysis for chloro-, bromo- and iodobenzene in the vapour phase has been studied using an excitation source emitting at 193 nm (Freedman et al., 1980). There appear to be two reaction pathways. One pathway involves intersystem crossing from a singlet $\pi\pi^*$ state to a triplet $\sigma\sigma^*$ which homolyzes. For chlorobenzene intersystem crossing occurs from the $S_1$ $\pi\pi^*$ state whereas for bromo- and iodobenzene $S_3$, $S_2$ and $S_1$ $\pi\pi^*$ states are involved. In the other pathway a distortion of the $S_1$ $\pi\pi^*$ state occurs as takes place in the formation of prefulvene. This distortion enables electronic energy to enter the phenyl ring as vibrational energy and so lead to homolysis. This concept of a distorted excited state being responsible for homolysis does not appear to have been considered as a possible reaction pathway for solution phase experiments. A study has been made of the vapour phase dechlorination of chlorotoluenes in the presence of ethene (Koso et al., 1979) and it was concluded the homolysis occurs from the triplet state.

There is much less precise information concerning the photophysical parameters for chlorobenzenes in solution. It is known that chlorobenzene and many of its derivatives homolyze on photolysis. Moreover, if benzene is used as solvent, the chlorine atom is replaced by a phenyl group, whereas the use of hydrogen-donating solvents such as methanol leads to dehalogenation. Irradiation of chlorobenzene in benzene leads to biphenyls and in this system the solvent gathers the light. Benzene can act both as a singlet and a triplet sensitizer (Dubois and Wilkinson, 1963). Thus, although the homolysis of chlorobenzenes is sensitized by benzene, one cannot draw any conclusion as to the excited state of chlorobenzene responsible for reaction. To elucidate the mechanism of homolysis of chlorobenzene and other chloroaromatics, sensitization and quenching experiments have been undertaken (Bunce et al., 1980). The results favour the view that the triplet state is responsible for reaction and, in support of this, it was pointed out that the quantum yield for dechlorination is high (0.54). However, the energy of the triplet state (85 kcal mol$^{-1}$) lies well below that of the aryl—chlorine bond strength (Table 1) and therefore the question arises as to how this energy deficiency is made up. As yet this question has not been satisfactorily answered. That alkyl ketones can sensitize the homolysis of chlorobenzene has been demonstrated (Augustyniak, 1980). Since it has recently been shown that triplet states of carbonyl compounds will form excited complexes with electron deficient alkenes (Maharaj and Winnik, 1982) one cannot be certain whether the sensitization observed by Augustyniak is due purely to energy transfer. A variety of substituted chlorobenzenes undergo homolysis on irradiation in benzene solution, e.g. 4-chloroaniline and 2-, and 4-chlorobenzonitriles (Robinson and Vernon, 1969, 1971). Interestingly 4-chloroacetophenone and 4-chlorobenzophenone display very low photoreactivity even though their triplet yields are close to unity. This lack of reactivity may be due to the lowest triplet states of these compounds being associated with an n → π* transition rather than a π → π* or σ → σ* transition.

In another study aimed at determining the mechanism whereby halobenzenes homolyze, compounds such as [12] were irradiated in methanol

$$Cl-\text{C}_6\text{H}_4-O(CH_2)_nBr \xrightarrow[\text{MeOH}]{h\nu} \text{C}_6\text{H}_5-O(CH_2)_nBr + PhOH$$

[12] n = 2,3

$$Cl-\text{C}_6\text{H}_4-\underset{\underset{\text{Bu}^t}{|}}{\overset{\overset{\text{Bu}^t}{|}}{C}}H\text{CH}-\text{C}_6\text{H}_4-Cl \xrightarrow{h\nu} \text{C}_6\text{H}_5-\underset{\underset{\text{Bu}^t}{|}}{\overset{\overset{\text{Bu}^t}{|}}{C}}H\text{CH}-\text{C}_6\text{H}_5$$

[13]

(Davidson et al., 1980b). It was found that the compounds dechlorinate rather than suffer homolysis of the carbon—bromine bond. This is rather surprising when one considers that the methylene—bromine bond is weaker than the aryl—chlorine bond by ~35 kcal mol$^{-1}$. The homolysis of the carbon—chlorine bond obviously competes very effectively with energy transfer to the methylene—bromine bond. The triplet state of [12] is of higher energy than the methylene—bromine bond and is of lower energy than that of the aryl—chlorine bond. Thus it was concluded that the excited singlet state of [12] is responsible for homolysis. If by any chance the triplet state of chlorobenzenes can undergo homolysis in solution then this process must have a very high rate constant to compete so effectively with energy transfer.

Another very interesting example of dechlorination competing effectively with other processes is shown by [13] (Eichin et al., 1980). It was found that irradiation of meso-[13] in decane gave the meso dechlorinated product, dechlorination having competed with homolysis of the benzylic carbon—carbon bond. Homolysis of this benzylic bond should be facile since its energy lies below that of both the excited singlet and triplet states of chlorobenzene. The photochemistry of the benzylic compounds [14a] and [14b] has also been investigated (Robinson and Vernon, 1969; Davidson et al., 1980b). Compound [14a] was found to be relatively stable when irradiated in benzene solution but irradiation of [14b] in methanol solution led to dechlorination. Thus in [14b] the dechlorination process competed with both singlet and triplet energy transfer. This result suggests that the excited singlet state of the chloroaromatics must be at least in part responsible for the homolysis.

The observation that benzene can sensitize the homolysis of chlorobenzene (Robinson and Vernon, 1969, 1971) raised the question as to whether a benzene—chlorobenzene exciplex could be involved in the process. This point has been examined by Bunce and Ravanal (1977) who found that [15]

ArCH$_2$CH$_2$—⟨benzene ring⟩—Cl            ⟨benzene ring⟩—(CH$_2$)$_3$—⟨benzene ring⟩—Cl

[14a] Ar = 4-Biphenylyl                               [15]
[14b] Ar = 1-Naphthyl

undergoes dechlorination relatively inefficiently. They found that [15] exhibits exciplex emission showing that the two aryl groups can interact intramolecularly. It would appear that exciplex formation leads to energy wastage rather than efficient chemical reactions. By extrapolation it seems reasonable to suggest that irradiation of chlorobenzenes in benzene solution leads to intermolecular exciplex formation. If high concentrations of benzene are employed, termolecular complex formation (one molecule of chlorobenzene and two molecules of benzene) is also likely. One really needs to know the extent to which these complexes lead to deactivation before one can exclude the excited singlet state of the chlorobenzene as being the reactive species. That the relative concentrations of chlorobenzene and benzene play an important part in determining the product distribution has been demonstrated by Kojima et al. (1981). Thus irradiation of either neat chlorobenzene or concentrated benzene solutions of chlorobenzene gives a different ratio of the o-, m- and p-chlorobiphenyls than when dilute benzene solutions of chlorobenzene are used. The concentration effect was attributed to the participation of chlorobenzene excimers at high chlorobenzene concentrations. However, Bunce *et al.* (1980) claim that the excited singlet excimer is an energy-wasting intermediate rather than a reactive one.

The formation of exciplexes between chlorobenzene and benzene and its derivatives has been looked at in other contexts. It has been found (Grimshaw and de Silva, 1980a) that the chloro-compound [16a] undergoes photocyclization in cyclohexane solution whereas the bromo- and iodo-compounds

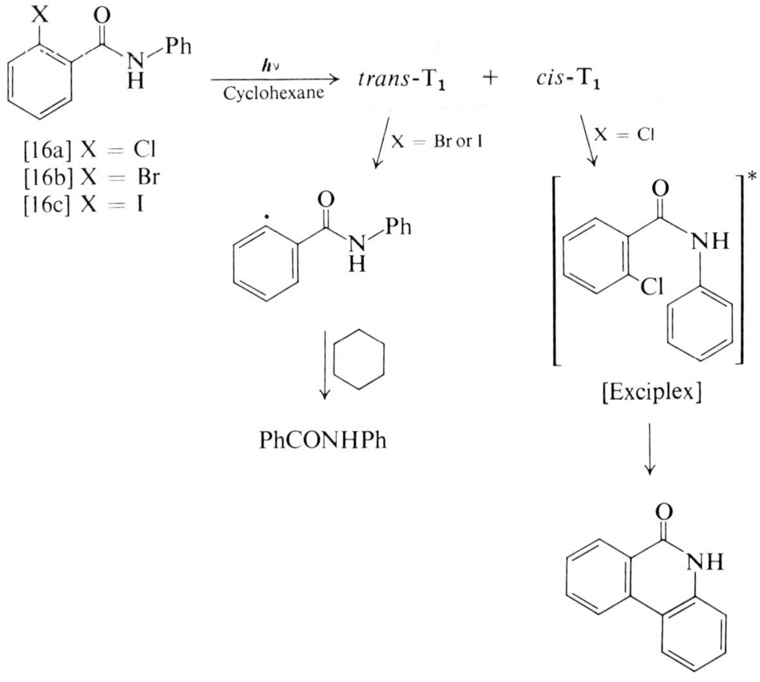

[16a] X = Cl
[16b] X = Br
[16c] X = I

$\xrightarrow{h\nu, \text{Cyclohexane}}$ *trans*-$T_1$ + *cis*-$T_1$

PhCONHPh

[Exciplex]

[16b,c] dehalogenate. To rationalize these observations it was suggested that interaction of the chlorine atom in the triplet state of the *cis*-conformer of [16a] with the other benzenoid group leads to an exciplex from which carbon—chlorine bond homolysis occurs with resultant cyclization. If N,N-diethylaniline is added to the reaction mixture only benzanilide is formed, suggesting that the greater electron-donating ability of the amine has negated the rather weak intramolecular complexing ability. For [16b] and [16c] the carbon—halogen bonds are of lower energy than the triplet state and therefore homolysis can occur in the triplet manifold of the *trans*-conformer, which is the most extensively populated ground-state conformer. For the chlorocompound [16a] the *trans*-conformer is unreactive since the carbon—chlorine bond

is of higher energy than the triplet state and the anilino group is not suitably situated spatially to assist dechlorination. Interestingly, the reactivity of [16a] is reduced to zero by the addition of polar solvents to the cyclohexane solution. This suggests that the initially created excited state has some charge-transfer character and that this can be solvated by polar solvents to such an extent that its energy is lowered below that of the intramolecular exciplex formed between the chlorophenyl and anilino groups. The excited state responsible for the reactivity of [16a–c] was assigned to the triplet state on the basis of sensitization and quenching studies. However, it is difficult to tell if some of the reactivity of [16a] is not due to the excited singlet state and whether dissociation of [16b,c] will not be so efficient as to compete with intersystem crossing.

A wide variety of substituted dichlorobenzenes has been shown to undergo dechlorination on irradiation in methanol solution (Mansour et al., 1980). Usually monochloro-compounds are the major products. Accurate quantum-yield measurements have been made and, unlike chlorobenzene, some of the reactions can be carried out by sensitization using either acetophenone or benzophenone. Quantum yields are temperature dependent and dechlorination does not occur below −70°C. This indicates that electronic excitation is insufficient to cause bond homolysis and that some energy from the surroundings (as heat) has to be supplied to overcome the activation energy barrier for homolysis.

1,4-Dichlorobenzene on irradiation in the presence of nitrogen oxides leads to the production of nitro-compounds with little evidence of dechlorination (Nojima and Kanno, 1980). This is rather surprising when one considers the relative ease with which chlorobenzenes dechlorinate.

The photochemistry of trichlorobenzenes has been examined (Åkermark et al., 1976; Choudhry et al., 1979). Irradiation of 1,2,4-trichlorobenzene [17] in either cyclohexane or propan-2-ol produces 1,3- and 1,4-dichlorobenzene; loss of a chlorine atom occurs so that the steric compression between the *ortho*-chlorine atoms is relieved. No trace of chlorine loss from the 4-position could be found. The relief of steric congestion must play some part in the energetics of the homolytic process and it is interesting to find that there is some evidence which supports the view that both the excited singlet and triplet states are reactive. Thus it is found that the ratio of the products formed in the reaction is sensitive to the presence of oxygen. In propan-2-ol the ratio of 1,3-dichlorobenzene to 1,4-dichlorobenzene changes from 0.65 in degassed solution to 0.5 in aerated solution. Furthermore the use of acetone as sensitizer leads to a ratio of 3 : 1. These results support the view that the triplet state is involved and, surprisingly, that the triplet state leads to a different product distribution to that from the singlet state. The question of the relative reactivity of excited singlet and triplet states is not

[Scheme showing 1,2,4-trichlorobenzene [17] → hv → 1,3-dichlorobenzene + 1,4-dichlorobenzene]

[Scheme showing 1,2,3-trichlorobenzene [18] → hv, Degassed MeOH, 22 h → 1,3-dichlorobenzene (72.3%) + 1,2-dichlorobenzene (13.3%) + chlorobenzene (8%)]

really understood and is the subject of current debate (Henne and Fischer, 1977). The sensitization and quenching experiments on [18] are in agreement but the problems associated with such experiments are highlighted by this example. The fact that the product distribution is unaffected by substrate concentration suggests that only triplet sensitization is occurring. Since the sensitizer also reacts with the solvent and therefore some is probably consumed during the irradiation period, it is difficult to make a quantitative appraisal of the results. In a related piece of work (Choudhry et al., 1979) it was shown that [18] and 1,3,5-trichlorobenzene dechlorinate on irradiation in methanol solution. The product distribution from [18] clearly shows that the relief in strain plays a major part in deciding which is the most favourable chlorine atom to lose. Sensitization experiments, using acetone, and quenching experiments, utilizing isoprene and oxygen, demonstrate that the triplet state of [18] is a reactive intermediate and that 1,3-dichlorobenzene is the favoured product from this species. The results indicate that formation of 1,3-dichlorobenzene is more favoured by reaction of the triplet than the excited singlet state. Why should this be so? The excited singlet state has approximately 15 kcal mol$^{-1}$ more energy than the triplet state and therefore should exhibit less regioselectivity. Since formation of 1,3-dichlorobenzene is the process leading to the greatest relief of strain, the process presumably has the lowest energy of activation and is consequently the more favoured pathway for the less energetic triplet state.

Measurements of the quantum yields for dechlorination of chloroanisoles and chlorotoluenes has been made (Mansour et al., 1979) (Table 2). If it is assumed that the quantum yields of formation of the excited singlet and triplet states are similar for all the compounds, it appears that chlorotoluenes are more reactive than chloroanisoles. This is what one would expect on the basis of the influence of OMe vs Me upon the polar contribution to the

TABLE 2

Quantum yields for dechlorination of chloroanisoles and chlorotoluenes in methanol[a]

| Starting compound | Products | Quantum yield |
|---|---|---|
| 2,6-Dichloroanisole | 2-Chloroanisole<br>Anisole | $5.24 \times 10^{-3}$ |
| 3,5-Dichloroanisole | 3-Chloroanisole<br>Anisole | $7.18 \times 10^{-2}$ |
| 2,4-Dichlorotoluene | 4-Chlorotoluene<br>Toluene | $7.59 \times 10^{-3}$ |
| 2,5-Dichlorotoluene | 3-Chlorotoluene<br>Toluene | $1.44 \times 10^{-1}$ |
| 3,4-Dichlorotoluene | 4-Chlorotoluene<br>Toluene | $1.10 \times 10^{-1}$ |

[a] Mansour et al., 1979

transition state for the homolytic reaction. It is, however, surprising to find that 3,5-dichloroanisole dechlorinates more efficiently than dichloroanisoles bearing a 2-chloro substituent. A similar situation arises with the chlorotoluenes. From the work on trichlorobenzenes (Åkermark et al., 1976; Choudhry et al., 1979) one would have expected that the chlorine atoms in the 2-position would be particularly prone to homolysis. Irradiation of 4-chloroanisole in methanol gives rise to both homolysis (producing anisole) and nucleophilic substitution (Soumillion and De Wolf, 1981). It has been suggested that the substitution reaction involves formation of a radical ion pair, the radical cation of which reacts with methanol to give 1,4-dimethoxybenzene.

The relevance of the strength of the carbon—halogen bond relative to the energy of the excited state undergoing reaction has been examined using compounds [19a–c] (Grimshaw and de Silva, 1980b). These compounds

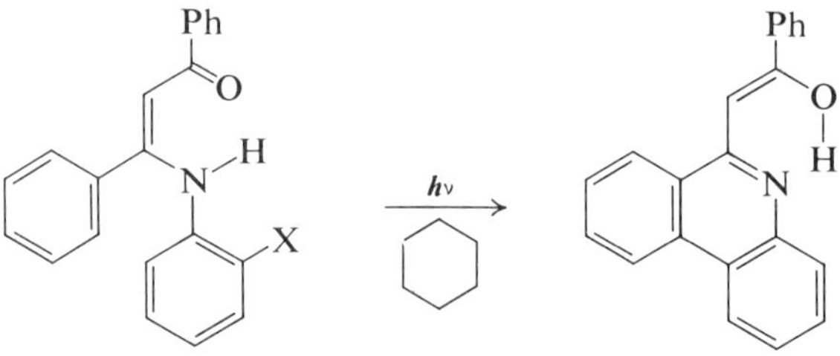

[19a] X = Cl
[19b] X = Br
[19c] X = I

undergo photocyclization on irradiation in cyclohexane and the quantum yields for this process show a strong wavelength dependence (Table 3). This wavelength effect suggests that dissociation is occurring from either a vibrationally excited first excited singlet state or an upper excited singlet state. The latter could take place if the energy entering the molecule *via* the $\sigma \to \sigma^*$ transition of the arylhalo group leads solely to dissociation rather than being channelled into the lower lying $\pi \to \pi^*$ transition.

TABLE 3

Quantum yields for the photocyclization of [19 a–c][a]

|  Irradiation wavelength/nm | Quantum yield ($\times 10^2$) | | |
| --- | --- | --- | --- |
| | X = Cl | X = Br | X = I |
| 245 | 1.2 | 110 | 460 |
| 315 | < 0.2 | 4 | 90 |
| 335 | < 0.4 | < 0.4 | 9 |
| 370 | < 0.1 | < 1 | < 4 |

[a] Grimshaw and de Silva, 1980b

CHLORINATED BIPHENYLS

The photophysics of these compounds is documented far better than that for the chlorobenzenes. Biphenyl itself is a twisted molecule due to the steric interaction between the hydrogen atoms at the 2,2'- and 6,6'-positions. This twisting leads to a decrease in conjugation. Excitation produces a non-planar singlet state which relaxes to an excited singlet state having a greater degree of planarity and a planar triplet state. Measurement of the quantum yields of phosphorescence and phosphorescence spectra gives data on the relaxed (planar) triplet state. This relaxed state may however not be a reactive state and one must use caution in using phosphorescence data to interpret chemical reactivity. Table 4 summarizes some of the known photophysical parameters of the chlorinated biphenyls and the quantum yields for dechlorination at room temperature (Bunce *et al.*, 1978; Ruzo *et al.*, 1974, 1975a). Inspection of Table 4 shows that for all the chlorinated biphenyls, intersystem crossing is efficient although they do have a measurable quantum yield of fluorescence. The photoreactive compounds contain a chlorine atom at the 2-position. This substituent will, because of its size, increase the energy of activation for the biphenyl system to achieve planarity. Consequently, its expulsion will lead to a far greater relief of strain than if the chlorine had been substituted at either a 3- or 4-position. Compounds containing a 2-chloro group have a shorter triplet lifetime (calculated from quenching data obtained from experiments using cyclohexa-1,3-diene as a triplet quencher) than those not having a chlorine substituent in an *ortho*-position, and this may be a manifestation of the importance of homolytic carbon—chlorine bond cleavage as a non-radiative decay route. Product studies (Ruzo *et al.*, 1975a; Ruzo and Zabik, 1975) show that it is the *o*-chloro substituents which are lost preferentially.

TABLE 4

Photophysical parameters and quantum yields for dechlorination ($\varphi_r$) of chlorinated biphenyls

| Positions of substitution | $\varphi_F{}^a$ | $\varphi_P{}^a$ | $\tau_T/10^{-8}$s | $\varphi_r$ | |
|---|---|---|---|---|---|
| 2 | 0.09 | 0.46 | | $0.39^a$ | |
| 3 | 0.07 | 0.27 | | $0.0011^a$ | |
| 4 | 0.07 | 0.60 | | $0.0006^a$ | |
| 2,4 | 0.02 | | | $0.62^a$ | |
| 2,5 | 0.06 | | | $0.13^a$ | |
| 2,21 | 0.005 | 0.95 | | $0.16^a$ | |
| 2,4,5 | 0.11 | | | $0.43^a$ | $0.05^c$ |
| 2,4,6 | 0.003 | | | $0.14^a$ | $0.02^c$ |
| 2,2′,4,4′ | 0.02 | 0.94 | 0.78 | $0.45^a$ | $0.10^b$ |
| 2,2′,3,3′ | | | 0.77 | $0.007^a$ | |
| 2,2′,5,5′ | 0.0007 | | 0.67 | $0.0053^a$ | $0.01^b$ |
| 2,2′,6,6′ | | | 0.70 | | $0.006^b$ |
| 3,3′,4,4′ | 0.08 | | 2.20 | $0.005^a$ | |
| 3,3′,5,5′ | 0.02 | 0.41 | 1.91 | $0.0003^a$ | $0.002^c$ |
| 2,2′,4,4′,6,6′ | 0.003 | ~1.0 | | | |

$^a$ Bunce *et al.*, 1978; $\varphi_r$ measured using 245 nm radiation with iso-octane as solvent
$^b$ Ruzo *et al.*, 1974; $\varphi_r$ measured using 300 nm radiation with cyclohexane as solvent
$^c$ Ruzo *et al.*, 1975a; $\varphi_r$ measured using 300 nm radiation with cyclohexane as solvent

The quenching studies (Bunce *et al.*, 1978; Ruzo *et al.*, 1974) utilizing cyclohexa-1,3-diene suggest that the triplet state is responsible for reaction. However, it has been pointed out (Bunce *et al.*, 1978) that the diene also quenches the excited singlet state of the chlorobiphenyls and it is therefore possible that quenching of both excited singlet and triplet states is occurring. If this had been the case one should have observed some curvature in Stern–Volmer plots and from the published data (Ruzo *et al.*, 1974) this does not seem to be the case. Recently it has been shown (Bunce and Gallacher, 1982) that the kinetics for the 4-chlorobiphenyl/diene system are not amenable to a simple kinetic analysis due to the fact that it is impossible to select conditions under which the whole of the incident radiation is absorbed by the biphenyl. Apparently the chloroaromatic reacts with the excited diene, and in the case of 1-chloronaphthalene, cycloadducts were isolated. As a consequence of these findings it is impossible to give a rigorous explanation for the effect of added dienes upon the photochemistry of chloroaromatic compounds. Attempts have been made to employ the heavy atom effect. This method relies upon the ability of an added compound containing a heavy atom (e.g. bromine or iodine) or a heavy atom itself (e.g. xenon) to enhance the rate of intersystem crossing between an excited singlet state and a triplet state. If this occurs, the triplet yield being thereby increased, and if the triplet lifetime is not appreciably shortened, then the yield of triplet product should be increased. The results (Bunce et al., 1978) are not totally convincing but do lend some credence to the view that the triplet state participates in the reaction. Sensitization studies (Bunce et al., 1978) provided very little convincing evidence.

From the fact that the quantum yields for dechlorination are often higher than the quantum yields of fluorescence (Table 4) it has been concluded that the triplet state is the reactive state. The quantum yields of reaction are greater than those of fluorescence only when there is an o-chloro substituent and the evidence presented suggests that the triplet states are involved in the reactions of these compounds. However, within the series of chlorobiphenyls containing an o-chloro substituent there is little correlation between quantum yields of reaction and of phosphorescence and this is not surprising when one considers that the quantum yield of phosphorescence is made up of many rate constants which include the rate constant for reaction. Given that the triplet state is the reactive species, the question arises as to how homolysis of a bond can occur in a compound whose triplet energy lies below the bond dissociation energy. What is not known with any certainty is the bond energy of the carbon—chlorine in bonds in *ortho*-substituted chlorobiphenyls. As previously pointed out the relief of steric strain caused by homolyzing such a bond may well bring the energy of activation for homolysis below the triplet energy of the biphenyl. Furthermore the triplet energy of a 2-chlorobiphenyl is higher than that of biphenyl itself by 8 kcal mol$^{-1}$ (Wagner and Scheve, 1977). The fact that the phosphorescence spectrum of 2-chlorobiphenyl (Wagner and Scheve, 1977) shows a broad structureless band whereas that of 4-chlorobiphenyl is structured and similar to that of biphenyl, shows that the o-chloro substituent does affect the geometry of both the ground and excited states. Thus the energetics for the loss of chlorine from the triplet state of an o-chlorobiphenyl may be favourable. For biphenyls bearing chlorine substituents at positions other than the *ortho* ones, the carbon—chlorine bond strength should be similar to the value given in

Table 1. Since the triplet energy of these compounds is around 64 kcal mol$^{-1}$ (Ruzo et al., 1975a) (slightly higher for the nonplanar triplets) the loss of chlorine is an endothermic process. This would account for the very low quantum yields for dehalogenation of the compounds, and it is possible that the reaction that does occur arises from the first excited singlet state. On irradiation in neutral or alkaline propan-2-ol solution, 2,4,6-trichlorobiphenyl undergoes preferential loss of an *ortho* chlorine (Nishiwaki et al., 1980). Polychlorinated biphenyls also undergo loss of chlorine upon irradiation in alkaline solutions of alcohols (Nishiwaki, 1981) and when they are adsorbed upon silica gel (Occhiucci and Patacchiola, 1982).

## CHLORINATED TERPHENYLS

Terphenyls, like biphenyls, are nonplanar compounds. Introduction of chlorine substituents decreases the quantum yield of fluorescence and presumably enhances intersystem crossing (Chittini et al., 1978). Irradiation of chlorinated terphenyls in cyclohexane and in methanol leads to dechlorination. Another significant reaction in methanol is replacement of the chlorine by a methoxyl group, that is to say, nucleophilic substitution is competing with dehalogenation. The quantum yields for dechlorination of compounds bearing an *o*-chloro substituent are considerably higher than those in which the substituent is in other positions. Evidence presented in favour of the reaction occurring from the triplet state included the finding that the presence of oxygen retards the reaction and the addition of xenon (a heavy atom quencher) enhances the quantum yield for reaction. Sensitization experiments using benzophenone gave inconclusive results. Thus there is some evidence for the participation of the triplet state in the dechlorination reaction although there is the possibility that some reaction does occur from the excited singlet state.

## 1-CHLORONAPHTHALENE AND RELATED COMPOUNDS

On irradiation in hydrogen-donating solvents such as methanol and cyclohexane, 1-chloronaphthalene gives naphthalene and binaphthyls with $\varphi$ (disappearance) in the range $2-20 \times 10^{-3}$ (Ruzo et al., 1975b). Photoinduced dechlorination of 2-chloronaphthalene and 1,2-dichloronaphthalene has also been reported. When methanol was used as a solvent formation of methoxynaphthalenes was also observed. Nucleophilic substitution increases in importance when solvents of poor hydrogen-donating ability (e.g. acetonitrile) containing a nucleophile such as water are used. When such solvents are used and nucleophiles are absent, the formation of binaphthyls is enhanced. The fact that the reactions are retarded by the presence of oxygen and that they can apparently be sensitized by benzophenone led to the initial suggestion that reaction occurs *via* the triplet state. However, in later papers (Bunce *et al.*, 1976, 1980) the reactivity was ascribed to the excited singlet state. If the value for the vinyl—chlorine bond strength from Table 1 is used, reaction from the excited singlet state appears to be endothermic. To overcome this difficulty it was suggested that reaction occurs *via* an excimer and spectroscopic evidence for such a species was obtained. The most convincing evidence for the participation of an excimer comes from the finding that the quantum yield for dechlorination of 1-chloronaphthalene increases as the concentration of the chloro-compound is increased. Furthermore, the addition of 4-chlorobiphenyl, a nonabsorbing additive, also increases the quantum yield of dechlorination (2).

$$1\text{-NpCl}^* + 4\,\text{Cl-Biph} \rightleftharpoons |1\text{-NpCl}/4\,\text{Cl-Biph}|^* \rightarrow \text{Np}\cdot + \text{Cl}\cdot + 4\,\text{Cl-Biph} \quad (2)$$
$$\text{Hetero-excimer}$$

The formation of binaphthyls from 1-chloronaphthalene is also indicative of excimeric intermediates, as indicated in (3). The breakdown of the excimeric

$$1\text{-NpCl}^* + \text{ArH} \rightarrow \text{Excimer} \rightarrow \text{Np}\cdot + \text{Cl}^- + \text{ArH}^{+\cdot}$$
$$\text{Np}\cdot + \text{R}\text{—}\text{H} \rightarrow \text{NpH} + \text{R}\cdot \quad (3)$$
$$\text{ArH}^{+\cdot} + \text{ArH} \rightarrow \text{Ar}\text{—}\text{Ar}$$
$$\text{ArH}^{+\cdot} + \text{CH}_3\text{OH} \rightarrow \text{Ar}\text{—}\text{OCH}_3$$

intermediates (formed between two molecules of the chloro-compound or between one molecule of chloro-compound and one molecule of dechlorinated aromatic hydrocarbon) to give radical cations explains the formation of methoxylated products when methanol is used as solvent.

That chloroaromatic hydrocarbons such as 1-chloronaphthalene can form excited complexes with aromatic hydrocarbons appears to offer an explanation for the observation that the addition of typical triplet quenchers such as cyclohexa-1,3-diene and piperylene in some cases accelerates the reaction rather than retards it and in other cases apparently leads to quenching efficiencies less than diffusion controlled (Ruzo and Bunce, 1975). Enhanced loss of chlorine can be explained if the dienes and the chloro-compound form a hetero-excimer which breaks up to give the radical cation of the diene, chloride ion and a naphthyl radical. However, use of high diene concentrations and a polar solvent alters the course of the reaction and cycloaddition products formed between the diene and the chloro-compound are produced (Bunce and Gallacher, 1982). It is also difficult to rationalize the finding that added electron acceptors such as biacetyl also increase the rate of dechlorination; for example, when the concentration of biacetyl in methanol is increased from $1.7 \times 10^{-3}$M to $5.10^{-2}$M a 2.27-fold increase in rate is observed.

The photodechlorination of polychlorinated naphthalene in methanol solution has been investigated (Ruzo et al., 1975c). Besides dechlorination, binaphthyl formation and to a lesser extent formation of methoxylated naphthalenes were observed. The formation of these products can be accounted for by the mechanisms given for 1-chloronaphthalene.

## 9,10-DICHLOROANTHRACENE

Irradiation of this compound in acetonitrile solution containing 2,5-dimethylhexa-2,4-diene or cyclohexa-1,3-diene leads to dechlorination and the formation of 9-chloroanthracene (Smothers et al., 1979). By way of contrast when benzene is used as solvent a cycloaddition reaction takes place to give

[20] (Yang et al., 1975). The formation of [20] has been rationalized in terms of an exciplex intermediate. In polar solvents, interaction of the excited anthracene with conjugated dienes and aromatic hydrocarbons may give rise to full electron transfer. The use of acetonitrile containing deuterium oxide as solvent and 2,5-dimethylhexa-2,4-diene as diene led to the formation of deuteriated 9-chloroanthracene [23]. Not only does the formation of this compound substantiate the view that interaction of the diene with the excited anthracene gives rise to extensive if not complete electron transfer but also demonstrates that the loss of chlorine is not from radical anion [21] but rather from the radical [22].

## REACTIONS OF OTHER CHLORINATED AND RELATED COMPOUNDS OF MECHANISTIC INTEREST

The photoinduced decomposition of di- and octachlorobenzofuran in methanol and hexane solutions has been investigated and not surprisingly dechlorination was observed (Hutzinger et al., 1973). Thus in the case of the octachloro compound, hepta-, hexa-, penta- and tetra-chlorobenzofurans were produced upon irradiation with light of wavelength 310 nm. Irradiation of halogenopyridines in benzene gives phenylpyridines (Terashima et al., 1981). The order of reactivity of the halogeno substituents is chlorine < bromine < iodine. Interestingly, the positional reactivity is dependent upon the halogen substituent. Thus for the bromo- and iodo-compounds the positional reactivity is 2 > 3 > 4 whereas for the chloropyridines it is 3 > 2 > 4. Pentachloropyridine undergoes dechlorination to give 2,3,4,6-tetra-chloropyridine on irradiation in ethers (Ager et al., 1972). The reason for this regiospecific loss of chlorine is not known. 4-Bromo- and 4-iodotetra-chloropyridines undergo exclusive loss of bromine and iodine respectively (Bratt et al., 1980) indicating homolysis of the weakest bond in the system.

Pentachloropyridine-N-oxide rearranges to the isocyanate [24]. Its formation was postulated as occurring *via* an intermediate oxaziridine.

A most remarkable reaction has been found to occur when chlorobenzene (Bryce-Smith *et al.*, 1980) and 3-chlorotetrafluoropyridines (Barlow *et al.*, 1979, 1980) are irradiated in the presence of alkenes such as cyclopentene. It was found that the alkene inserted into the carbon—halogen bond to give 1,2-addition products. Use of bromo- and iodobenzene led to the production of unstable 1,2-addition products. Available evidence suggests that aryl radicals are not involved. Thus in the reaction of 3-chlorotetrafluoropyridine with ethylene (Barlow *et al.*, 1979, 1980), telomerization of the ethylene was not observed. Fluorobenzenes, in which homolysis of the carbon—halogen bond does not occur on photolysis because of the high bond dissociation energy, also form 1,2-addition products with alkenes (Bryce-Smith *et al.*, 1980). It is possible that these reactions occur *via* an exciplex with the alkene acting as an electron donor. 2,3,4,5,6-Pentafluorobiphenyl [25] is formed in a related reaction which appears to take place when a mixture of benzene and hexafluorobenzene is irradiated (Bryce-Smith *et al.*, 1979). The yield of this

product is dependent upon the concentration of benzene, the presence of proton donors, and the polarity of the solvent. Increasing the polarity of the solvent leads to an increase in product yield which suggests that an exciplex intermediate is involved. However, use of deuteriated solvents, such as $CH_3OD$, does not lead to incorporation of deuterium in the product and therefore free radical ions cannot be involved. Quenching studies indicate that triplet hexafluorobenzene is the excited species responsible for reaction. In this reaction homolysis of the carbon—halogen does not appear to take place, but rather to occur from an intermediate *via* an ionic process. This is somewhat reminiscent of the electrocyclic reactions which produce compounds such as [3] and which give products by elimination of a molecule of hydrogen halide.

The reaction of tetrahalophthalonitriles with aryl ethers is a most interesting reaction because in some cases (e.g. 1,4-dimethoxybenzene) competition between dehalogenation and biaryl formation occurs (Al-Fakhri et al., 1980). The formation of these products is nicely rationalized in terms of the radical ion pair [26]. Thus the formation of the dehalogenated product appears to follow a similar mechanism to that postulated for the reaction of, for example, 1-chloronaphthalene (Bunce et al., 1976). Irradiation of tetrahalobenzonitriles (the halogen being either chlorine or fluorine) in the presence of ethers leads to [27] (Al-Fakhri and Pratt, 1976) and here again it seems necessary to invoke reaction via radical ions in order to explain the loss of fluorine as in the earlier example (Bryce-Smith et al., 1980). The tetrachloropyridazine [28] undergoes a series of rearrangements in preference to homolysis of the carbon—chlorine bond (Fox et al., 1982).

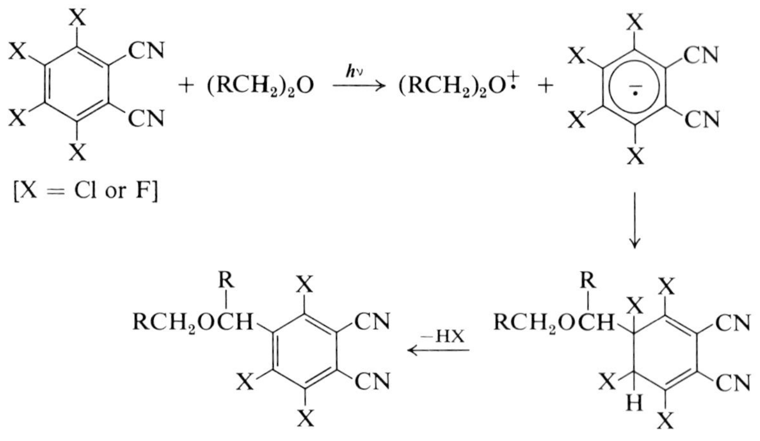

[27]

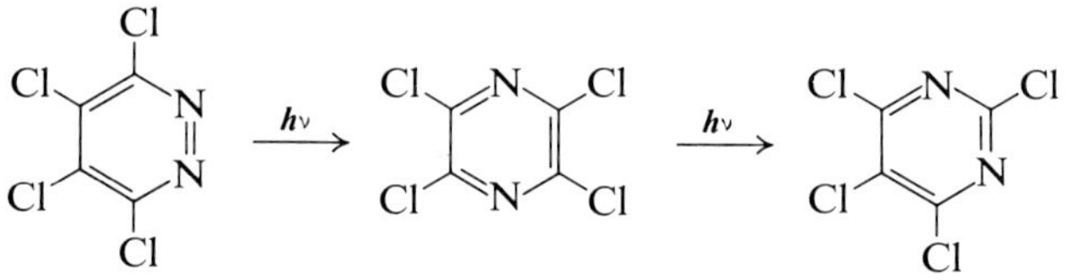

[28]

## 3  Bromo- and iodoaromatics

BROMO- AND IODOBENZENES

As can be seen from Table 1, the carbon—halogen bond strength in these compounds is substantially lower than the aryl—chlorine bond strength and falls below the energy of the excited singlet state for benzenoid compounds. Furthermore the $\sigma \to \sigma^*$ absorption bands of carbon—bromine bonds have a tail which extends beyond 254 nm. Thus irradiation at short wavelengths such as 254 nm can lead to reaction *via* direct population of the $\sigma \to \sigma^*$ transition as well as *via* the more classical route involving the $\pi \to \pi^*$ transition. Which of the two transitions is of lower energy in the singlet manifold has in most cases not been defined. Epr studies of the radical anions derived from simple halogenobenzenes (Andrieux *et al.*, 1979) show that the lowest anti-bonding orbital is $\sigma^*$. Introduction of electron-withdrawing groups having —R effects, such as cyano and nitro groups, reduces the energy of the $\pi^*$ orbital and brings it below the $\sigma^*$ orbital. It is therefore likely that direct irradiation of bromo- and iodobenzenes leads to cleavage in the excited singlet state, this process competing with intersystem crossing. In the case of iodo-compounds where the bond-dissociation energy is less than the energy of the triplet state, bond homolysis from this state is also likely.

By examination of the excitation spectrum of phosphorescence for a number of iodoaromatics it has proved possible to locate the $S_0 \rightarrow T_n$ absorption band (Marchetti and Kearns, 1967). That irradiation into this band leads to homolysis with comparable efficiency to excitation in the $S_0 \rightarrow S_n$ absorption bands suggest that the triplet state is primarily responsible for homolysis of these compounds.

There is also some evidence for the triplet state of 1,4-dibromobenzene being responsible for homolysis (Pedersen and Lohse, 1979). By means of the technique of flash photolysis excited 1,4-dibromobenzene was shown to react in benzene to give a transient which was identified as a phenylcyclohexadienyl radical, a species generated by attack of a 4-bromophenyl radical upon benzene. The addition of the classical triplet quencher penta-1,3-diene quenches the formation of the cyclohexadienyl radical. This suggests that the diene quenches the triplet state of the bromo-compound so reducing the amount of 4-bromophenyl radicals available for reaction.

Examination of the products of photolysis of a series of 4-bromophenoxyalkyl bromides (Davidson et al., 1980b) showed that the aryl—bromine bond is cleaved in preference to the weaker methylene—bromine bond. It was shown that the homolysis reaction competed with total success against energy transfer to the weaker bond. For this reason it was postulated that homolysis occurred from the singlet manifold. The correctness of this mechanism was verified by the finding that [29] underwent homolysis of the

ArCH$_2$CH$_2$—⟨C$_6$H$_4$⟩—Br        PhO(CH$_2$)$_n$Hal        PhO(CH$_2$)$_{n-2}$CH=CH$_2$

[29]                    [30] Hal = Br              [32]
(Ar = 1-Naphthyl)       [31] Hal = I

aryl—bromine bond even though a naphthalene group, capable of deactivating the excited singlet and triplet state of the bromophenyl group, is ideally situated within the molecule.

The competition that exists between excitation into the $\sigma \rightarrow \sigma^*$ transition of the carbon—iodine bond and the $\pi \rightarrow \pi^*$ transition of the aryl group when 254 nm light is used is illustrated by the reactions of [30] and [31] (Davey et al., 1982). In the case of [30], reaction to give [34] is very dependent upon the value of $n$, and, when $n > 6$, little if any reaction occurs. It was shown that when $n$ is as large as 6, energy transfer to the carbon—bromine bond becomes inefficient due to the inability of the methylene chain to coil so as to allow the bromine atom to approach close enough to the aryl group to allow energy transfer to take place. In contrast, the efficiency of the reactions of [31] seems to be far less dependent upon the value of $n$. Examination of the absorption spectra of [31] shows that a measurable percentage of the light absorbed at 254 nm enters the $\sigma \rightarrow \sigma^*$ transition of the carbon—iodine bond.

Since the aryl—iodine bond is so much weaker than the aryl—bromine bond (Table 1), one may expect that the quantum yields of products formed *via* photolytic cleavage of the halo-compounds should be the greatest for iodo-compounds. This is not found in practice. Irradiation of [33] gives [34] and the quantum yields for reaction of the iodo-compounds are lower than those of the bromo-compound (Grimshaw and de Silva, 1979, 1980c). Certainly the irradiation of iodo-compounds in neat solvents with which aryl radicals can react often leads to very high yields of products *via* this process (Sharma and Kharasch, 1968; Wolf and Kharasch 1961, 1965; Kharasch, *et al.*, 1966; Matsuura and Omura, 1966; Pinhey and Rigby, 1969), and these reactions usually occur more readily than those for the corresponding bromo-compounds. However, the quantum yield of product formation is not a true reflection of the efficiency of the photoinduced homolysis since the recombination of geminate radicals will reduce the efficiency of the product-forming reaction. In principle the extent to which cage recombination occurs can be probed by examining the effect of increasing the macroscopic solvent viscosity upon the reaction. In the case of [33], increasing the solvent viscosity caused a decrease in quantum yield of product formation (Grimshaw and de Silva, 1979, 1980c), e.g. for the iodo-compound in hexane ($\eta = 0.28$ cp) $\varphi = 0.35$, whereas in cyclohexane ($\eta = 0.80$ cp) $\varphi = 0.21$. Change in solvent polarity also affects the quantum yields but the fact that within a given type of solvent the quantum yield decreases as the viscosity is increased indicates that recombination of geminate aryl radicals and halogen atoms is an important energy-wasting process.

It has also been shown that the cyclization of compounds such as [33] can be triplet sensitized. The chloro-compound undergoes reaction and if this occurs from the triplet state, the process is endothermic. To account for reactivity of the compounds [33] it was suggested that the triplet state of the arylamine portion of the compound forms an excited complex with the haloarene group and it is this complex which undergoes reaction. A similar argument was used to explain the reactivity of [35] which contrasts with the stability of [36] (Grimshaw and de Silva, 1980d). In the complex formed by [35] the aryl—iodine bond can interact with the arylamino group. This is not the case in [36] and it is believed that this is the reason for the stability of this compound. It is interesting that compound [37] is also photostable. This stability was attributed to light-induced charge transfer occurring as with [35] but with back electron transfer effectively competing with aryl—bromine bond homolysis.

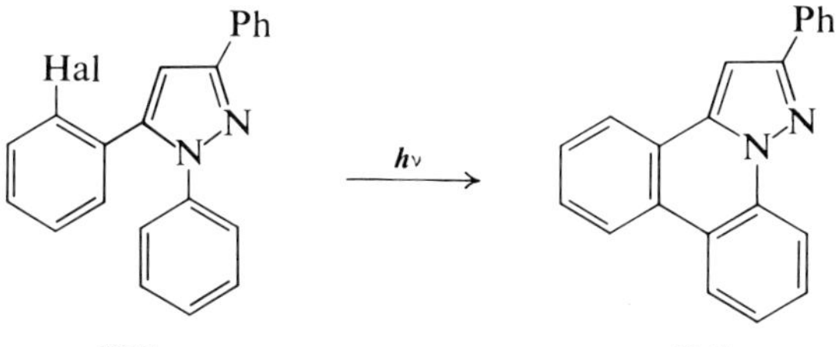

[33] → hν → [34]

[35] → hν → [34] + (pyrazoline with Ph, Ph, Ph)

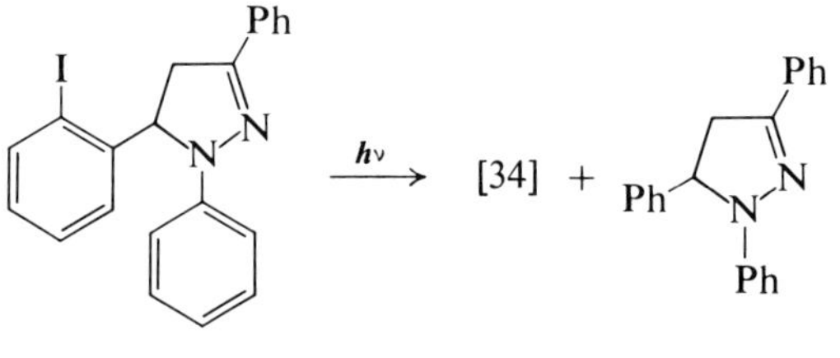

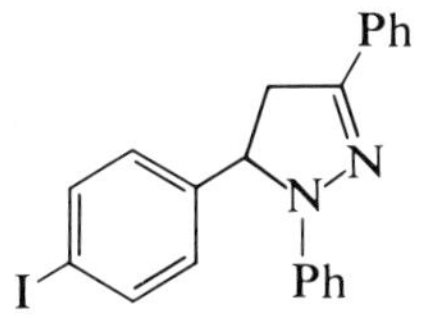

[36]

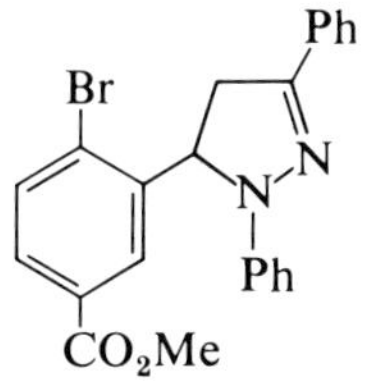

[37]

The reported reactions of the isomeric methyl bromobenzoates (Nikishin and Cheltsova, 1968) are of interest in that for these compounds there is a possibility of reaction from an n → π* excited state. Photolysis of the esters in benzene solution containing acetone gave low yields of methyl benzoate and high yields of phenylated methyl benzoates. Thus homolysis of the carbon—bromine bond does occur in these compounds. The rate of reaction was greatest for methyl-2-bromobenzoate. In this compound there is considerable steric interaction between the ester group and the bromine substituent. Expulsion of the bromine atom not only relieves the steric congestion but also allows better overlap of the π-system of the ester group with that of the benzene nucleus. As with the chlorobiphenyls (Ruzo et al., 1974, 1975a) the isomer which homolyzes leading to the greatest relief of strain is the most reactive. Surprisingly, the efficiency of the reaction is dependent upon the concentration of acetone. It is hard to tell whether the acetone is acting as a singlet or a triplet sensitizer. One would expect the reaction to occur predominantly from the singlet manifold and to be favoured by population of the $\pi \to \pi^*$ transition rather than the $n \to \pi^*$ transition since for a $\pi \to \pi^*$ transition, energy can be dissipated into vibrational modes one of which is the stretching of the carbon—bromine bond. Photolysis of methyl-4-bromobenzoate in methanol gives methyl benzoate (Davidson and Goodin 1982) but the rate of formation of the product is far from linear. The acceleration observed as the reaction proceeded was attributed to the accumulation of hydrogen bromide which may possibly hydrogen bond to the ester group so raising the energy of the $n \to \pi^*$ transition relative to the $\pi \to \pi^*$ transitions. It was shown that the photolysis of methyl-4-bromobenzoate in methanol containing hydrogen bromide is very much more efficient than in the absence of hydrogen bromide. Furthermore the production of methyl benzoate was found to be almost linear with time.

The photocyclization of [38] to give [39] is an interesting reaction (Paramasivam et al., 1979) because not only does the thiocarbamate group intramolecularly scavenge the aryl radicals in competition with the reaction of the

[Structures: compound [38] (dichloro-bromo aryl with NHCSCH₃) → hv, CH₃OH → compound [39] (chloro-benzothiazole with Me)]

radicals with solvent methanol but also reaction occurs solely via the homolytic cleavage of the aryl—bromine bond, the weakest carbon—halogen bond. Cleavage of aryl—bromine bonds in preference to aryl—chlorine bonds has also been shown to occur on photolysis of 4-bromo-4'-chlorodiphenylmethane in benzene (Robinson and Vernon, 1970).

Irradiation of 2-bromobenzoyldiphenylphosphine [40] gives products which indicate that there is competition between acyl—phosphorus bond cleavage and aryl—bromine bond homolysis (Dankowski et al., 1979). This gives some indication as to the lability of the acyl—phosphorus bond and in view of the use of acylphosphines and related compounds as photoinitiators for polymerization this would appear to warrant further attention. The

[Structures: compound [40] (2-bromo COPPh₂ benzene) → hv → CHO-Br benzene + CHO-Ph benzene + HOCHPPh₂(=O)-Br benzene]

photoaddition of bromobenzene to alkenes carrying electron-withdrawing groups has been accomplished with the aid of cuprous chloride as a catalyst (Mitani *et al.*, 1980). A similar type of reaction, but uncatalyzed, occurs upon irradiation of chlorobenzene (Bryce-Smith *et al.*, 1980) and 3-chlorotetrafluoropyridine (Barlow *et al.*, 1979, 1980) in the presence of alkenes. It has been reported that irradiation of bromobenzene in the presence of nitrogen oxides leads to the formation of a number of nitrophenols but as yet there is little information concerning the mechanism of this reaction (Nojima *et al.*, 1980).

### BROMO- AND IODOPOLYCYCLIC AROMATIC HYDROCARBONS

It was noted earlier that the addition of compounds such as cyclohexa-1,3-diene and biacetyl accelerated the dechlorination of chloronaphthalene and this was attributed to the operation of a charge-transfer mechanism (Bunce *et al.*, 1976; Ruzo and Bunce, 1975). The presence of such compounds in the reactions of 1-bromonaphthalene and 4-iodobiphenyl did not cause any acceleration and in some cases caused a slight retardation. The observation that carbonyl compounds quench the excited singlet state of aromatic hydrocarbons by an exciplex mechanism makes the use of 1,2-dicarbonyl compounds as triplet quenchers open to question (Busch *et al.*, 1977). The lack of acceleration of dehalogenation of bromo- and iodo-compounds by added dienes etc., suggests that the rate of dehalogenation is very high and this is no doubt a reflection of the relatively weak carbon—halogen bond strengths.

The question of whether the dehalogenation of 1-iodonaphthalene, 4-iodobiphenyl and 1-bromonaphthalene occurs from the singlet or triplet manifold has been probed by examining the effect of added lithium halides upon the reactions with methanol as solvent (Ruzo *et al.*, 1977). In all cases the efficiency of retardation by the ions was $I^- > Br^- > Cl^-$. Halide ions are known to quench the excited singlet states of aromatic hydrocarbons (Beer *et al.*, 1970; Davidson and Lewis, 1973; Watkins, 1973). The efficiency of quenching is not related to the heavy atom effect but rather to the ease of oxidation of the anions. The quenching does lead to triplet production (Watkins, 1973) and therefore the observed retardation may indicate that the added ions quench the excited singlet state and that reaction occurs from this state. However, it is well known that heavy atom effects can increase the efficiency of intersystem crossing between $T \leadsto S_0$. It is known that halide ions do accelerate this process (Najbar and Rodakiewicz-Nowak, 1978), and therefore the quenching effect of the ions may indicate that reaction is occurring from the triplet state. This example illustrates the problems of using external heavy atom effects as a diagnostic tool for investigating reaction mechanisms. In many cases the relative efficiency of $S_1 \leadsto T_1$, and

T $\rightsquigarrow$ $S_0$ are largely unknown. For some ω-naphthylalkyl halides it has been found that the presence of the halogeno groups leads to an increase in triplet yield (Davidson et al., 1980a) but has little effect upon triplet lifetime. Grieser and Thomas (1980) have used a much more direct method for determining the lability of triplet bromo- and iodonaphthalenes and also biphenyls. They studied the triplet lifetime and yield of these compounds as a function of temperature and in this way were able to obtain energies of activation for homolysis from the triplet state. For 1- and 2-iodonaphthalenes a value of ∼5 kcal mol$^{-1}$ was obtained and for 4-bromobiphenyl a value of 16 kcal mol$^{-1}$. The energy of activation for homolysis of triplet 1-bromonaphthalene could not be obtained due to the high temperatures that are required to assist the homolysis of this species. This particularly simple but elegant way of studying triplet reactivity opens up possibilities for assessing the reality of the involvement of triplet states in the dehalogenation of haloaromatics. By use of nanosecond flash photolysis (Pineault et al., 1981) it was shown that internal conversion and intersystem crossing compete with homolysis of the excited singlet state of 2-iodoanthracene. Little homolysis occurs from the lowest triplet state of this compound. Similarly homolysis occurs from upper triplet states of 9-iodoanthracene (produced by intersystem crossing) but not from the lowest triplet state which has insufficient energy.

The debromination of monobrominated and dibrominated biphenyls has been examined (Bunce et al., 1975) and it is not surprising to find that the 2-bromo-compound is more reactive than the 4-bromo-compound. On irradiation 2,5-dibromobiphenyl loses the bromine which gives the largest release of steric strain to give 3-bromobiphenyl.

Irradiation of 9-bromophenanthrene in hexane solution gives [41] (Weiss et al., 1971). The formation of this product may be rationalized as occurring

[41]

by attack of 9-phenanthryl radicals upon the starting material. However, the reaction was carried out in hexane solution and one would expect the 9-phenanthryl radicals to abstract hydrogen from the solvent. The absence of this reaction may be due to the production of the phenanthryl radicals from an excimer composed of two molecules of the bromo-compound.

## 4 Assisted dehalogenation of halogenoaromatics

AMINE- AND SULPHIDE-ASSISTED DEHALOGENATION OF HALOGENOAROMATICS

There is an extensive literature concerned with the formation of excited complexes between aromatic hydrocarbons and tertiary amines (Davidson, 1975, 1983; Gordon and Ware, 1975; Mataga and Ottolenghi, 1979). In many cases the reduction potentials of haloaromatic hydrocarbons are similar to the parent hydrocarbons (Andrieux et al., 1980) and therefore the excited singlet state of these compounds may well form excited complexes or radical ions with electron donors. The presence of a halogen atom may affect the probability of such reactions occurring by enhancing intersystem crossing to such a degree that the excited singlet state has a very short life-time and therefore little chance of participating in a bimolecular reaction. If the triplet state does not react with the amine, the addition of amine will have no effect. Another possibility is that the photoinduced homolysis has a very large rate constant thus attenuating the singlet life-time, and consequently reaction with an amine will be inefficient. This appears to be the case for bromobenzene and iodobenzene and related compounds since the dehalogenation reactions are unaffected by the presence of tertiary amines (Davidson and Goodin, 1981). 4-Chlorobenzonitrile, which does not dechlorinate on irradiation in methanol, does dechlorinate when triethylamine is added; the inefficiency of the homolytic reaction allows the bimolecular reaction with the amine to compete. The dechlorination of many chlorinated biphenyls is accelerated by the addition of triethylamine (Bunce et al., 1978; Ruzo et al., 1975a) and it is those compounds with a low quantum yield for dechlorination which show the most marked acceleration. Other compounds which undergo triethylamine-assisted dehalogenation include chlorinated terphenyls (Chittini et al., 1978), where again only the relatively light stable compounds showed the enhancement, 1-chloronaphthalene (Ruzo et al., 1975b; Bunce et al., 1976) and bromobiphenyls (Bunce et al., 1975). In the latter reactions, the addition of triethylamine leads to a much cleaner reaction and, if the quantum yield of biphenyl formation is being measured, the apparent acceleration by the amine is due to the suppression of side reactions.

So far terms such as "amine-assisted" and "acceleration of reaction" have been used and taken to indicate that some form of complex is formed between the haloaromatic and the amine. Fluorescence from complexes formed between 1-chloronaphthalene and triethylamine has been observed (Bunce et al., 1976). In many cases the fluorescence of haloaromatics (Bunce et al., 1976, Ohashi et al., 1973) and haloheterocycles (Nasielski et al., 1972) is quenched by triethylamine and, since this cannot be a classic energy-transfer process, it is reasonable to assume that the quenching is due to excited complex formation. Thus the observation that the use of highly polar solvents increases the accelerating effect of added amines is often taken as evidence for the participation of complexes or radical ions in these reactions as shown in (4) (Bunce et al., 1976; Davidson and Goodin, 1981; Ohashi et al., 1973). In order to gain more substantial evidence for the participation of

$$ArHal\ (S_1) + Et_3N \rightarrow (ArHal)^{\cdot -} + Et_3N^{\cdot +}$$

$$ArHal^{\cdot -} \rightarrow Ar\cdot + Hal^{-} \tag{4}$$

$$Ar\cdot \xrightarrow[\text{or Amine}]{\text{Solvent}} ArH$$

radical ions, various dehalogenation reactions have been carried out using a mixture of acetonitrile and deuterium oxide as solvent. If the radical ions (e.g. [42]) have a reasonable life-time, and some of these have been determined (Andrieux et al., 1980), then they may be deuteriated which can lead to deuterium incorporation into the product. In many cases deuterium incorporation

has been observed e.g. 1-chloro and 1-bromonaphthalene, 4-chlorobiphenyl and 9-chloroanthracene (Davidson and Goodin, 1981; Bunce, 1982). Somewhat surprisingly, deuterium incorporation into benzonitrile produced from 4-chlorobenzonitrile was not observed even though the addition of triethylamine is essential for reaction to occur (Davidson and Goodin, 1981). In none of the reactions studied was deuterium incorporation into the starting material observed. Thus in these cases, once the radical ion of the halocarbon has been formed, it does not revert to starting material. This contrasts with the finding that N,N-dimethyl-4-chlorophenylpropylamine undergoes dechlorination more slowly than 4-chlorophenylpropane (Bunce and Ravanal, 1977) which suggests that the intramolecular excited complex formed between the amine and the chlorophenyl group undergoes efficient back electron transfer. Because of the efficiency of this type of reaction it is often very difficult to obtain spectroscopic evidence for intramolecular exciplexes and radical ions.

Since the formation of excited complexes between aromatic hydrocarbons and amines appears to involve most commonly the excited singlet rather than the triplet state of the hydrocarbon, it has been tacitly assumed that the amine-assisted dechlorination reactions occur *via* the singlet state of the chloroaromatic. Bunce (1982) has made a careful kinetic study of these reactions and reaches the conclusion that the excited singlet complex formation is an energy-wasting process and that it is the triplet complex which leads to reaction. Flash photolysis studies have now confirmed that the triplet states of chloroaromatic hydrocarbons are quenched by tertiary amines in acetonitrile (Beecroft *et al.*, 1983).

Addition of triethylamine accelerates the dechlorination of 4-chlorobiphenyl (Bunce *et al.*, 1978; Tsujimoto *et al.*, 1975), but surprisingly 3-chlorobiphenyl is reduced to give 1,4-dihydro-3-chlorobiphenyl. This reaction is without precedent, although it is known that many polycyclic aromatic hydrocarbons are photoreduced by tertiary amines (Davidson, 1975).

The debromination of bromopyrimidines is reported to be accelerated by the addition of secondary aliphatic amines as well as tertiary amines (Nasielski *et al.*, 1973; Parkányi, 1981) and radical ions have been postulated as intermediates. The higher oxidation potential of secondary amines compared with tertiary amines makes them less likely to participate in electron-transfer reactions. In the case of the reaction question, the ease of reduction of pyrimidines may help the thermodynamics of the process to be balanced.

The fluorescence of aromatic amines such as N,N-dimethylaniline is quenched by haloaromatics such as chloro- and bromobenzene and exciplex fluorescence has been observed (Tosa *et al.*, 1969; Pac *et al.*, 1972; Grodowski and Latowski, 1974; Bunce and Gallacher, 1982). Irradiation of the amine in the presence of halobenzenes produces benzene, biphenyl, phenylated N,N-dimethylanilines and N-methylaniline. These products are formed as a result of the reactions generating aryl radicals.

Sulphides quench the fluorescence of aromatic hydrocarbons and this is attributed to the occurrence of a charge-transfer rather than an energy-transfer process. Not surprisingly sulphides such as diethyl sulphide accelerate the dehalogenation of haloaromatic hydrocarbons (Davidson *et al.*, 1982). Furthermore, if the reactions are carried out in the presence of deuterium oxide, deuterium is incorporated into products thus indicating that either radical ions or exciplexes are intermediates.

DEHALOGENATION OF HALOAROMATIC COMPOUNDS ASSISTED BY SODIUM BOROHYDRIDE

The addition of sodium borohydride has been shown to accelerate the photo-induced dechlorination of 3-chloro- and 4-chlorobiphenyl (Tsujimoto *et al.*, 1975). Available evidence suggests that sodium borohydride photoreduces excited 1-cyanonaphthalene by hydride attack (Barltrop and Owers, 1972). Presumably chlorinated aromatic hydrocarbons are reduced by a similar mechanism.

## 5 Photoinduced nucleophilic substitution haloaromatics

HALOAROMATICS

This vast area has been recently reviewed (Havinga *et al.*, 1975; Havinga and Cornelisse, 1976). There are many examples of the displacement of the halo group by the hydroxide ion and amines (Nijhoff and Havinga, 1965; Joschek and Miller, 1966; Omura and Matsuura, 1971) although in some of these reactions radical processes play a part, e.g. formation of [44]–[46] from [43] (Omura and Matsuura, 1971).

Solvent polarity also seems to play a part in determining the extent of radical *vs* ionic reaction. Irradiation of aqueous solutions of 3,4-dichloro-aniline produces 2-chloro-5-aminophenol and it has been suggested that the dichloro-compound undergoes heterolysis on irradiation to give a phenyl cation (Miller *et al.*, 1979). By way of contrast, irradiation of the dichloroaniline in methanol generates 3-chloroaniline as the exclusive product; thus not only does homolysis occur but reaction takes place at a different position in the nucleus. The fact that methanol is a good hydrogen-atom donating solvent, whereas water is very poor in this respect, may also be an important factor. Thus it was noted with 1-chloronaphthalenes that the use of polar solvents that are poor hydrogen atom donors favours nucleophilic substitution (Ruzo *et al.*, 1975b). 4-Chloroanisole is also photosolvolyzed on irradiation in methanol (Soumillion and De Wolf, 1981) but in this case the solvolysis was envisaged as involving the radical cation of the chlorocompound rather than an aryl cation.

A variety of halo-9,10-anthraquinones undergo photosolvolysis to give products which are of use in the synthesis of dyes. Irradiation of 1-chloro-9,10-anthraquinone in concentrated sulphuric acid produces 1-hydroxy-9,10-anthraquinone (Seguchi and Ikeyama, 1980). The 2-chloroquinone is unreactive. However, the 2-chloro-compound does react with pyridine (Loskutov *et al.*, 1981). The mechanism of these reactions is far from clear. A mechanistic study has been made of the aminolysis, by primary amines, of 1-bromo-4-acetylamino-9,10-anthraquinone to give 1,4-diamino-9-10-anthraquinones (Tajima *et al.*, 1979). A triplet charge-transfer state of the quinone was shown to be responsible for reaction. 1-Bromoquinones which do not possess a charge-transfer triplet state as the lowest triplet state are unreactive. Primary amines react with 1-amino-4-bromo-9,10-anthraquinone-2-sulphonate to give a 1,4-diamino-quinone (Inoue *et al.*, 1973, 1975). The reaction is only successful if oxygen is present. This led to the suggestion that the product is formed by reaction of the amine with a "quinone-oxygen" excited complex. There is the possibility that the function of oxygen is to suppress photoreduction of the quinone. This particular quinone does not photoreact with aromatic amines. However, the case is different for 1-amino-2,4-dibromo-9,10-anthraquinone which reacts with both aliphatic and aromatic amines to give 1,4-diamino-2-bromoquinones (Inoue and Hida, 1978; Inoue *et al.*, 1980). The reaction occurs in both the presence and absence of oxygen. In the absence of oxygen, reaction is proposed as occurring *via* the triplet charge-transfer state and in the presence of oxygen by reaction of the amine with the excited singlet and triplet complexes of the quinone with oxygen.

Many bromoquinolines and isoquinolines, but not their chloro counterparts, dehalogenate on irradiation in methanol solution containing hydroxide ions (Parkányi and Lee, 1974). The rate of reduction is dependent upon hydroxide ion concentrations but surprisingly the use of a mixture of $CH_3OD/D_2O$ as solvent did not lead to any deuterium incorporation into the product. It was suggested that the hydroxide ion acts as an electron donor so producing the bromoquinoline radical ion which dehalogenates. Hydroxide ions acting as electron donors to excited states is not unknown (Phillips et al., 1969), but it is strange that, if this process does occur, chloroquinolines are unreactive since their reduction potentials are likely to be similar to bromoquinolines.

REACTIONS INVOLVING VINYL CATIONS

Since the photosolvolysis of a number of haloaromatics appears to involve phenyl cations it is instructive to consider the evidence for the formation of vinyl cations in analogous reactions. McNeeley and Kropp (1976) have provided convincing evidence for the formation of vinyl cations *via* the triplet sensitized heterolysis of vinyl halides. A number of arylalkenes (e.g. [47]) give products upon irradiation which can be conveniently rationalized

$$\underset{\underset{Ar}{\overset{Ar}{\diagdown}}}{\overset{}{C}}=\underset{\underset{Hal}{\diagup}}{\overset{CH_3}{\diagup}}C \quad \xrightarrow[MeOH]{h\nu} \quad \underset{\underset{Ar}{\overset{Ar}{\diagdown}}}{\overset{}{C}}=\underset{}{\overset{CH_3}{\diagup}}\dot{C} + Hal\cdot \quad \xrightarrow{MeOH} \quad \underset{\underset{Ar}{\overset{Ar}{\diagdown}}}{\overset{}{C}}=\underset{\underset{H}{\diagup}}{\overset{CH_3}{\diagup}}C$$

[47]

$$\underset{\underset{Ar}{\overset{Ar}{\diagdown}}}{\overset{}{\phantom{C}}}C=CH_2 \quad \longleftarrow \quad \underset{\underset{Ar}{\overset{Ar}{\diagdown}}}{\overset{}{C}}=\overset{+}{C}-CH_3 + Hal^- \quad \longrightarrow \quad Ar-\overset{+}{C}=\underset{\underset{Ar}{\diagdown}}{\overset{CH_3}{\diagup}}C$$

$$\downarrow MeOH$$

$$\underset{\underset{MeO}{\diagup}}{\overset{Ar}{\diagdown}}C=\underset{\underset{Ar}{\diagdown}}{\overset{CH_3}{\diagup}}C$$

as arising *via* vinyl cations (Kitamura et al., 1982). The formation of ionic products is favoured by the presence of electron-donating groups in the aryl moieties. The particular mechanism shown for product formation from [47] illustrates the dilemma with regard to the mechanism involving vinyl cations. The cations can be viewed as arising directly from photoinduced heterolysis or as involving electron transfer between geminate radicals produced by photoinduced homolysis. As yet there is no clear indication for a unique pathway. Compelling evidence for the intermediacy of ionic species comes from laser flash photoconductivity experiments with [48] (Schnabel et al., 1980). The closely-related compound [49] undergoes photocyclization to give [50] and this again can best be rationalized as involving vinyl cations (Suzuki *et al.*, 1981). It appears that the formation of [51] and [52] may well represent one of the few cases of trapping aryl cations by alkenes (Maruyama *et al.*, 1980), and this suggests that it may be worth exploring the reactions with alkenes of those compounds which readily undergo solvolysis.

## $S_{RN}1$ REACTIONS

In the earlier section on the amine- and sulphide-assisted dehalogenation of haloaromatics it was suggested that the radical anion of the haloaromatic is formed and that this cleaves to give an aryl radical and a halide ion. Many anions can act as electron donors for the excited states of haloaromatics and this can set up a chain reaction, as shown in (5). This type of reaction is

$$X^- + (ArHal)^* \to X\cdot + (ArHal)^{\bar{\cdot}}$$
$$(ArHal)^{\bar{\cdot}} \to Ar\cdot + Hal^- \quad (5)$$
$$Ar\cdot + X^- \to (ArX)^{\bar{\cdot}}$$
$$(ArX)^{\bar{\cdot}} + (ArHal)^* \to ArX + (ArHal)^{\bar{\cdot}}$$

known as a substitution (S) – radical (R) – nucleophilic (N) reaction, i.e. $S_{RN}1$. Much of the earlier work has been reviewed by Bunnett (1978). A variety of nucleophiles can be used, e.g. enolates, thiolate, arsenate, etc. If the haloaromatic contains more than one type of halogen atom then not surprisingly the most easily lost halogen is iodide followed by bromide with chloride being quite the most stable. Many of the reactions have proved of synthetic value, for example, in the synthesis of isoquinolines (Beugelmans et al., 1982).

Recently an important point of mechanistic interest has been discussed, relating to reaction (6). In some cases it has been found that the anion $(ArX)^{\bar{\cdot}}$

$$(ArX)^{\bar{\cdot}} + (ArHal)^* \to ArX + (ArHal)^{\bar{\cdot}} \quad (6)$$

can undergo fragmentation in competition with the electron-transfer process; thus in the reaction with thiolate ions [53] there is competition between

formation of the arenethiolate ion [54] and the sulphide [55] (Rossi and Palacios, 1981). Which route takes preference is determined by the nature of R. Thus if R forms a stable radical (benzyl or t-butyl) formation of the arenethiolate ion [54] takes preference. The reactions of haloaromatics with phosphide, arsenide and stibide anions have also been examined (Alonso and

Rossi, 1982). It was found that in accordance with the bond strengths which are in the order aryl—phosphorus > aryl—arsenic > aryl—antimony, the likelihood of the abnormal reaction taking preference increased on going from phosphide to stibide anions. Thus, in the reaction (7) of diphenylarsenide anions with an aryl halide, a variety of mixed triarylarsenides were formed. In general, if the σ* molecular orbital of the aryl—metal bond is of lower energy than that of the radical anion, then the abnormal reaction occurs.

$$ArX + Ph_2As^- \rightarrow Ph_3As + Ph_2ArAs + PhAr_2As + Ar_3As \qquad (7)$$

REACTIONS OF VINYLIC HALIDES WITH AROMATIC COMPOUNDS

A number of aromatic and heterocyclic compounds have been found to react with vinylic halides on irradiation; for example methoxynaphthalene reacts with the uracil [56] (Matsuura et al., 1980). It has been proposed that the

bromouracil [56] reacts with the excited aromatic hydrocarbon to give a charge-transfer complex as indicated in (8). Other hydrocarbons, e.g. pyrene,

$$ArH^* + BrU \rightarrow [ArH^{\overset{+}{\cdot}} \; BrU^{\overset{-}{\cdot}}]^*$$
$$[ArH^{\overset{+}{\cdot}} \; BrU^{\overset{-}{\cdot}}]^* \rightarrow Br^- + U\cdot + ArH^{\overset{+}{\cdot}} \qquad (8)$$
$$U\cdot + ArH^{\overset{+}{\cdot}} \rightarrow Ar\text{—}U + H^+$$

react with 5-halouridines to give adducts such as [57] (Saito et al., 1980). Such adducts are highly fluorescent and are potentially useful fluorescence probes. Uracils and uridines also react with tryptophan in the presence of a triplet sensitizer such as acetone. In this way, products such as [58] and [59] can be obtained (Ito et al., 1980). It was suggested (Matsuura et al., 1980) that these reactions occur *via* the triplet state of the uracil and uridine. The mechanism of the reaction is dependent upon the type of halouracil or uridine

used. Use of iodo-compounds leads to reaction *via* homolysis of the carbon—iodine bond followed by attack of the uracyl and uridyl radicals so produced upon the tryptophan. The bromo-compounds react by forming a charge-transfer complex with the tryptophan whereas the chloro-compounds react *via* a heterolytic process. The electron transfer can be sensitized by incorporating electron donors such as 2-methoxynaphthalene into the reaction (Ito et al., 1979) giving the reaction sequence (9).

$(BrU)^*T_1 + NpOMe \rightarrow BrU^{\cdot -} + NpOMe^{\cdot +}$

$NpOMe^{\cdot +} + Try \rightarrow Try^{\cdot +} + NpOMe$  (9)

$Try^{\cdot +} + BrU^{\cdot -} \rightarrow Products$

NpOME = 1-Methoxynaphthalene; Try = Tryptophan

Dihalogenomaleimides (e.g. [60]) react with indoles (Matsuo et al., 1976; Wald et al., 1980) and with furans and thiophens (Wamhoff and Hupe, 1978). These reactions are thought to involve excited charge-transfer complexes.

## References

Ager, E., Chivers, G. E. and Suschitzky, H. (1972). *J. Chem. Soc. Chem. Commun.* 505

Åkermark B., Baeckström P., Westlin U. E., Göthe, R. and Wachtmesiter C. (1976). *Acta. Chem. Scand.* **B30**, 49

Al-Fakhri, K. A. K. and Pratt, A. C. (1976). *J. Chem. Soc. Chem. Commun.* 484

Al-Fakhri, K. A. K., Mowatt, A. C. and Pratt, A. C. (1980). *J. Chem. Soc. Chem. Commun.* 566

Alonso, R. A. and Rossi, R. A. (1982). *J. Org. Chem.*, **47**, 77

Andrieux, C. P., Blocman, C., Dumas-Bouchiat, J. M. and Savéant, J. M. (1979). *J. Am. Chem. Soc.* **101**, 3431

Andrieux, C. P., Blocman, C., Dumas-Bouchiat, J. M., M'Halla, F. and Savéant, J. M. (1980). *J. Am. Chem. Soc.* **102**, 3806

Augustyniak, W. (1980). *Ser. Chem.*, **39**, 72

Barlow, M. G., Haszeldine, R. N. and Langridge, J. R. (1979). *J. Chem. Soc. Chem. Commun.* 608

Barlow, M. G., Haszeldine, R. N. and Langridge, J. R. (1980). *J. Chem. Soc. Perkin Trans. 1*, 2520

Barltrop, J. A. and Owers, R. J. (1972). *J. Chem. Soc. Chem. Commun.* 592

Beecroft, R. A., Davidson, R. S. and Goodwin, D. C. (1983). *Tetrahedron Lett.* 5673

Beer, R., Davis, K. M. C. and Hodgson, R. (1970). *J. Chem. Soc. Chem. Commun.* 840
Begley, W. J. and Grimshaw, J. (1977). *J. Chem. Soc. Perkin Trans. 1*, 2324
Beugelmans, R., Chastanet, J. and Roussi, G. (1982). *Tetrahedron Lett.* 2313
Bratt, J., Iddon, B., Mack, A. G., Suschitzky, H., Taylor, J. A. and Wakefield, B. J. (1980). *J. Chem. Soc. Perkin Trans. 1*, 648
Bryce-Smith, D., Gilbert, A. and Twitchett, P. (1979). *J. Chem. Soc. Perkin Trans. 1*, 558
Bryce-Smith, D., Dadson, W. M. and Gilbert A. (1980). *J. Chem. Soc. Chem. Commun.* 112
Bunce, N. J. (1982). *J. Org. Chem.* **47**, 1948
Bunce, N. J. and Gallacher, J. C. (1982). *J. Org. Chem.* **47**, 1955
Bunce, N. J. and Ravanal, L. (1977). *J. Am. Chem. Soc.* **99**, 4150
Bunce, N. J., Safe, S., and Ruzo, L. O. (1975). *J. Chem. Soc. Perkin Trans. 1*, 1607
Bunce, N. J., Pilon, P., Ruzo, L. O. and Sturch, D. J. (1976). *J. Org. Chem.* **41**, 3023
Bunce, N. J., Kumar, Y., Ravanal, L. and Safe, S. (1978). *J. Chem. Soc. Perkin Trans. 2*, 880
Bunce, N. J., Bergsma, J. P., Bergsma, M. D., De Graaf, W., Kumar, Y. and Ravanal, L. (1980). *J. Org. Chem.* **45**, 3708
Bunnett, J. F. (1978). *Acc. Chem. Res.* **11**, 413
Busch, D., Dahm, L. L., Suvicke, B. and Ricci, R. W. (1977). *Tetrahedron Lett.* 4489
Chittim, B., Safe, S., Bunce, N. J., Ruzo, L., Olie, K. and Hutzinger, O. (1978). *Can. J. Chem.* **56**, 1253
Choudhry, G. G., Roof, A. A. M. and Hutzinger, O. (1979). *Tetrahedron Lett.* 2059
Dankowski, M., Praefcke, K., Nyburg, S. C. and Wong-Ng, W. (1979). *Phosphorus and Sulfur* **7**, 275
Davey, L., Davidson, R. S., Goodin, J. W. and Taylor, C. (1982). Unpublished results
Davidson, R. S. (1975). *In* "Molecular Association" (R. Foster, ed.) Vol. I. Academic Press, London and New York
Davidson, R. S. (1983). *Adv. Phys. Org. Chem.* **19**, 1
Davidson, R. S. and Goodin, J. W. (1981). *Tetrahedron Lett.* 163
Davidson, R. S. and Goodin, J. W. (1982). Unpublished results
Davidson, R. S. and Lewis, A. (1973). *J. Chem. Soc. Chem. Commun.* 262
Davidson, R. S., Bonneau, R., Joussot-Dubien, J. and Trethewey, K. R. (1980a). *Chem. Phys. Lett.* **74**, 3181
Davidson, R. S., Goodin, J. W. and Kemp, G. (1980b). *Tetrahedron Lett.* 2911
Davidson, R. S., Goodin, J. W. and Pratt, J. E. (1982). *Tetrahedron Lett.* 2225
Dubois, J. T. and Wilkinson, F. (1963). *J. Chem. Phys.* **38**, 2541
Egger, K. W. and Cocks, A. T. (1973). *Helv. Chim. Acta.* **56**, 1516
Eichin, K. H., Beckhaus, H. D. and Ruchardt, C. (1980). *Tetrahedron Lett.* 269
Fox, M. A., Lemal, D. M., Johnson, D. W. and Hohman. J. R. (1982). *J. Org. Chem.* **47** 398
Freedman, A., Yang, S. C., Kawasaki, M. and Bersohn, R. (1980). *J. Chem. Phys.* **72**, 1028.
Gordon, M. and Ware, W. R. (eds.) (1975). "The Exciplex". Academic Press, London and New York
Grieser, F. and Thomas, J. K. (1980). *J. Chem. Phys.* **73**, 2115
Grimshaw, J. and de Silva, A. P., (1979). *J. Chem. Soc. Chem. Commun.* 193
Grimshaw, J. and de Silva, A. P. (1980a). *J. Chem. Soc. Chem. Commun.* 302

Grimshaw, J. and de Silva, A. P. (1980b). *J. Chem. Soc. Chem. Commun.* 301
Grimshaw, J. and de Silva, A. P. (1980c). *Can. J. Chem.* **58**, 1880
Grimshaw, J. and de Silva, A. P. (1980d). *J. Chem. Soc. Chem. Commun.* 1236
Grimshaw, J. and de Silva, A. P. (1981). *Chem. Soc. Rev.* **10**, 181
Grimshaw, J. and Haslett, R. J. (1980). *J. Chem. Soc. Perkin Trans 1*, 657
Grodowski, M. and Latowski, T. (1974). *Tetrahedron*, **30**, 767
Havinga, E. and Cornelisse, J. (1976). *Pure. Appl. Chem.* **47**, 1
Havinga, E., Cornelisse, J. and de Gunst, G. P. (1975). *Adv. Phys. Org. Chem.* **11**, 225
Henne, A. and Fischer, H. (1977). *J. Am. Chem. Soc.* **99**, 300
Hey, D. H., Jones, G. H. and Perkins, M. J. (1971). *J. Chem. Soc. (C)* 116
Hey, D. H., Jones, G. H. and Perkins, M. J. (1972). *J. Chem. Soc. Perkin Trans. 1*, 1150
Hutzinger, O., Safe, S., Wentzell, B. R. and Zitko, V. (1973). *Environ. Health Perspectives* **5**, 267
Inoue, H. and Hida, M. (1978). *Bull Chem. Soc. Jpn.* **51**, 1793
Inoue, H., Hida, M., Tuong, T. D. and Murata, T. (1973). *Bull. Chem. Soc. Jpn.* **46**, 1759
Inoue, H., Nakamura, K., Kato, S. and Hida, M. (1975). *Bull. Chem. Soc. Jpn.* **48**, 2872
Inoue, H., Shinoda, T. and Hida, M. (1980). *Bull. Chem. Soc. Jpn.* **53**, 154
Ito, S., Saito, I. and Matsuura, T. (1979). *Tetrahedron Lett.* 4067
Ito, S., Saito, I. and Matsuura, T. (1980). *J. Am. Chem. Soc.* **102**, 7535
Joschek, H. I. and Miller, S. I. (1966). *J. Am. Chem. Soc.* **88**, 3269
Kharasch, N. (1969). *Intra-Sci. Chem. Rep.* **3**, 203
Kharasch, N., Sharma, R. K. and Lewis, H. B. (1966). *J. Chem. Soc. Chem. Commun.* 418
Kitamura, T., Kobayashi, S. and Taniguchi, H. (1982). *J. Org. Chem.* **47**, 2323
Kojima, M., Sakuragi, H. and Tokumaru, K. (1981). *Chem. Lett.* 1539
Koso, Y., Ichimura, T., Hikida, T. and Mori, Y. (1979). *Kokagaku Toronkai Koen Yoshishu* **176** (*Chem. Abs.* **92**, 197578y)
Loskutov, V. A., Lukonina, S. M., Konstantinova, A. V. and Fokin, E. P. (1981). *Zh. Org. Khim.* **17**, 584
Maharaj, U. and Winnik, M. A. (1982). *Tetrahedron Lett.* 3035
Mansour, M., Parlar, H. and Korte, F. (1979). *Naturwissenschaften* **66**, 579
Mansour, M., Wawrik, S., Parlar, H. and Korte, F. (1980). *Chem. -Ztg.* **104**, 339
Marchetti, A. and Kearns, D. R. (1967). *J. Am. Chem. Soc.* **89**, 5335
Maruyama, K., Tojo, M. and Otsuki, T. (1980). *Bull. Chem. Soc. Jpn.* **53**, 567
Mataga, N. and Ottolenghi, M. (1979). *In* "Molecular Association" (R. Foster, ed.) Vol. II. Academic Press, London and New York, p. 1
Matsuo, T., Mihara, S. and Ueda, I. (1976). *Tetrahedron Lett.* 4581
Matsuura, T. and Omura, K. (1966). *Bull. Chem. Soc. Jpn.* **39**, 944
Matsuura, T., Saito, I., Ito, S. and Shinimura, T. (1980). *Pure and Appl. Chem.* **52**, 2705
McNeeley, S. A. and Kropp, P. J. (1976). *J. Am. Chem. Soc.* **98**, 4319
Miller, G. C., Mille, M. J., Crosby, D. G., Sontum, S. and Zepp. R. G. (1979). *Tetrahedron* **35**, 1797
Mitani, M., Nakayama, M. and Koyama, K. (1980). *Tetrahedron Lett.* 4457
Najbar, J. and Rodakiewicz-Nowak, J. (1978). *Chem. Phys. Lett.* **58**, 545
Nasielski, J., Kirsch-Demesmaeker, A. and Nasielski-Hinkens, R. (1972). *Tetrahedron* **28**, 3767

Nasielski, J., Kirsch-Demesmaeker, A., Kirsch, P. and Nasielski-Hinkens, R. (1973). *Tetrahedron* **29**, 3767
Nijhoff, D. F. and Havinga, E. (1965). *Tetrahedron Lett.* 4199
Nikishin, G. I. and Cheltsova, M. A. (1968). *Izv. Akad. Nauk. S.S.S.R. Ser. Khim.* **157** (*Chem. Abs.* **69**, 6690f)
Nishiwaki, T. (1981). *Yuki Gosei Kagaku Kyokai Shi* **39**, 226
Nishiwaki, T., Usui, M. and Anda, K., (1980). *Tokyo-Toritsu Kogyo Gijutsu Senta Kenkyu Hokoku* **133** (*Chem. Abs.* **93**, 113514j)
Nojima, K. and Kanno, S. (1980). *Chemosphere* **9**, 437
Nojima, K., Ikarigawa, T. and Kanno, S. (1980). *Chemosphere* **9**, 421
Occhiucci, G. and Patacchiola, A. (1982). *Chemosphere* **11**, 255
Ohashi, M., Tsujimoto, K. and Seki, K. (1973). *J. Chem. Soc. Chem. Commun.* 384
Omura, K. and Matsuura, T. (1971). *Tetrahedron* **27**, 3101
Pac, C., Tosa, T. and Sakurai, H. (1972). *Bull. Chem. Soc. Jpn.* **45**, 1169
Paramasivam, R., Palaniappan, R. and Ramakrishnan, V. T. (1979). *J. Chem. Soc. Chem. Commun.* 260
Parkányi, C. (1981). *Bull. Soc. Chim. Belg.* **90**, 599
Parkányi, C. and Lee, Y. J. (1974). *Tetrahedron Lett.* 1115
Pedersen, C. L. and Lohse, C. (1979). *Acta Chem. Scand.* **B33**, 649
Phillips, G. O., Worthington, N. W., McKellar, J. F. and Sharpe, R. R. (1969). *J. Chem. Soc. (A)* 767
Pineault, R. L., Morgante, G. C. and Struve, W. S. (1981). *J. Photochem.* **17**, 435
Pinhey, J. T. and Rigby, R. D. G. (1969). *Tetrahedron Lett.* 1267
Robinson, G. E. and Vernon, J. M. (1969). *J. Chem. Soc. Chem. Commun.* 977
Robinson, G. E. and Vernon, J. M. (1970). *J. Chem. Soc. (C)* 2586
Robinson, G. E. and Vernon, J. M. (1971). *J. Chem. Soc. (C)* 3363
Rossi, R. A. and Palacios, S. M. (1981). *J. Org. Chem.* **46**, 5300
Ruzo, L. O. and Bunce, N. J. (1975). *Tetrahedron Lett.* 511
Ruzo, L. O. and Zabik, M. J. (1975). *Bull. Environ. Contam. Toxicol.* **13**, 181
Ruzo, L. O., Zabik, M. J. and Schuetz, R. D. (1974). *J. Am. Chem. Soc.* **96**, 3809
Ruzo, L. O., Safe, S. and Zabik, M. J. (1975a). *J. Agric. Food. Chem.* **23**, 594
Ruzo, L. O., Bunce, N. J. and Safe, S., (1975b). *Can. J. Chem.* **53**, 688
Ruzo, L. O., Bunce, N. J., Safe, S. and Hutzinger, O. (1975c). *Bull. Environ. Contam. Toxicol.* **14**, 341
Ruzo, L. O., Sunström, G., Hutzinger, O. and Safe, S. (1977). *Rec. Trav. Chim. Pays-Bas* **96**, 249
Saito, I., Ito, S., Shinimura, T. and Matsuura, T. (1980). *Tetrahedron Lett.* 2813
Sammes, P. G. (1973). In "Chemistry of the Carbon-Halogen Bond" (S. Patai, ed.) Part II. J. Wiley and Sons, New York, Ch. 11
Schnabel, W., Naito, I., Kitamura, T., Kobayashi, S. and Taniguchi, H. (1980). *Tetrahedron* **36**, 3229
Seguchi, K. and Ikeyama, H. (1980). *Chem. Lett.* 1493
Sharma, R. K. and Kharasch, N. (1968). *Angew. Chem. Int. Ed. Engl.* **7**, 36
Shimoda, A., Hikida, T. and Mori, Y. (1979). *J. Phys. Chem.* **83**, 1309
Smothers, W. K., Schanze, K. S. and Saltiel, J. (1979). *J. Am. Chem. Soc.* **101**, 1895
Soumillion, J. P. and De Wolf B. (1981). *J. Chem. Soc. Chem. Commun.* 436
Suzuki, T., Kitamura, T., Sonoda, T., Kobayashi, S. and Taniguchi, H. (1981). *J. Org. Chem.* **46**, 5324
Tajima, M., Inoue, H. and Hida, M. (1979). *Nippon Kagaku Kaishi* **1728** (*Chem. Abs.* **92**, 214527g)

Terashima, M., Seki, K., Yoshida, C. and Kanaoka, Y. (1981). *Heterocycles* **15**, 1075
Tosa, T., Pac, C. and Sakurai, H. (1969). *Tetrahedron Lett.* 3635
Tsujimoto, K., Tasaka, S. and Ohashi, M. (1975). *J. Chem. Soc. Chem. Commun.* 758
Wagner, P. J. and Scheve, B. J. (1977). *J. Am. Chem. Soc.* **99**, 2888
Wald, K. M., Nada, A. A., Szilagyi, G. and Wamhoff, H. (1980). *Chem. Ber.* **113**, 2884
Wamhoff, H. and Hupe, H. J. (1978). *Tetrahedron Lett.* 125
Watkins, A. R. (1973). *J. Phys. Chem.* **77**, 1207
Weiss, U., Ziffer, H. and Edwards, J. M. (1971). *Aust. J. Chem.* **24**, 657
Wolf, W. and Kharasch, N. (1961). *J. Org. Chem.* **26**, 284
Wolf, W. and Kharasch, N. (1965). *J. Org. Chem.* **30**, 2493
Yang, N. C., Srinivasachar, K., Kim, B. and Libman, J. (1975). *J. Am. Chem. Soc.* **97**, 5006

# Author Index

*Numbers in italic refer to the pages on which references are listed at the end of each article*

Aalstad, B., 58, 59, 62, 65, 68, 87, *180, 187*
Achord, J. M., 162, *180*
Adams, R. N., 60, 61, 71, 76, *180, 185, 188*
Ager, E., 209, *229*
Augustyniak, W., 197, *229*
Ahlberg, E., 76, 135, 146, *180, 181*
Akaba, R., 109, 110, *186*
Åkermark, B., 200, 202, *229*
Albini, A., 141, *180*
Alder, R. W., 106, *180*
Al-Fakhri, K. A. K., 211, *229*
Almgren, C. W., 109, 110, *182*
Al-Omran, F., 166, *180*
Alonso, R. A., 226, *229*
Amatore, C., 58, 59, *180*
Ambrose, J. F., 61, *180*
Amouyal, E., 95, *180*
Anda, K., 206, *232*
Anderson, D. R., 21, 44, *53*
Ando, T., 149, *186*
Andrade, J. G., 57, *189*
Andrieux, C. P., 101, 103, 104, 212, 219, 220, *180, 181, 229*
Andrulis, Jr., P. J., 136, *181*
Anson, F. C., 104, *181*
Arnold, D. R., 124, 125, 126, 128, 141, *180, 186, 189*
Artes, R., 161, *184*
Asmus, K.-D., 56, 124, *181, 186*
Atkins, P. W., 1, 2, 13, 25, *51, 52*

Baciocchi, E., 137, 138, 139, *181*
Baeckström, P., 200, 202, *229*
Baertschi, P., 116, 117, *184*
Bailey, S. I., 127, *181*
Bally, T., 116, 117, *184*
Banthorpe, D. V., 151, *181*

Bard, A. J., 56, 90, 95, *181*
Barek, J., 135, *181*
Baretz, B. H., 23, *52*
Bargon, J., 129, 172, *181, 184*
Barlow, M. G., 210, 217, *229*
Barltrop, J. A., 222, *229*
Barnes, G. E., 162, 163, *181*
Barth, W., 116, *184*
Barton, D. H. R., 95, 109, *181*
Basselier, J. J., 31, *52*
Bauld, N. L., 120, 123, *181*
Bawn, C. E. H., 100, *181*
Bechgaard, K., 57, 58, 148, *187*
Becker, W. G., 30, 116, *53, 184*
Beckhaus, H. D., 198, *230*
Beecroft, R. A., 221, *229*
Beer, R., 217, *230*
Begley, W. J., 193, *230*
Belchenko, O. I., 45, *53*
Bell, F. A., 95, 100, 120, *181*
Bellamy, F., 27, *52*
Bellville, D. J., 120, 123, *181*
Belyaeva, S. G., 25, 29, *52*
Belyakov, V. A., 49, *52*
Benati, L., 112, 114, *181*
Bennett, J. E., 49, *52*
Bergsma, J. P., 197, 199, 207, *230*
Bernadi, R., 173, *185*
Berndt, A., 31, 45, *52*
Berneth, H., 116, 117, *183*
Bernstein, R. B., 42, *52*
Bersohn, R., 196, *230*
Berson, J. A., 115, *185*
Bertini, F., 173, *185*
Bethell, D., 87, 112, 113, 114, *181, 187*
Beugelmans, R., 226, *230*
Bewick, A., 134, 147, *181, 182*
Bilevich, K. A., 152, *182*
Billups, W. E., 177, *183*
Blankespoor, R. L., 159, *182*

Blocman, C., 101, 103, *181*, 212, 219, 220, *229*
Blount, H. N., 76, 77, 80, 81, 86, *182*, *183*, 188
Blucher, W. G., 165, *187*
Blum, Z., 90, 91, 94, *182*, *183*
Bock, H., 56, 116, 117, 118, 148, *182*, *184*
Bonneau, R., 218, *230*
Boyd, J. W., 144, *182*
Bramley, R., 151, *181*
Bratt, J., 209, *230*
Braun, A. M., 26, *53*
Brechbiel, M., 151, *188*
Brien, D. J., 130, *186*
Brixius, D. W., 141, *189*
Brown, R. D., 156, *182*
Brownstein, S., 152, *182*
Bryce-Smith, D., 210, 211, 217, *230*
Buchachenko, A. L., 2, 12, 13, 33, 34, 49, *52*, *53*
Bull, H. G., 68, *183*
Bunce, N. J., 197, 198, 199, 203, 204, 205, 206, 207, 208, 211, 217, 218, 219, 220, 221, 223, *230*, *232*
Bunnett, J. F., 107, 226, *182*, *230*
Bunton, C. A., 152, *182*
Bünzli, J.-C. G., 115, *185*
Busch, D., 217, *230*
Bussey, R. J., 160, *188*
Buttrill, Jr., S. E., 161, *188*

Campbell, K. A., 116, 117, *183*
Caronna, T., 173, 174, *182*
Carpenter, A. K., 146, *184*
Carpenter, L. L., 61, *180*
Carrington, A., 5, 33, *52*
Cedheim, L., 94, 120, *182*
Chambers, J. Q., 84, *185*
Chanon, M., 106, *182*
Chastanet, J., 226, *230*
Chelsky, R., 123, *181*
Cheltsova, M. A., 215, *232*
Chen, S. C., 178, *182*
Cheng, H. Y., 80, 83, *182*
Cherry, W. R., 18, *53*
Chittini, B., 206, 219, *230*
Chivers, G. E., 209, *229*
Choudhry, G. G., 200, 201, 202, *230*
Chow, M. F., 19, 20, 31, 32, 39, 42, 44, 47, 48, *53*

Chow, Y. L., 129, 159, 174, 175, 177, 178, *182*
Chu, S.-Y., 117, *182*
Chung, C.-J., 19, 20, 21, 30, 39, 40, 42, 44, *52*, *53*
Citterio, A., 56, 140, 172, 173, 174, *182*, *185*
Clemens, A. H., 168, *182*
Clennan, E. L., 109, 110, *182*
Closs, G. L., 2, 34, *52*
Cocks, A. T., 194, 195, *230*
Coe, D. E., 147, *182*
Cogswell, G., 123, *181*
Colon, C., 178, *182*
Comi, R., 144, *189*
Conlin, R. T., 100, 112, 113, *184*
Connon, H. A., 115, *189*
Cook, G. B., 42, *52*
Coombes, R. G., 163, *182*
Cordes, E. H., 68, *183*
Corey, R. J., 177, *182*
Cornelisse, J., 107, 222, *184*, *231*
Cramer, J., 141, *189*
Creason, S. C., 61, *182*
Crellin, R. A., 120, *181*, *182*
Crolla, T., 173, 174, *182*
Crosby, D. G., 223, *231*

Dadson, W. M., 210, 211, 217, *230*
Dahm, L. L., 217, *230*
Danen, W. C., 129, 174, 177, *182*
Dankowski, M., 216, *230*
Dapperheld, S., 98, *183*
Darwent, J. R., 95, *183*
Dauphin, G., 111, *185*
Davey, L., 213, *230*
Davidson, R. S., 56, *183*, 197, 198, 213, 216, 217, 218, 219, 220, 221, 222, *229*, *230*
De Graaf, W., 197, 199, 207, *230*
de Gunst, G. P., 222, *231*
Dehnicke, K., 115, *185*
Deno, N. C., 174, 176, 177, *183*
Deronzier, A., 95, *183*
de Silva, A. P., 192, 199, 202, 203, 214, *230*, *231*
Dewar, M. J. S., 136, 151, *181*, *183*
De Wolf, B., 202, 223, *232*
Dietz, R., 136, *181*
Dinghra, O. P., 57, *185*
Dixon, W. T., 124, 172, *183*

# AUTHOR INDEX

do Amaral, L., 68, *183*
Dobaeva, N. M., 159, 160, *185*, *186*
Dobson, B. C., 172, *187*
Doyle, M. P., 159, *182*
Draper, M. R., 168, 169, *183*
Dreher, E.-L., 134, *187*
Dresswick, W. J., 95, *188*
Dubois, J. T., 196, *230*
Dumas-Bouchiat, J. M., 101, 103, 212, 219, 220, *180*, *181*, *229*
Duncan, J. F., 42, *52*

Eberhardt, M. K., 128, *183*
Eberson, L., 56, 60, 73, 80, 90, 91, 92, 93, 94, 101, 107, 108, 120, 128, 136, 139, 140, 161, *182*, *183*, *187*
Ecker, P., 161, *184*
Edwards, J. M., 218, *233*
Edwards, G. J., 134, *181*
Eeles, M. F., 113, *181*
Egger, K. W., 194, 195, *230*
Eichin, K. H., 198, *230*
Elliott, I. W., 57, *183*
Engel, P. S., 17, *52*
Epling, G. A., 144, *189*
Eriksen, J., 109, *183*
Esposito, F., 95, *183*
Evans, D. H., 56, 134, *183*, *187*
Evans, G. T., 1, 33, *52*
Evans, J. F., 76, 77, 80, 81, 86, *183*
Evans, T. R., 93, 116, *183*

Fairhurst, S. A., 112, 113, 114, *181*
Farid, S., 120, *183*
Fedorov, A. V., 49, *52*
Feldberg, S. W., 61, *186*
Fendler, E. J., 18, *52*
Fendler, J. H., 18, *52*
Fentiman, A., 154, *189*
Fessenden, R. W., 124, 128, 175, *183*, *186*
Field, K. W., 174, *185*
Fischer, H., 31, 45, 201, *52*, *231*
Fishbein, R., 177, *183*
Fokin, E. P., 223, *231*
Foote, C. S., 109, *183*
Fox, M. A., 116, 117, 211, *183*, *230*
Freedman, A., 196, *230*
Friedrich, P., 161, *184*
Frith, P. G., 2, *52*
Fritsch, J. M., 61, *188*

Fritz, F., 161, *184*
Fuentes-Aponte, A., 128, *183*
Fujii, H., 120, *181*
Fujiwara, K., 166, *180*
Fukuzumi, S., 156, 157, 158, 159, *184*,
Fuller, G. B., 147, *182*
Fung, L. W.-M., 109, *188*

Galimov, E. M., 49, *52*
Gallacher, J. C., 204, 207, 221, *230*
Galli, R., 173, *185*
Gannett, P. M., 130, *186*
Gardini, G. P., 129, *184*
Gardner, S. A., 123, *181*
Gaspar, P. P., 100, 112, *184*
Gassman, P. G., 117, 119, *184*
Gebauer, H., 161, *184*
Geisel, H., 134, *187*
Genies, M., 103, *184*
Ghirardini, M., 174, *182*
Giffney, J. C., 164, 165, 166, *180*, *184*
Gilbert, A., 210, 211, 217, *230*
Giordan, J., 148, *184*
Giordano, C., 56, 140, *185*
Göbl, M., 124, *186*
Goldanskii, V. I., 50, *52*
Golden, J. T., 154, *189*
Golding, J. G., 163, *182*
Goodin, J. W., 197, 198, 213, 216, 219, 220, 221, 222, *230*
Goodwin, D. C., 221, *229*
Gordon, M., 219, *230*
Göthe, R., 200, 202, *229*
Grand, D., 95, *180*
Grätzel, M., 26, *53*
Green, G., 123, *181*
Greenberg, A., 115, *184*
Grieser, F., 218, *230*
Grimshaw, J., 192, 193, 199, 202, 203, 214, *230*, *231*
Grodowski, M., 221, *231*
Gupta, A., 25, *52*

Hadjigeorgiou, P., 163, *182*
Hammerich, O., 56, 57, 67, 70, 71, 72, 73, 74, 78, 79, 80, 81, 82, 83, 84, 85, 86, 87, 127, 148, *184*, *187*, *189*
Hammond, G. S., 25, *52*
Hammond, W. B., 115, *185*
Hand, R., 146, *184*

Handoo, K. L., 100, 112, 113, 114, *181*, *184*
Handoo, S. K., 100, *184*
Hanotier, J., 139, *184*
Hanotier-Bridoux, M., 139, *184*
Hanson, P., 80, *184*
Harris, R. H., 50, *52*
Haselbach, E., 116, 117, *184*
Haslett, R. J., 193, *231*
Haszeldine, R. N., 210, 217, *229*
Hata, N., 27, 28, *52*
Hathaway, C., 154, *189*
Havinga, E., 107, 222, *184*, *231*, *232*
Hayashi, H., 31, *53*
Haynes, R. K., 95, 109, *181*
Hayon, E., 124, 175, *188*
Heighway, C. J., 172, *184*, *187*
Heijer, J. Den, 107, *184*
Helgée, B., 90, 91, 94, 146, *180*, *183*
Helsby, P., 167, *184*
Henne, A., 201, *231*
Hertler, W. R., 177, *182*
Hewgill, F. R., 127, *181*
Hey, D. H., 194, *231*
Hida, M., 223, *231*, *232*
Hikada, T., 196, *231*, *232*
Hiller, K.-O., 124, *186*
Hisahara, H., 31, *53*
Hisamitsu, K., 116, *186*
Ho, C.-T., 100, 112, 113, *184*
Ho, T.-I., 129, *185*
Hodgson, R., 217, *230*
Hoffmann, R. W., 116, *184*
Hohman, J. R., 211, *230*
Hokawa, M., 28, *52*
Holcman, J., 128, 129, *188*
Holton, D. M., 124, *184*
Horgan, A. G., 151, *188*
Houk, K. N., 156, *188*
House, D. B., 172, *187*
Howard, J. A., 49, *52*
Hughes, E. D., 151, 152, *181*, *182*, *184*
Hull, V. J., 72, *185*
Hum, G. P., 165, *187*
Hünig, S., 95, 116, 117, *183*, *184*
Hunt, R. L., 136, *181*
Hupe, H. J., 229, *233*
Hurysz, L. F., 93, *183*
Hussey, C. L., 162, *180*
Huston, 154, *189*

Hutzinger, O., 200, 201, 202, 206, 208, 209, 217, 219, *230*, *231*, *232*

Ichimura, T., 196, *231*
Iddon, B., 209, *230*
Ikarigawa, T., 217, *232*
Ikeyama, H., 223, *232*
Ingold, C. K., 151, *181*
Inoue, H., 223, *231*, *232*
Inoue, K., 129, *188*,
Inozemtsev, A. N., 159, *185*
Ito, S., 227, 228, *231*, *232*
Iwai, K., 159, *182*

Jempty, T. C., 67, *185*
Jenson, B. S., 78, *184*
Johnson, D. W., 211, *230*
Jones, C. R., 99, *185*
Jones, G., 30, *53*
Jones, G. H., 194, *230*
Jones, G. T., 152, *184*
Jones, G., II, 116, *185*
Jönsson, L., 107, 108, 128, 140, 161, *183*, *185*
Joschek, H. I., 222, *231*
Joussot-Dubien, J., 218, *230*
Jugelt, W., 112, 113, *185*, *187*

Kabakoff, D. S., 115, *185*
Kaim, W., 56, *182*
Kalinowski, H. O., 115, *185*
Kameswaran, V., 57, *185*
Kamha, M. A., 16, 45, *52*, *53*
Kamiya, Y., 139, *187*
Kanaoka, Y., 209, *233*
Kanno, S., 200, 217, *232*
Kapovits, I., 84, *185*
Kaptein, R., 2, 9, 10, 11, *52*
Kashimura, S., 129, *188*
Kaskar, B., 109, *188*, *189*
Kato, S., 223, *231*
Kawasaki, M., 196, *230*
Kearns, D. R., 213, *231*
Keller, P., 95, *180*
Kemp, G., 197, 198, 213, *230*
Kerr, J. B., 67, *185*
Kessel, C. R., 130, *186*
Khalifa, M. H., 134, *187*
Kharasch, N., 192, 214, *231*, *232*, *233*
Khiznyi, V. A., 152, 159, *187*
Kim, C.-K., 57, *185*

# AUTHOR INDEX

Kim, B., 208, *233*
Kim, K., 72, *185*
Kim-Thuan, N., 95, *186*
Kirsch, P., 221, *232*
Kirsch-Demesmaeker, A., 219, 221, *231, 232*
Kitamura, T., 224, 225, *231, 232*
Klink, J. R., 154, *189*
Kobayashi, S., 156, 224, *186, 231, 232*
Kochi, J. K., 156, 157, 158, 159, *184*
Kojima, M., 198, *231*
Konstantinova, A. V., 223, *231*
Korte, F., 200, 201, 202, *231*
Koser, G. F., 117, *184*
Koshechko, V. G., 152, 159, *185, 187*
Koso, Y., 196, *231*
Kossai, R., 111, *185*
Kovacic, P., 174, *185*
Koyama, K., 217,
Kraeutler, B., 1, 2, 17, 19, 21, 27, 30, 31, 34, 39, 42, 44, *53*
Kricka, L. J., 100, *185*
Krishnamurthy, V. V., 171, 172, *186*
Kropp, P. J., 224, *231*
Kumar, Y., 197, 199, 203, 204, 205, 207, 219, 221, *230*
Kupchan, S. M., 57, *185*
Kuwana, T., 128, *186*
Kwart, H., 151, *188*

Lambert, M. C., 120, *182*
Lambert, T. P., 1, 2, *52*
Land, E. J., 124, *185*
Langridge, J. R., 210, 217, *229*
Lanyiova, Z., 116, 117, *184*
Latowski, T., 221, *231*
Lawesson, S.-O., 156, 157, 160, *187*
Lawler, R. G., 1, 2, 33, 39, 40, *52, 53*
Lazdins, D., 154, *189*
Leclerc, G., 95, 109, *181*
Ledwith, A., 56, 90, 95, 100, 120, *181, 182, 185*
Lee, T.-S., 117, *182*
Lee, Y. J., 223, *232*
Leedy, D. W., 61, *188*
Lemal, D. M., 211, *230*
Leshina, T. V., 16, 25, 29, 45, 50, *52, 53*
Lewis, A., 217, *230*
Lewis, F. D., 129, *185*
Lewis, H. B., 214, *231*
Liao, C.-S., 84, *185*

Libert, M., 147, *182*
Libman, J., 208, *233*
Liebman, J. F., 115, *184*
Liepa, A. J., 57, *185*
Line, L. L., 96, *187*
Lines, R., 144, *185*
Lohse, C., 213, *232*
Loskutov, V. A., 223, *231*
Lowery, M. K., 174, *185*
Lukonina, S. M., 223, *231*
Lynn J. T., 57, *185*
Mack, A. G., 209, *230*
Madhavan, V., 128, *186*
Maeda, H., 170, *188*
Magnus, P. D., 95, 109, *181*
Maharaj, U., 197, *231*
Maier, G., 115, 116, 117, 118, *182, 183, 185*
Makhon'kov, D. I., 152, *189*
Malsch, K.-D., 115, 116, 117, *183, 185*
Maltsev, V. I., 34, 49, *52, 53*
Mandolini, L., 137, 138, 139, *181*
Mann, B. E., 50, *52*
Manning, G., 71, 76, *185*
Mansour, M., 200, 201, 202, *231*
Marchand, A. P., 151, *183*
Marchetti, A., 213, *231*
Marcoux, L. S., 61, 72, 76, *188*
Martigny, P., 104, *185*
Martinez, G. A., 128, *183*
Maruyama, K., 225, *231*
Maryasova, V. I., 25, 29, *52*
Mataga, N., 219, *231*
Matsumura, Y., 129, *188*
Matsuo, T., 229, *231*
Matsuura, T., 214, 222, 227, 228, *231, 232*
Mattay, J., 23, 45, *53*
Mattes, S. L., 120, *185*
Matusch, R., 115, *185*
Mayausky, J. S., 80, 83, 84, 85, *185*
Mayeda, E. A., 141, 144, *185*
McCreery, R. L., 80, 83, 84, 85, *182, 185, 186, 187*
McCubbin, I., 95, *183*
McKellar, J. F., 224, *232*
McKillop, A. 57, *185, 189*
McLauchlan, K. A., 2, 5, 33, *52*
McNeeley, S. A., 224, *231*
Melicharek, M., 146, *184*
Mellor, J. M., 134, 147, *181, 182*

Menzies, I. D., 95, 109, *181*
Meyer, T. J., 95, *188*
M'Halla, F., 101, 103, 219, 220, *181*, *229*
Migron, Y., 123, *181*
Mihara, S., 229, *231*
Mille, M. J., 223, *231*
Miller, G. C., 223, *231*
Miller, H. M., 172, *187*
Miller, L. L., 67, 141, 144, *182*, *185*
Miller, S. I., 222, *231*
Minisci, F., 56, 140, 172, 173, 174, 177, 179, *182*, *185*
Mitani, M., 217, *231*
Mo, Y. K., 155, *186*
Molin, Y. N., 1, 16, 17, 25, 29, 45, 50, 52, *53*
Montevecchi, P. C., 112, 114, *181*
Moradpour, A., 95, *180*
Morgante, G. C., 218, *232*
Mori, Y., 196, *231*, *232*
Morkovnik, A. S., 159, 160, *185*, *186*
Moutet, J.-C., 103, 105, *184*, *186*
Mowatt, A. C., 211, *229*
Mukai, T., 116, *186*
Murata, T., 223, *231*
Murata, Y., 71, 72, 73, 80, *186*, *188*
Murphy, D., 124, *183*, *184*
Musker, W. K., 56, *186*
Myhre, P. C., 162, 163, *181*

Nada, A. A., 229, *233*
Nagae, K., 170, *188*
Nagai, T., 112, 114, *187*
Nagakura, S., 31, *53*, 156, *186*
Naito, I., 224, *232*
Najbar, J., 217, *231*
Nakagawa, F., 28, *52*
Nakamura, K., 223, *231*
Nakayama, M., 217, *231*
Nasielski, J., 219, 221, *231*, *232*
Nasielski-Hinkens, R., 219, 221, *231*, *232*
Neale, R. S., 174, *186*
Nelsen, S. F., 56, 109, 110, 124, 129, 130, 145, 174, 177, *182*, *183*, *186*
Nelson, R. F., 57, 61, 96, 146, *180*, *182*, *184*, *186*, *187*, *188*
Neptune, M., 80, *186*
Neta, P., 124, 128, 175, *183*, *186*
Nicholas, A. M. de P., 124, 125, 126, 128, *186*
Nijhoff, D. F., 222, *232*

Nikishin, G. I., 215, *232*
Nilsson, A., 134, *186*
Nishiwaki, T., 206, *232*
Nojima, K., 200, 217, *232*
Nojima, M., 149, *186*
Norman, R. O. C., 172, *183*
Nyberg, K., 57, 60, 90, 91, 92, 93, 94, 107, 140, *183*, *186*
Nyburg, S. C., 216, *230*

Occhiucci, G., 206, *232*
Ohashi, M., 219, 220, 221, 222, *232*, *233*
Ohmizu, H., 129, *188*
Ohnesorge, W. E., 63, *188*
Ohta, N., 139, *187*
Oivers, R. J., 222, *229*
Okada, K., 116, *186*
Okhlobystin, O. Yu., 152, 159, 160, *182*, *186*
Okura, I., 95, *186*
Okuyama, T., 160, *188*
Olah, G. A., 155, 171, 172, *186*
Olie, K., 206, 219, *230*
Omura, K., 214, 222, *231*, *232*
Ono, M., 148, *189*
Ono, Y., 28, *52*
Osa, T., 128, *186*
Oshima, T., 112, 114, *187*
Osuka, A., 170, *188*
Oth, J. F. M., 115, *185*
Otsuki, T., 225, *231*
Ottolenghi, M., 219, *231*
Owers, R. J., 222, *229*

Pabon, R., 123, *181*
Pac, C., 221, *232*, *233*
Packer, J. E., 172, *184*, *187*
Palacios, S. M., 226, *232*
Palaniappan, R., 216, *232*
Palmquist, U., 67, 134, *186*, *187*
Panov, V. B., 159, *186*
Paramasivam, R., 216, *232*
Park, K. H., 151, *188*
Parkányi, C., 107, 221, 223, *187*, *232*
Parker, D. G., 155, *186*
Parker, V. D., 56, 57, 58, 59, 60, 62, 63, 64, 65, 67, 68, 70, 71, 72, 73, 74, 75, 76, 78, 79, 80, 81, 82, 83, 84, 85, 86, 87, 127, 132, 134, 135, 146, 148, 149, *180*, *181*, *184*, *185*, *186*, *187*, *188*, *189*

# AUTHOR INDEX

Parlar, H., 200, 201, 202, *231*
Parmelee, W. P., 124, 145, *186*
Patacchiola, A., 206, *232*
Paul, H., 31, 45, *52*
Pedersen, C. L., 213, *232*
Pedersen, E. B., 156, 157, 160, *187*
Perchinunno, M., 173, *185*
Perkins, M. J., 194, *231*
Perrin, C. L., 156, 160, *188*
Peterson, T. E., 156, 157, 160, *187*
Pettersson, T., 134, *186*
Pfriem, S., 115, *185*
Phillips, G. O., 224, *232*
Pierson, C., 177, *183*
Pilon, P., 207, 211, 217, 219, 220, *230*
Pineault, R. L., 218, *232*
Pines, A., 34, *53*
Pinhey, J. T., 214, *232*
Podoplelov, A. V., 50, *52*
Pokhudenko, V. D., 152, 159, *185*, *187*
Pons, S., 134, *181*, *182*
Porta, O., 174, *182*
Porter, G., 95, 124, *183*, *185*
Praefcke, K., 216, *230*
Pragst, F., 112, 113, 142, 143, *185*, *187*
Pratt, A. C., 211, *229*
Pratt, J. E., 222, *230*
Purtov, P. A., 25, 29, *52*

Rabai, J., 84, *185*
Radner, F., 161, *183*
Radzitzky, P. de, 139, *184*
Ramachandran, V., 57, *185*
Rasburn, E. J., 172, *187*
Ravanal, L., 197, 198, 199, 203, 204, 205, 207, 219, 221, *230*
Reitano, M., 144, *189*
Reverdy, G., 103, 105, *184*, *186*
Reynolds, R., 96, *187*
Rezvukhin, A. L., 45, *53*
Ricci, R. W., 217, *230*
Richards, J. A., 134, *187*
Richardson, R. K., 172, *184*, *187*
Ridd, J. H., 154, 155, 164, 165, 166, 167, 168, 169, *180*, *182*, *183*, *184*, *187*
Rieker, A., 131, 132, 134, *187*, *188*
Rigaudy, J., 31, *52*, *53*
Rigby, R. D. G., 214, *232*
Ristagno, C. V., 159, *187*
Ritchie, I. M., 127, *181*
Rittmeyer, P., 148, *182*

Rivera, J. I., 128, *183*
Robinson, G. E., 197, 198, 216, *232*
Robinson, S. R., 166, *180*
Rodakiewicz-Nowak, J., 217, *231*
Rol, C., 137, 138, *181*
Ronis, D., 34, *53*
Ronlán, A., 57, 58, 59, 62, 63, 64, 65, 67, 68, 87, 127, 134, *180*, *184*, *186*, *187*, *189*
Roof, A. A. M., 200, 201, 203, *230*
Rosenblatt, D. H., 129, 174, 177, *182*
Ross, D. S., 161, 165, *187*, *188*
Rossi, R. A., 226, *229*, *232*
Roth, B., 116, 117, 118, *182*
Roth, H. D., 116, *187*
Roussi, G., 226, *230*
Rüchardt, C., 172, 198, *189*, *230*
Ruzo, L. O., 203, 204, 206, 207, 208, 211, 215, 217–220, 223, *230*, *232*
Rys, P., 156, *187*

Sackett, P. H., 80, 83, *182*, *187*
Safe, S., 203, 204, 205, 206, 208, 209, 215, 217, 218, 219, 221, 223, *230*, *231*, *232*
Sagdeev, R. Z., 1, 16, 17, 25, 29, 45, 50, *52*, *53*
Saito, I., 227, 228, *231*, *232*
Sakaguchi, Y., 31, *53*
Sakota, K., 139, *187*
Sakuragi, H., 198, *231*
Sakurai, H., 221, *232*, *233*
Salikhov, K. M., 1, 16, 17, 25, 29, 45, *52*, *53*
Saltiel, J., 208, *232*
Sammes, P. G., 192, *232*
Sandall, J. P. B., 154, 165, 167, 168, *182*, *184*, *187*
Santiago, C., 156, *188*
Sato, K., 116, *186*
Savéant, J. M., 58, 59, 101, 103, 104, 106, 212, 219, 220, *180*, *181*, *188*, *229*
Schaap, A. P., 109, *188*, *189*
Schäfer, H., 57, 62, *182*, *188*
Schäfer, U., 115, *185*
Schanze, K. S., 208, *232*
Scheve, B. J., 205, *233*
Schilling, M. L. M., 116, *187*
Schmalzl, P. W., 144, *182*
Schmid Baumberger, R., 135, *188*

Schmidt, W., 96, 97, 101, 102, *188*
Schmitt, R. J., 161, *188*
Schnabel, W., 224, *232*
Schnieder, K.-A., 116, 117, *183*
Schubert, U., 161, *184*
Schuetz, R. D., 203, 204, 215, *232*
Scoggin, D. I., 146, *184*
Seguchi, K., 223, *232*
Sehested, K., 128, 129, *188*
Seifert, K.-G., 172, *181*
Seki, K., 209, 219, 220, *232*, *233*
Seo, E. T., 61, *188*
Sesana, G., 174, *182*
Shadid, O. B., 107, *184*
Shaik, S. S., 159, *188*
Shang, D. T., 76, *188*
Sharma, R. K., 192, 214, *231*, *232*
Sharpe, R. R., 224, *232*
Shealer, S. E., 120, *183*
Shein, S. M., 16, 45, *53*
Sherrington, D. C., 95, *181*
Shigehara, K., 104, *181*
Shimoda, A., 196, *232*
Shine, H. J., 56, 71, 72, 73, 80, 90, 95, 151, 159, 160, *181*, *185*, *186*, *187*, *188*
Shinimura, T., 227, 228, *231*, *232*
Shinoda, T., 223, *231*
Shkrebtii, O. I., 152, 159, *187*
Shono, T., 129, *188*
Sifain, M. M., 116, *183*
Silber, J. J., 72, 73, 160, *188*
Simic, M., 124, 175, *188*
Simmons, W., 109, 110, *182*
Simonet, J., 104, 111, *185*
Simpson, J. T., 129, *185*
Siskin, M., 142, *188*
Skrabal, P., 156, *187*
Smith, D. J., 159, *182*
Smith, W. A., 39, 40, *53*
Smothers, W. K., 208, *232*
Sonoda, T., 224, 225, *232*
Sontum, S., 223, *231*
Soole, P. J., 172, *187*
Soumillion, J. P., 202, 223, *232*
Spagnolo, P., 112, 114, *181*
Speiser, B., 131, 132, *188*
Srinivasachar, K., 208, *233*
Steckhan, E., 62, 96, 97, 98, 101, 102, *183*, *188*
Stein, U., 148, *182*

Stella, L., 175, 177, 178, *188*
Stenerup, H., 60, *183*
Sterna, L., 34, *53*
Stotz, R., 146, *184*
Strachan, E., 124, *185*
Streith, J., 27, *52*
Struve, W. S., 218, *232*
Stuart, J. D., 63, *188*
Sturch, D. J., 207, 211, 217, 219, 220, *230*
Sullivan, B. P., 95, *188*
Sundholm, C., 57, *187*
Sunström, G., 217, *232*
Surzur, J. M., 175, 178, *188*
Suschitzky, H., 209, *229*, *230*
Sutcliffe, L. H., 112, 113, 114, *181*
Suttie, A. B., 134, *188*
Suvicke, B., 217, *230*
Suzuki, H., 170, 224, 225, *188*, *232*
Svanholm, U., 57, 63, 64, 65, **73**, **74**, 75, 76, 78, 79, 80, 86, 148, 149, *187*, *188*, *189*
Szilagyi, G., 229, *233*

Tajima, M., 223, *232*
Takeuchi, M., 95, *186*
Tanaka, J., 156, *186*
Taniguchi, H., 224, *231*, *232*
Tanimoto, Y., 31, *53*
Tarasov, V. F., 34, *52*, *53*
Tasaka, S., 221, 222, *233*
Tashiro, M., 156, *186*
Taylor, C., 213, *230*
Taylor, E. C., 57, *185*, *189*
Taylor, J. A., 209, *230*
Terashima, M., 209, *233*
Thelen, P. J., 175, *189*
Thomas, J. A., 151, *181*
Thomas, J. K., 218, *230*
Tobe, M. L., 106, *182*
Todres, Z. V., 152, *189*
Tojo, M., 225, *231*
Tokumaru, K., 31, *53* 198, *231*
Tokura, N., 149, *186*
Tordo, P., 175, 178, *188*
Torii, S., 148, *189*
Torsell, K., 156, 157, 160, *187*
Tosa, T., 221, *232*, *233*
Trahanovsky, W. S., 141, *189*
Trethewey, K. R., 218, *230*
Trojanek, A., 60, *186*

Tsujimoto, K., 219, 220, 221, 222, *232, 233*
Tung, C. H., 19, 20, *53*
Tuong, T. D., 223, *231*
Turrell, A. C., 57, *185*
Turro, N. J., 1, 2, 17, 18, 19, 20, 21, 23, 25, 26, 27, 30, 31, 32, 34, 39, 40, 42, 44, 45, 47, 48, *52, 53*
Twitchett, P., 210, *230*

Ueda, I., 229, *231*
Uneyama, K., 148, *189*
Usui, M., 206, *232*
Utley, J. H. P., 144, *185*

Van Dyke, D. A., 159, *182*
Veltwisch, D., 124, *186*
Vernon, J. M., 197, 198, 216, *232*

Wachtmeister, C., 200, 202, *229*
Wagner, P. J., 205, *233*
Wake, R. W., 116, *183*
Wakefield, B. J., 209, *230*
Wald, K. M., 229, *233*
Waldyke, M. J., 159, *182*
Walter, R. I., 95, *189*
Walton, D. J., 147, *182*
Wamhoff, H., 229, *233*
Waters, W. A., 172, *189*
Watkins, A. R., 217, *233*
Ware, W. R., 219, *230*
Wawrik, S., 200, *231*
Wawzonek, S., 175, *189*
Wayner, D. D. M., 141, *189*
Webster, D. R., 172, *187*
Weed, G. C., 19, *53*
Weinreb, S. M., 144, *189*
Weiss, U., 218, *233*
Wentzell, B. R., 209, *231*
Werner, R., 172, *189*
Westlin, U. E., 200, 202, *229*
Whalen, R., 177, *183*

Wheeler, J., 61, *182*
White, H. S., 154, *189*
White, W. N., 154, *189*
Whitson, P. E., 134, *187*
Wiberg, K. B., 115, *189*
Wilkinson, F., 196, *230*
Williams, D. K., 57, *185*
Winnik, M. A., 197, *231*
Wirth, D. D., 120, 123, *181*
Wistrand, L.-G., 107, 108, 128, 139, 140, *183, 185, 186*
Wolf, J. F., 141, 144, *185*
Wolf, W., 214, *233*
Wolfe, S., 34, *53*
Wolff, M. E., 174, 176, *189*
Wong-Ng, W., 216, *230*
Worthington, N. W., 224, *232*
Wu, S.-M., 160, *188*
Wyckoff, J. C., 177, *183*

Yamaguchi, R., 117, 119, *184*
Yamashita, Y., 116, *186*
Yang, N. C., 208, *233*
Yang, S. C., 196, *231*
Yasina, L. L., 49, *52*
Yildiz, A., 128, *186*
Yoshida, C., 209, *233*
Yoshioka, A., 112, 114, *187*
Yu, S. H., 155, *186*

Zabik, J. M., 203, 204, 206, 215, 219, *232*
Zaklika, K. A., 109, *188, 189*
Zefirov, N. S., 152, *189*
Zemel, H., 128, *186*
Zepp, R. G., 223, *231*
Zhidomirov, F. M., 2, *52*
Ziebig, R., 142, 143, *187*
Ziffer, H., 218, *233*
Zitko, V., 209, *231*
Zmuda, H., 151, *188*
Zollinger, H., 156, 172, *187, 189*

# Cumulative Index of Authors

Ahlberg, P., **19**, 223
Albery, W. J., **16**, 87
Allinger, N. L., **13**, 1
Anbar, M., **7**, 115
Arnett, E. M., **13**, 83
Bard, A. J., **13**, 155
Bell, R. P., **4**, 1
Bennett, J. E., **8**, 1
Bentley, T. W., **8**, 151; **14**, 1
Berger, S., **16**, 239
Bethell, D., **7**, 153: **10**, 53
Blandamer, M.J., **14**, 203
Brand, J. C. D., **1**, 365
Brändström, A., **15**, 267
Brinkman, M. R., **10**, 53
Brown, H. C., **1**, 35
Buncel, E., **14**, 133
Cabell-Whiting, P. W., **10**, 129
Cacace, F., **8**, 79
Carter, R. E., **10**, 1
Collins, C. J., **2**, 1
Cornelisse, J., **11**, 225
Crampton, M. R., **7**, 211
Davidson, R. S., **19**, 1; **20**, 191
Desvergne, J. P., **15**, 63
de Gunst, G. P., **11**, 225
de Jong, F., **17**, 279
Eberson, L., **12**, 1: **18**, 79
Engdahl, C., **19**, 223
Farnum, D. G., **11**, 123
Fendler, E. J., **8**, 271
Fendler, J. H., **8**, 271; **13**, 279
Ferguson, G., **1**, 203
Fields, E. K., **6**, 1
Fife, T. H., **11**, 1
Fleischmann, M., **10**, 155
Frey, H. M., **4**, 147
Gilbert, B. C., **5**, 53
Gillespie, R. J., **9**, 1

Gold, V., **7**, 259
Goodin, J. W., **20**, 191
Gould, I. R., **20**, 1
Greenwood, H. H., **4**, 73
Hammerich, O., **20**, 55
Havinga, E., **11**, 225
Hine, J., **15**, 1
Hogen-Esch, T. E., **15**, 153
Hogeveen, H., **10**, 29, 129
Ireland, J. F., **12**, 131
Johnson, S. L., **5**, 237
Johnstone, R. A. W., **8**, 151
Jonsäll, G., **19**, 223
Kemp, G., **20**, 191
Kice, J. L., **17**, 65
Kirby, A. J., **17**, 183
Kohnstam, G., **5**, 121
Kramer, G. M., **11**, 177
Kreevoy, M. M., **6**, 63; **16**, 87
Kunitake, T., **17**, 435
Ledwith, A., **13**, 155
Liler, M., **11**, 267
Long, F. A., **1**, 1
Maccoll, A., **3**, 91
McWeeny, R., **4**, 73
Melander, L., **10**, 1
Mile, B., **8**, 1
Miller, S. I., **6**, 185
Modena, G., **9**, 185
More O'Ferrall, R. A., **5**, 331
Morsi, S. E., **15**, 63
Neta, P., **12**, 223
Norman, R. O. C., **5**, 33
Nyberg, K., **12**, 1
Olah, G. A., **4**, 305
Parker, A. J., **5**, 173
Parker, V. D., **19**, 131; **20**, 55
Peel, T. E., **9**, 1

Perkampus, H. H., **4**, 195
Perkins, M. J., **17**, 1
Pittman, C. U. Jr., **4**, 305
Pletcher, D., **10**, 155
Pross, A., **14**, 69
Ramirez, F., **9**, 25
Rappoport, Z., **7**, 1
Reinhoudt, D. N., **17**, 279
Ridd, J. H., **16**, 1
Reeves, L. W., **3**, 187
Robertson, J. M., **1**, 203
Rosenthal, S. N., **13**, 279
Samuel, D., **3**, 123
Schaleger, L. L., **1**, 1
Scheraga, H. A., **6**, 103
Schleyer, P. von R., **14**, 1
Schmidt, S. P., **18**, 187
Schuster, G. B., **18**, 187
Scorrano, G., **13**, 83
Shatenshtein, A. I., **1**, 156
Shine, H. J., **13**, 155
Shinkai, S., **17**, 435
Silver, B. L., **3**, 123
Simonyi, M., **9**, 127
Stock, L. M., **1**, 35
Symons, M. C. R., **1**, 284
Tedder, J. M., **16**, 51
Thomas, A., **8**, 1
Thomas, J. M., **15**, 63
Tonellato, U., **9**, 185
Toullec, J., **18**, 1
Tüdös, F., **9**, 127
Turner, D. W., **4**, 31
Turro, N. J., **20**, 1
Ugi, I., **9**, 25
Walton, J. C., **16**, 51
Ward, B., **8**, 1
Whalley, E., **2**, 93
Williams, D. L. H., **19**, 381
Williams, J. M. Jr., **6**, 63
Williams, J. O., **16**, 159
Williamson, D. G., **1**, 365

Wilson, H., **14,** 133
Wolf, A. P., **2,** 201
Wyatt, P. A. H., **12,** 131
Zimmt, M. B., **20,** 1
Zollinger, H., **2,** 163
Zuman, P., **5,** 1

# Cumulative Index of Titles

Abstraction, hydrogen atom, from O–H bonds, **9,** 127
Acid solutions, strong, spectroscopic observation of alkylcarbonium ions in, **4,** 305
Acid-base properties of electronically excited states of organic molecules, **12,** 131
Acids, reactions of aliphatic diazo compounds with, **5,** 331
Acids, strong aqueous, protonation and solvation in, **13,** 83
Activation, entropies of, and mechanisms of reactions in solution, **1,** 1
Activation, heat capacities of, and their uses in mechanistic studies, **5,** 121
Activation, volumes of, use for determining reaction mechanisms, **2,** 93
Addition reactions, gas-phase radical, directive effects in, **16,** 51
Aliphatic diazo compounds, reactions with acids, **5,** 331
Alkylcarbonium ions, spectroscopic observation in strong acid solutions, **4,** 305
Ambident conjugated systems, alternative protonation sites in, **11,** 267
Ammonia, liquid, isotope exchange reactions of organic compounds in **1,** 156
Aqueous mixtures, kinetics of organic reactions in water and, **14,** 203
Aromatic photosubstitution, nucleophilic, **11,** 225
Aromatic substitution, a quantitative treatment of directive effects in, **1,** 35
Aromatic substitution reactions, hydrogen isotope effects in, **2,** 163
Aromatic systems, planar and non-planar, **1,** 203
Aryl halides and related compounds, photochemistry of, **20,** 191
Arynes, mechanisms of formation and reactions at high temperatures, **6,** 1
A-$S_E2$ reactions, developments in the study of, **6,** 63

Base catalysis, general, of ester hydrolysis and related reactions, **5,** 237
Basicity of unsaturated compounds, **4,** 195
Bimolecular substitution reactions in protic and dipolar aprotic solvents, **5,** 173

$^{13}$C N.M.R. spectroscopy in macromolecular systems of biochemical interest, **13,** 279
Carbene chemistry, structure and mechanism in, **7,** 163
Carbanion reactions, ion-pairing effects in, **15,** 153
Carbocation rearrangements, degenerate, **19,** 223
Carbon atoms, energetic, reactions with organic compounds, **3,** 201
Carbon monoxide, reactivity of carbonium ions towards, **10,** 29
Carbonium ions (alkyl), spectroscopic observation in strong acid solutions, **4,** 305
Carbonium ions, gaseous, from the decay of tritiated molecules, **8,** 79
Carbonium ions, photochemistry of, **10,** 129
Carbonium ions, reactivity towards carbon monoxide, **10,** 29
Carbonyl compounds, reversible hydration of, **4,** 1
Carbonyl compounds, simple, enolisation and related reactions of, **18,** 1
Catalysis by micelles, membranes and other aqueous aggregates as models of enzyme action, **17,** 435
Catalysis, enzymatic, physical organic model systems and the problem of, **11,** 1
Catalysis, general base and nucleophilic, of ester hydrolysis and related reactions, **5,** 237

Catalysis, micellar, in organic reactions; kinetic and mechanistic implications, **8**, 271
Catalysis, phase-transfer by quaternary ammonium salts, **15**, 267
Cation radicals in solution, formation, properties and reactions of, **13**, 155
Cation radicals, organic, in solution, kinetics and mechanisms of reaction of, **20**, 55
Cations, vinyl, **9**, 135
Charge density—N.M.R. chemical shift correlations in organic ions, **11**, 125
Chemically induced dynamic nuclear spin polarization and its applications, **10**, 53
Chemiluminescence of organic compounds, **18**, 187
CIDNP and its applications, **10**, 53
Conduction, electrical, in organic solids, **16**, 159
Conformations of polypeptides, calculations of, **6**, 103
Conjugated, molecules, reactivity indices, in, **4**, 73
Crown-ether complexes, stability and reactivity of, **17**, 279

$D_2O$–$H_2O$ mixtures, protolytic processes in, **7**, 259
Degenerate carbocation rearrangements, **19**, 223
Diazo compounds, aliphatic, reactions with acids, **5**, 331
Diffusion control and pre-association in nitrosation, nitration, and halogenation, **16**, 1
Dimethyl sulphoxide, physical organic chemistry of reactions, in, **14**, 133
Dipolar aprotic and protic solvents, rates of bimolecular substitution reactions in, **5**, 173
Directive effects in aromatic substitution, a quantitative treatment of, **1**, 35
Directive effects in gas-phase radical addition reactions, **16**, 51

Effective molarities of intramolecular reactions, **17**, 183
Electrical conduction in organic solids, **16**, 159
Electrochemical methods, study of reactive intermediates by, **19**, 131
Electrochemistry, organic, structure and mechanism in, **12**, 1
Electrode processes, physical parameters for the control of, **10**, 155
Electron spin resonance, identification of organic free radicals by, **1**, 284
Electron spin resonance studies of short-lived organic radicals, **5**, 23
Electron-transfer reactions in organic chemistry, **18**, 79
Electronically excited molecules, structure of, **1**, 365
Electronically excited states of organic molecules, acid-base properties of, **12**, 131
Energetic tritium and carbon atoms, reactions of, with organic compounds, **2**, 201
Enolisation of simple carbonyl compounds and related reactions, **18**, 1
Entropies of activation and mechanisms of reactions in solution, **1**, 1
Enzymatic catalysis, physical organic model systems and the problem of, **11**, 1
Enzyme action, catalysis by micelles, membranes and other aqueous aggregates as models of, **17**, 435
Equilibrium constants, N.M.R. measurements of, as a function of temperature, **3**, 187
Ester hydrolysis, general base and nucleophilic catalysis, **5**, 237
Exchange reactions, hydrogen isotope, of organic compounds in liquid ammonia, **1**, 156
Exchange reactions, oxygen isotope, of organic compounds, **2**, 123
Excited complexes, chemistry of, **19**, 1
Excited molecules, structure of electronically, **1**, 365

Force-field methods, calculation of molecular structure and energy by, **13**, 1
Free radicals, identification by electron spin resonance, **1**, 284
Free radicals and their reactions at low temperature using a rotating cryostat, study of, **8**, 1

Gaseous carbonium ions from the decay of tritiated molecules, **8**, 79
Gas-phase heterolysis, **3**, 91
Gas-phase pyrolysis of small-ring hydrocarbons, **4**, 147
General base and nucleophilic catalysis of ester hydrolysis and related reactions, **5**, 237

$H_2O$–$D_2O$ mixtures, protolytic processes in, **7**, 259
Halogenation, nitrosation, and nitration, diffusion control and pre-association in, **16**, 1
Halides, aryl, and related compounds, photochemistry of, **20**, 191
Heat capacities of activation and their uses in mechanistic studies, **5**, 121
Heterolysis, gas-phase, **3**, 91
Hydrated electrons, reactions of, with organic compounds, **7**, 115
Hydration, reversible, of carbonyl compounds, **4**, 1
Hydrocarbons, small-ring, gas-phase pyrolysis of, **4**, 147
Hydrogen atom abstraction from O—H bonds, **9**, 127
Hydrogen isotope effects in aromatic substitution reactions, **2**, 163
Hydrogen isotope exchange reactions of organic compounds in liquid ammonia, **1**, 156
Hydrolysis, ester, and related reactions, general base and nucleophilic catalysis of, **5**, 237

Intermediates, reactive, study of, by electrochemical methods, **19**, 131
Intramolecular reactions, effective molarities for, **17**, 183
Ionization potentials, **4**, 31
Ion-pairing effects in carbanion reactions, **15**, 153
Ions, organic, charge density–N.M.R. chemical shift correlations, **11**, 125
Isomerization, permutational, of pentavalent phosphorus compounds, **9**, 25
Isotope effects, hydrogen, in aromatic substitution reactions, **2**, 163
Isotope effects, magnetic, magnetic field effects and, on the products of organic reactions, **20**, 1
Isotope effects, steric, experiments on the nature of, **10**, 1
Isotope exchange reactions, hydrogen, of organic compounds in liquid ammonia, **1**, 150
Isotope exchange reactions, oxygen, of organic compounds, **3**, 123
Isotopes and organic reaction mechanisms, **2**, 1

Kinetics and mechanisms of reactions of organic cation radicals in solution, **20**, 55
Kinetics, reaction, polarography and, **5**, 1
Kinetics of organic reactions in water and aqueous mixtures, **14**, 203

Least nuclear motion, principle of, **15**, 1

Macromolecular systems of biochemical interest, $^{13}$C N.M.R. spectroscopy in, **13**, 279

Magnetic field and magnetic isotope effects on the products of organic reactions, **20**, 1

Mass spectrometry, mechanisms and structure in: a comparison with other chemical processes, **8**, 152

Mechanism and structure in carbene chemistry, **7**, 153

Mechanism and structure in mass spectrometry: a comparison with other chemical processes, **8**, 152

Mechanism and structure in organic electrochemistry, **12**, 1

Mechanisms and reactivity in reactions of organic oxyacids of sulphur and their anhydrides, **17**, 65

Mechanisms, nitrosation, **19**, 381

Mechanisms, organic reaction, isotopes and, **2**, 1

Mechanisms of reaction in solution, entropies of activation and, **1**, 1

Mechanisms of solvolytic reactions, medium effects on the rates and, **14**, 10

Mechanistic applications, the reactivity–selectivity principle, **14**, 69

Mechanistic studies, heat capacities of activation and their use, **5**, 121

Medium effects on the rates and mechanisms of solvolytic reactions, **14**, 1

Meisenheimer complexes, **7**, 211

Methyl transfer reactions, **16**, 87

Micellar catalysis in organic reactions: kinetic and mechanistic implications, **8**, 271

Micelles, membranes and other aqueous aggregates, catalysis by, as models of enzyme action, **17**, 435

Molecular structure and energy, calculation of, by force-field methods, **13**, 1

Nitration, nitrosation, and halogenation, diffusion control and pre-association in, **16**, 1

Nitrosation mechanisms, **19**, 381

Nitrosation, nitration, and halogenation, diffusion control and pre-association in, **16**, 1

N.M.R. chemical shift–charge density correlations, **11**, 125

N.M.R. measurements of reaction velocities and equilibrium constants as a function of temperature, **3**, 187

N.M.R. spectroscopy, $^{13}$C, in macromolecular systems of biochemical interest, **13**, 279

Non-planar and planar aromatic systems, **1**, 203

Norbornyl cation: reappraisal of structure, **11**, 179

Nuclear magnetic relaxation, recent problems and progress, **16**, 239

Nuclear magnetic resonance, *see* N.M.R.

Nuclear motion, principle of least, **15**, 1

Nucleophilic aromatic photosubstitution, **11**, 225

Nucleophilic catalysis of ester hydrolysis and related reactions, **5**, 237

Nucleophilic vinylic substitution, **7**, 1

OH—bonds, hydrogen atom abstraction from, **9**, 127

Oxyacids of sulphur and their anhydrides, mechanisms and reactivity in reactions of organic, **17**, 65

Oxygen isotope exchange reactions of organic compounds, **3**, 123

Permutational isomerization of pentavalent phosphorus compounds, **9**, 25

Phase-transfer catalysis by quaternary ammonium salts, **15**, 267
Phosphorus compounds, pentavalent, turnstile rearrangement and pseudorotation in permutational isomerization, **9**, 25
Photochemistry of aryl halides and related compounds, **20**, 191
Photochemistry of carbonium ions, **9**, 129
Photosubstitution, nucleophilic aromatic, **11**, 225
Planar and non-planar aromatic systems, **1**, 203
Polarizability, molecular refractivity and, **3**, 1
Polarography and reaction kinetics, **5**, 1
Polypeptides, calculations of conformations of, **6**, 103
Pre-association, diffusion control and, in nitrosation, nitration, and halogenation, **16**, 1
Products of organic reactions, magnetic field and magnetic isotope effects on, **30**, 1
Protic and dipolar aprotic solvents, rates of bimolecular substitution reactions in, **5**, 173
Protolytic processes in $H_2O$–$D_2O$ mixtures, **7**, 259
Protonation and solvation in strong aqueous acids, **13**, 83
Protonation sites in ambident conjugated systems, **11**, 267
Pseudorotation in isomerization of pentavalent phosphorus compounds, **9**, 25
Pyrolysis, gas-phase, of small-ring hydrocarbons, **4**, 147

Radiation techniques, application to the study of organic radicals, **12**, 223
Radical addition reactions, gas-phase, directive effects in, **16**, 51
Radicals, cation in solution, formation, properties and reactions of, **13**, 155
Radicals, organic application of radiation techniques, **12**, 223
Radicals, organic cation, in solution, kinetics and mechanisms of reaction of, **20**, 55
Radicals, organic free, identification by electron spin resonance, **1**, 284
Radicals, short-lived organic, electron spin resonance studies of, **5**, 53
Rates and mechanisms of solvolytic reactions, medium effects on, **14**, 1
Reaction kinetics, polarography and, **5**, 1
Reaction mechanisms, use of volumes of activation for determining, **2**, 93
Reaction mechanisms in solution, entropies of activation and, **1**, 1
Reaction velocities and equilibrium constants, N.M.R. measurements of, as a function of temperature, **3**, 187
Reactions of hydrated electrons with organic compounds, **7**, 115
Reactions in dimethyl-sulphoxide, physical organic chemistry of, **14**, 133
Reactive intermediates, study of, by electrochemical methods, **19**, 131
Reactivity indices in conjugated molecules, **4**, 73
Reactivity-selectivity principle and its mechanistic applications, **14**, 69
Rearrangements, degenerate carbocation, **19**, 223
Refractivity, molecular, and polarizability, **3**, 1
Relaxation, nuclear magnetic, recent problems and progress, **16**, 239

Short-lived organic radicals, electron spin resonance studies of, **5**, 53
Small-ring hydrocarbons, gas-phase pyrolysis of, **4**, 147
Solid-state chemistry, topochemical phenomena in, **15**, 63
Solids, organic, electrical conduction in, **16**, 159
Solutions, reactions in, entropies of activation and mechanisms, **1**, 1

Solvation and protonation in strong aqueous acids, **13**, 83
Solvents, protic and dipolar aprotic, rates of bimolecular substitution-reactions in, **5**, 173
Solvolytic reactions, medium effects on the rates and mechanisms of, **14**, 1
Spectroscopic observations of alkylcarbonium ions in strong acid solutions, **4**, 305
Spectroscopy, $^{13}$C N.M.R., in macromolecular systems of biochemical interest, **13**, 279
Spin trapping, **17**, 1
Stability and reactivity of crown-ether complexes, **17**, 279
Stereoselection in elementary steps of organic reactions, **6**, 185
Steric isotope effects, experiments on the nature of, **10**, 1
Structure and mechanisms in carbene chemistry, **7**, 153
Structure and mechanism in organic electrochemistry, **12**, 1
Structure of electronically excited molecules, **1**, 365
Substitution, aromatic, a quantitative treatment of directive effects in, **1**, 35
Substitution, nucleophilic vinylic, **7**, 1
Substitution reactions, aromatic, hydrogen isotope effects in, **2**, 163
Substitution reactions, bimolecular, in protic and dipolar aprotic solvents, **5**, 173
Sulphur, organic oxyacids of, and their anhydrides, mechanisms and reactivity in reactions of, **17**, 65
Superacid systems, **9**, 1

Temperature, N.M.R. measurements of reaction velocities and equilibrium constants as a function of, **3**, 187
Topochemical phenomena in solid-state chemistry, **15**, 63
Tritiated molecules, gaseous carbonium ions from the decay of, **8**, 79
Tritium atoms, energetic, reactions with organic compounds, **2**, 201
Turnstile rearrangements in isomerization of pentavalent phosphorus compounds, **9**, 25

Unsaturated compounds, basicity of, **4**, 195

Vinyl cations, **9**, 185
Vinylic substitution, nucleophilic, **7**, 1
Volumes of activation, use of, for determining reaction mechanisms, **2**, 93

Water and aqueous mixtures, kinetics of organic reactions in, **14**, 203